Foundations of Statistical Mechanics

Volume II:
Nonequilibrium Phenomena

Fundamental Theories of Physics

An International Book Series on The Fundamental Theories of Physics: Their Clarification, Development and Application

Editor: ALWYN VAN DER MERWE
University of Denver, U.S.A.

Foundations of Statistical Mechanics

Volume II: Nonequilibrium Phenomena

by

Walter T. Grandy, Jr.

Department of Physics and Astronomy,
The University of Wyoming, U.S.A.

D. Reidel Publishing Company

A MEMBER OF THE KLUWER ACADEMIC PUBLISHERS GROUP

Dordrecht / Boston / Lancaster / Tokyo

Library of Congress Cataloging in Publication Data

Grandy, Walter T., 1933–
    Foundations of statistical mechanics, vol. II.

    (Fundamental theories of physics)
    Includes bibliographies and indexes.
    Contents: v. 1. Equilibrium theory — v. 2. Non-equilibrium phenomena.
    1. Statistical mechanics. I. Title. II. Series.
QC174.8.G73 1987   530.1′3   87–4881
ISBN 90–277–2649–3 (v. 2)

Published by D. Reidel Publishing Company,
P.O. Box 17, 3300 AA Dordrecht, Holland.

Sold and distributed in the U.S.A. and Canada
by Kluwer Academic Publishers,
101 Philip Drive, Norwell, MA 02061, U.S.A.

In all other countries, sold and distributed
by Kluwer Academic Publishers Group,
P.O. Box 322, 3300 AH Dordrecht, Holland.

Printed in The Netherlands

For Pat—
who is never far from equilibrium

Contents

Preface

In this volume we continue the logical development of the work begun in Volume I, and the equilibrium theory now becomes a very special case of the exposition presented here. Once a departure is made from equilibrium, however, the problems become deeper and more subtle—and unlike the equilibrium theory, many aspects of nonequilibrium phenomena remain poorly understood.

For over a century a great deal of effort has been expended on the attempt to develop a comprehensive and sensible description of nonequilibrium phenomena and irreversible processes. What has emerged is a hodgepodge of *ad hoc* constructs that do little to provide either a firm foundation, or a systematic means for proceeding to higher levels of understanding with respect to ever more complicated examples of nonequilibria. Although one should rightfully consider this situation shameful, the amount of effort invested testifies to the degree of difficulty of the problems. In Volume I it was emphasized strongly that the traditional exposition of equilibrium theory lacked a certain cogency which tended to impede progress with extending those considerations to more complex nonequilibrium problems. The reasons for this were adduced to be an unfortunate reliance on ergodicity and the notions of kinetic theory, but in the long run little harm was done regarding the treatment of *equilibrium* problems. On the nonequilibrium level the potential for disaster increases enormously, as becomes evident already in Chapter 1.

Let us cite two observations which serve to underscore the present state of nonequilibrium theory. At the close of a recent historical account [*Statistical Physics and the Atomic Theory of Matter From Boyle and Newton to Landau and Onsager*, Princeton Univ. Press, Princeton, 1983], S.G. Brush solicits observations from a number of current workers in the field regarding the pressing contemporary problems in statistical mechanics. "According to [R.B.] Griffiths, a basic difficulty is the lack of anything comparable to Gibbs' 'canonical' distribution for nonequilibrium situations." We are happy to report that something 'comparable' will indeed be found within the following pages—in fact, it has been known for about 25 years.

In a talk given at the 16th IUPAP Conference on statistical mechanics held in Boston in August 1986 [*Statphys 16*, H.E. Stanley (ed.), North-Holland, Amsterdam, 1986], J.L. Lebowitz observes the following: "It is no secret that there does not exist at present anything resembling a rigorous derivation of the hydrodynamic equations governing the time evolution of macroscopic variables from the laws governing their microscopic constituents." Although certainly lamentable, perhaps the situation is not quite as bad as first thought, and something 'resembling' such a derivation is provided here in Chapter 8.

This volume attempts to establish a coherent and rigorous foundation for nonequilibrium statistical mechanics, based on the principle of maximum entropy. The result is a straightforward extension of the equilibrium partition-sum algorithm of Gibbs and, as knowledgeable workers in the field might guess, its linear approximation is a correlation-function description of the system. It is, however, based on first principles and does not attempt to push the theory of linear *dynamical* response, for example, beyond its logical limits. Indeed, one finds that notion turned around: dynamical response theory emerges as a very special case of the general formulation of statistical mechanics.

After all is said and done, the result is by no means a closed theory, for it requires further articulation at various levels, and pragmatic calculational difficulties remain. The evaluation of space-time correlation functions remains one of the great computational challenges in the field.

It must be re-emphasized that this is a work focused on *foundations*, and thus numerous interesting topics such as Brownian motion are not included, for they are discussed at great length in many other sources. (This example is somewhat ironic, though, because Brownian motion was at the core of foundational discussions many years ago!) The enclosed material has been used with some success on numerous occasions in advanced statistical mechanics courses, but must be supplemented with many more detailed applications. One excellent source, particularly for macroscopic problems, is the recent translation of the work by Yu. L. Klimontovich [*Statistical Physics*, Harwood, Chur, 1986.]

As was the case with Volume I, both this volume and I owe a great deal to a long and fruitful association with Professor E.T. Jaynes. Despite the clear evidence of his influence, however, any shortcomings of the present volume must rest entirely with the author.

W.T. Grandy, Jr.

Laramie, Wyoming
September 1987

Prologue

Equilibrium statistical mechanics is known to provide an extraordinarily accurate description of many physical systems. When only a few macroscopic measurements are available for defining the system, and these correspond to constants of the motion, then the prescription advocated by Gibbs yields truly impressive results. Our summary of these methods has been given in Volume I. In particular, development of sophisticated mathematical techniques has enabled one to calculate in great detail the contributions of particle interactions to the thermodynamic functions, so that many of the essential features of equilibrium systems are now well understood.

The same can *not* be said for nonequilibrium systems, despite a history that is equally long. Part of the difficulty with describing in a general way systems out of equilibrium is the open-ended meaning of the term 'nonequilibrium'. Such phenomena can range from the swirl of cigarette smoke to the incredible violence of a supernova explosion— from the 'randomness' of dendritic growth to the raging torrent of a mountain stream in spring—thereby placing great burdens on attempts at comprehensive treatments. Further complications arise when we find that many fundamentally nonequilibrium processes are also irreversible. For this reason we often attempt to confine our description to steady-state phenomena, say, which are more readily controlled in the laboratory.

It is legitimate to inquire as to why, after all these years, we do not have a widely-accepted nonequilibrium theory, even for simple processes, and that brings us to a second major difficulty. Much of the theoretical work in this area has been based on the Boltzmann transport equation. This equation, however, is of dubious utility for all but the lowest-density gases, as we shall discuss in some detail presently. Alternatively, in order to circumvent these difficulties, it has been common to turn to the age-old desideratum of obtaining equations determining probabilities in the form

$$\frac{dP(t)}{dt} = F[P(t)],$$

where F is a functional of particle dynamics, as well as of probabilities. These equations, including Boltzmann's, go under the generic heading of Fokker-Planck, or 'master' equations, and theories of this type have been reviewed by Chester (1963), for example. But problems with these techniques emerge also, particularly with regard to their dependence on some kind of coarse-graining or forward-time-integration procedures. It has never been clear just why such procedures should be necessary on a fundamental level. In fact, most theoretical approaches to the nonequilibrium problem seem to involve some kind of premature approximation

1

rather early in their development.

In order to clarify these points further, consider the general problem of energy transfer, which might be typified by heating a kettle of water on an electric stove. In principle one can construct the Hamiltonians for the heating element, the kettle walls, and the body of water, as well as the interaction terms for element-kettle and kettle-water. But in actuality the complete solution of this (apparently simple!) problem is mathematically intractable, and several kinds of approximation are introduced. An often used, and perfectly acceptable way to treat a problem of this type consists of constructing a model that may or may not describe the actual physical situation well. One approach is to consider the various subsystems initially in different states of equilibrium, turn on the perturbations slowly, and then drastically simplify the Hamiltonians and interactions so that they may be handled mathematically. Another is to study the behavior of a single water molecule in a heat bath. Although some approximations are always necessary in treating particle-particle interactions, it might be argued that approximations which distort the essential features of a problem should be avoided whenever possible. In the present example it would seem more reasonable to focus on what can actually be measured in the physical situation; namely, the energy current entering the liquid. We then recognize that it is the temperature gradient between kettle and water that, in the final analysis, describes the heating we wish to understand, and which should appear in a *fundamental* way in the description.

As emphasized elsewhere—in Volume I, say—we are primarily interested in describing experimentally reproducible phenomena (ERP), at least during theory-construction stages of understanding. The example above surely falls into this category, and so eventually one wishes to focus on those features of such a process that relate to ERP. Quite generally, a theory of nonequilibria must address the question of how a many-body system responds to external stimuli, the application of which should be reproducible in the sense that the macroscopic initial conditions can be reproduced.

Generalization from one's introspection concerning such problems leads to the conclusion that a detailed understanding of arbitrary irreversible processes must necessarily pass through three distinct stages of calculation (e.g., Andrews, 1965). These are:

(i) Construction of the initial 'ensemble', or statistical operator $\hat{\rho}(t_0)$, describing the initial state of the system of interest;

(ii) Solution of the microscopic dynamical problem so as to obtain the time-evolved operator $\hat{\rho}(t)$;

(iii) Prediction of the final macroscopic physical quantities of interest using $\hat{\rho}(t)$.

(We shall employ the circumflex to denote quantum-mechanical operators.) In addition, there exists a stage (0) consisting of some kind of measurement or observation defining both the system and the problem. This is an essential procedure preceding any calculation, without which any real problem remains ill defined. Though perhaps obvious, the point is often ignored.

Stage (iii) has never presented any difficulties as a matter of principle—one merely calculates expectation values of the operators of interest via the prescription $\langle \hat{F} \rangle \simeq \text{Tr}(\hat{\rho}\hat{F})$—but its interpretation and justification by means of ergodic theory has been widely discussed and reviewed (e.g., Lebowitz, 1974), at least in the equilibrium case. Predictions of time-varying quantities, however, can not rely on ergodic theory to support their validity, and the reliability of such recipes must stem from a more general principle. As an aside, if one wishes to choose a number f as an estimate of the physical quantity represented by the operator $\hat{F}$, based on the statistical operator $\hat{\rho}$, then a reasonable criterion is that the expected square of the error $\langle (\hat{F} - f)^2 \rangle$ should be a minimum. The solution to this variational problem is just $f = \langle \hat{F} \rangle$, independent of ergodic theory.

As noted above, most recent progress has been made with stage (ii), although this seems to be the most difficult stage with which to cope in actual physical problems. Nevertheless, there exists an enormous literature on the needed techniques, and in a very real sense this problem is in principle solved. In one way or another, and most often to some degree of approximation, one need only solve the equation of motion

$$i\hbar \frac{\partial \hat{\rho}(t)}{\partial t} = [\hat{H}, \hat{\rho}(t)], \qquad (\text{P-1})$$

or the equivalent for open systems. The actual solution of this equation is practicably impossible to effect in all but the most trivial problems, of course, which is why over the years one has tried to replace this calculation by a simpler one through reduction to some kind of irreversible Boltzmann or master equation. But mathematical difficulties, however great, have little to do with matters of principle, and the fundamental aspect of stage (ii) remains that of solving the equations of motion subject to the initial conditions describing the physical situation. This last stipulation brings us back to stage (i).

No matter what the initial state of a system, its character can only be determined from observation or direct measurement, or in some cases by deliberate preparation. If, in addition, this initial state is reproducible, then the possibility of understanding the system is greatly enhanced. In the equilibrium situation certain quantities are taken to be constants of the motion, and verification of this presumption leads to construction of the equilibrium ensemble via the partition-sum algorithm of Gibbs. The problem of stage (i) is thus solved immediately, and almost trivially in this case. Perhaps for this reason stage (i) has been given little attention in the study of more complicated applications of statistical mechanics, with the notable exceptions of Tolman (1938), Andrews (1965), and Richardson (1974). From the historical view this is rather unfortunate, because a careful reading of Gibbs suggests that he recognized the existence of the general problem of ensemble construction and that the surface had only been scratched. (As in Volume I, we use the term 'ensemble' to denote a description provided by a definite statistical operator, rather than as the artifice introduced and recognized as such by Gibbs.)

For the nonequilibrium case, and for equilibrium problems in which the known quantities are not related to intrinsic constants of the motion, a much more general form of Gibbs' algorithm is required. Moreover, such a generalization should result in a prescription admitting in principle all many-body problems and all possible initial conditions. Indeed, one might envision the existence of a rule whose roots transcend its applications to physical problems, and whose foundations are simple enough to be tested easily.

The required generalization of the Gibbs algorithm has been known for some time, but both conceptual hangups and the lack of detailed applications in the open literature have mitigated against its general recognition. In his 1962 Brandeis Lectures, and subsequently, Jaynes (1963, 1967) reported his discovery of this general principle which, though elegant and simple, has eluded workers in the field for over eighty years. Usually the initial data defining the nonequilibrium state of a system can be put into the form of expectation values of a number of Heisenberg operators $\hat{F}_i(\mathbf{x},t)$. The variables $\mathbf{x}$ and t vary over some information-gathering space-time region R_i, which may be different for different operators $\hat{F}_i$. For arbitrary processes, therefore, neither present nor local expectation values will suffice to predict the future behavior of the system. Rather, one maximizes the entropy, $S = -\kappa \operatorname{Tr}(\hat{\rho} \ln \hat{\rho})$, subject to the constraints imposed by the values $\langle \hat{F}_i(\mathbf{x},t) \rangle$ over the regions R_i. The justification for this procedure has been discussed at length elsewhere by Jaynes (1957), Baierlein (1971), Hobson (1971), and Grandy (1980), and is merely a special case of a more general method of plausible reasoning. It is presented in some detail in Volume I.

As a solution to the above variational problem one obtains the optimal statistical operator describing the initial state of the system, as determined from what is known about the system, and only that. Hence, the problem of stage (i) is solved completely by writing the initial statistical operator as

$$\hat{\rho} = \frac{1}{Z} \exp\left[\sum_k \int_{R_k} \lambda_k(\mathbf{x},t) \hat{F}_k(\mathbf{x},t) \, d^3x \, dt \right], \qquad (\text{P--2})$$

where $Z = Z[\lambda_1(\mathbf{x},t), \ldots, \lambda_k(\mathbf{x},t), \ldots]$ is the *partition functional*:

$$Z = \operatorname{Tr} \exp\left[\sum_k \int_{R_k} \lambda_k(\mathbf{x},t) \hat{F}_k(\mathbf{x},t) \, d^3x \, dt \right]. \qquad (\text{P--3})$$

The *Lagrange-multiplier functions* $\lambda_k(\mathbf{x},t)$ are identified as usual from the initial data in the form of equations of constraint:

$$\langle \hat{F}_k(\mathbf{x},t) \rangle \equiv \operatorname{Tr}[\hat{\rho}\hat{F}_k(\mathbf{x},t)] = \frac{\delta}{\delta\lambda_k(\mathbf{x},t)} \ln Z, \qquad (\mathbf{x},t) \in R_k. \qquad (\text{P--4})$$

These form a set of coupled functional differential equations sufficient to determine the integrals in Eqs.(P--2) and (P--3). Outside the regions R_k the predicted expectation value of an operator $\hat{F}_k$, or of an arbitrary Heisenberg operator $\hat{J}$ at any point $(\mathbf{x},t)$, is just

$$\langle \hat{J}(\mathbf{x},t) \rangle = \operatorname{Tr}[\hat{\rho}\hat{J}(\mathbf{x},t)] = \operatorname{Tr}[\hat{\rho}(t)\hat{J}(\mathbf{x})], \qquad (\text{P--5})$$

a prescription completely justified within the theory itself. The second equality in Eq.(P–5) follows from cyclic invariance of the trace and the presumed unitary time development of the operators.

It is important to emphasize that $\hat{\rho}$ in Eq.(P–2) is nothing more than the initial statistical operator describing only what is known about the initial state of the system. Some reflection indicates that, aside from clearly specified driving, this is generally all one can hope to know regarding any experimentally reproducible situation. The time evolution of $\hat{\rho}$ is completely described by Eq.(P–1), in principle. Because the expression (P–2) is strikingly simple, although rather subtle, it is difficult to appreciate immediately either its full content or its broad consequences. In Chapter 2 we shall review its derivation and begin extracting some of its content.

In order to provide a perspective for the nonequilibrium theory to come, Chapter 1 is devoted to a review of classical hydrodynamics and the Boltzmann transport theory. We shall develop the general theory in Chapter 2 and then spend the remainder of the book applying it to both theoretical and practical problems. Just as we have spent eighty-some years fruitfully exploring the original algorithm of Gibbs, one may be sure that a complete understanding of its generalization to Eqs.(P–2)-(P–4) will only have begun in the following pages.

REFERENCES

Andrews, F.C.: 1965, 'Statistical Mechanics and Irreversibility', *Proc. Natl. Acad. Sci. (U.S.A)* **54**, 13.

Baierlein, R.: 1971, *Atoms and Information Theory*, Freeman, San Francisco..

Chester, G.V.: 1963, 'The Theory of Irreversible Processes', *Repts. Prog. Physics* **26**, 411.

Grandy, W.T., Jr.: 1980, 'Principle of Maximum Entropy and Irreversible Processes', *Phys. Repts.* **62**, 175.

Hobson, A.: 1971, *Concepts in Statistical Mechanics*, Gordon and Breach, New York.

Jaynes, E.T.: 1957, 'Information Theory and Statistical Mechanics', *Phys. Rev.* **106**, 620.

Jaynes, E.T.: 1963, 'Information Theory and Statistical Mechanics', in W.K. Ford (ed.), *Statistical Physics*, Benjamin, New York.

Jaynes, E.T.: 1967, 'Foundations of Probability Theory and Statistical Mechanics', in M. Bunge (ed.), *Delaware Seminar in the Foundations of Physics*, Springer-Verlag, New York.

Lebowitz, J.L: 1974, 'Ergodic Theory and Statistical Mechanics', in G. Kirczenow and J. Marro (eds.), *Transport Phenomena (Lecture Notes in Physics, Vol.31)*, Springer-Verlag, Berlin.

Richardson, J.M.: 1974, 'The Initial Distribution in Classical Statistical Mechanics', *J. Stat. Phys.* **11**, 323.

Tolman, R.C.: 1938, *The Principles of Statistical Mechanics*, Oxford Univ. Press, London.

Chapter 1

The Equations of Classical Hydrodynamics

Because it is the best known scenario and has been worked out in the most detail, we shall outline at some length what is loosely referred to as classical hydrodynamics in order to provide concrete illustrative examples of nonequilibrium phenomena. The point is to review the phenomenology so as to obtain a clear idea of just what the role of statistical mechanics must be, at least with respect to a definite class of problems. It is then useful to review the traditional approach to relating these macroscopic relations to the behavior of the microscopic constituents of the system, and to assess both the triumphs and shortcomings of what has been the conventional view for most of this century.

A. Rheology

One can define *rheology* quite broadly as the study of flow and deformation of matter under a system of applied forces— the Greek root *rheos* connotes 'current'. Almost every area of macroscopic phenomena is touched by the subject, from hydrodynamics to elastic vibrations in solids. The theory is well developed and a number of comprehensive surveys exist (e.g., Reiner, 1958; Fredrickson, 1964), all of which begin with a somewhat common set of definitions. A *volume strain* is a deformation in which the general shape of a body is unaltered, such as in compression and dilatation, whereas a *shearing strain* alters the shape of a body but not its size. By a 'body' we shall mean in what follows a constrained solid, liquid, or gas, and encompass the latter two under the general category of fluids.

Strains are described mathematically in terms of a *strain tensor*, $\sigma_{ij}(\mathbf{x}, t)$, and the *rate-of-strain tensor*, $\gamma_{ij}(\mathbf{x}, t) \equiv \dot{\sigma}_{ij}(\mathbf{x}, t)$. The principal invariants of the latter are usually designated by

$$\lambda_1 \equiv \gamma^m{}_m \,, \tag{1--1a}$$

$$\lambda_2 \equiv \frac{1}{2}(\lambda_1^2 - \gamma^n{}_m \gamma^m{}_n) \,, \tag{1--1b}$$

$$\lambda_3 \equiv \det \gamma \,, \tag{1--1c}$$

and when no confusion can arise we shall employ the summation convention over repeated indices. The *stress tensor* for the medium is denoted S^{ij}, with the following correspondences:

$$S^{ii} : \text{normal or volume stresses} \,, \tag{1--2a}$$

$$S^{ij} : \text{shear stresses} \quad (i \neq j) \,. \tag{1--2b}$$

6

As an example, a fluid is usually *defined* as a substance whose stress tensor has the form

$$S^{ij} = P g^{ij} - p^{ij} \,, \qquad (1\text{--}3)$$

where P is the scalar pressure, p^{ij} is called the *shear tensor*, and g^{ij} is the metric tensor describing the fixed coordinate system to which the body is referred. One usually adopts a general curvilinear coordinate system, but this will not be necessary in the ensuing development. In a fluid at rest the stresses are isotropic and, as we shall see, Eq.(1–3) reduces to

$$S^{ij} = P_0 \, g^{ij} \,, \qquad (1\text{--}4)$$

which defines the hydrostatic pressure, P_0.

A theory of rheology seeks to establish rheological equations of state, which are phenomenological constitutive relations of the form

$$S_{ij} = S_{ij}(\gamma) \,, \qquad (1\text{--}5)$$

for example. A number of material parameters will arise in such relations and are referred to as rheological coefficients. One concern of statistical mechanics is to describe these quantities characteristic of the substance in terms of the basic constituent particles and their interactions, as well as to predict eventually the appropriate macroscopic relations themselves.

We presume that the single-component collection of particles of mass m can be considered on the macroscopic scale as a continuous medium, in which the existence of a velocity field $\mathbf{v}(\mathbf{x}, t)$ is asserted. This is sometimes called the local velocity and, with respect to a fixed coordinate system, is the velocity of an infinitesimal volume element of the medium centered at point $\mathbf{x}$ at time t. Alternatively, we could define a displacement vector σ_i such that $\dot{\sigma}_i = v_i$, and all displacements refer to displacements of the element, which may contain many particles.

One can see how the velocity field arises quite naturally by recalling that in the system at rest the appropriate thermodynamic parameters are the number density $n \equiv N/V$, energy density $h \equiv E/V$, and pressure P, in the volume V. These macroscopic quantities are actually equilibrium expectation values of the microscopic operators to be discussed later. If the system is not in equilibrium, and small elements are moving with velocity $\mathbf{v}$ relative to the laboratory frame, then the above thermodynamic parameters can provide a valid description only in the elemental rest frame. We shall always work in the laboratory frame, and if we require the description to be invariant under Galilean transformations one obtains immediately the desired quantities:

$$n(\mathbf{x}, t) = n \,, \qquad (1\text{--}6\text{a})$$

$$h(\mathbf{x}, t) = h + \tfrac{1}{2} n m v^2 \,, \qquad (1\text{--}6\text{b})$$

$$\mathbf{J}(\mathbf{x}, t) = n \mathbf{v} \,, \qquad (1\text{--}6\text{c})$$

$$T_{ij}(\mathbf{x}, t) = S_{ij}(\mathbf{x}, t) + m n v_i v_j \,, \qquad (1\text{--}6\text{d})$$

$$q_i(\mathbf{x}, t) = (h + \tfrac{1}{2} m n v^2) v_i + v_j S_{ji} \,. \qquad (1\text{--}6\text{e})$$

Thus, Galilean invariance alone implies the existence in the laboratory frame of a particle current density $\mathbf{J}$, an energy current density $\mathbf{q}$, and a kinetic-energy tensor $t_{ij} \equiv mnv_iv_j$, in terms of which we write the total energy-momentum tensor as

$$T_{ij} = S_{ij} + t_{ij} \, . \tag{1-7}$$

Note that all quantities on the left-hand sides of Eqs.(1–6) must be considered eventually as expectation values of microscopic quantities, and that here $S_{ij}(\mathbf{x}, t)$ is just the local equilibrium pressure $P_0(\mathbf{x}, t)$ as specified by Eq.(1–4). In addition, it must be emphasized that these results are purely kinematic, for dissipation can not be introduced by transformation alone.

Let us describe the laboratory frame by means of a Cartesian coordinate system and write the total differential of the velocity field as

$$dv_i = \frac{\partial v_i}{\partial x_j} dx^j \equiv v_{i,j} \, dx^j \, . \tag{1-8}$$

Now write the partial derivatives as sums of symmetric and antisymmetric parts:

$$v_{i,j} = \frac{1}{2}\left(v_{i,j} + v_{j,i}\right) + \frac{1}{2}\left(v_{i,j} - v_{j,i}\right) . \tag{1-9}$$

Then the symmetric part is just the rate-of-strain tensor

$$\gamma_{ij} = \frac{1}{2}\left(v_{i,j} + v_{j,i}\right) , \tag{1-10}$$

whereas the antisymmetric part is called the *rotation tensor*:

$$\omega_{ij} \equiv \tfrac{1}{2}\left(v_{i,j} - v_{j,i}\right) . \tag{1-11}$$

This latter quantity describes vorticity in the medium, which can also be described by a vorticity vector

$$\Omega^r \equiv \tfrac{1}{2}\varepsilon^{rji} \, v_{i,j} \, , \tag{1-12}$$

where ε^{rji} is the Levi-Cività symbol. For a *pure shear* $v_{i,j} = v_{j,i}$, so that there is no vorticity and γ_{ij} reduces to the velocity gradient. In the other extreme of a *rigid rotation* about the axis perpendicular to the (i, j)-plane $v_{i,j} = -v_{j,i}$, and the vorticity is proportional to the velocity gradient. It will be presumed that angular momentum is always conserved in the system, so that ω_{ij} is taken to be zero in the sequel and the energy-momentum tensor is symmetric (see, e.g., Appendix A of Martin, *et al*, 1972). Finally, note that the strain tensor can be written in terms of derivatives of the displacement,

$$\sigma_{ij} = \tfrac{1}{2}\left(\sigma_{i,j} + \sigma_{j,i}\right) , \tag{1-13}$$

and a time differentiation reproduces Eq.(1–10).

The basic conservation laws lead to the continuity equation

$$\frac{\partial n}{\partial t} + J^k{}_{,k} = 0, \tag{1-14}$$

as well as equations of motion. Let F_i represent a component of any external force that may be imposed on the medium. Then the equations of motion are

$$m\frac{\partial J^i}{\partial t} + T^{ik}{}_{,k} = mnF^i, \tag{1-15a}$$

$$\frac{\partial h}{\partial t} + q^i{}_{,i} = -S^{kj}\, v_{j,k}, \tag{1-15b}$$

where the right-hand side of the last equation describes energy dissipation as heat owing to internal viscous forces, and we employ the laboratory frame.

Rheological parameters are introduced with the aid of elementary ideas from kinetic theory, along with phenomenological laws constructed long ago. Consider, for example, an imaginary plane normal to the z-direction in a gas, and denote by p_{zx} the mean force per unit area of the plane exerted by the fluid below the plane on the fluid above. This component of the shear tensor arises from a local shearing force. The x-component of the fluid velocity as a function of z is $v_x(z)$, but if in fact v_x does not vary with z, then $p_{zx} = 0$. Intuition thus suggests a first-order linear relation

$$p_{zx} = -\eta\frac{\partial v_x}{\partial z}, \tag{1-16}$$

where we choose the sign so as to make the proportionality factor η positive. This constant is called the *coefficient of shear viscosity*, and characterizes an internal friction between layers of the fluid.

In a similar manner, suppose the absolute temperature in a dilute fluid varies as $T = T(z)$, and that q_z is the thermal energy crossing unit area of a plane normal to the z-direction per unit time. Then to first order

$$q_z = -\lambda\frac{\partial T}{\partial z}, \tag{1-17}$$

which is known as Fourier's law (Fourier, 1822). We refer to λ as the *coefficient of thermal conductivity*.

One can also consider arrangements like those above in which we imagine it possible to 'label' a number of particles so that the general number density n remains constant but that of the labeled particles varies as $n_1 = n_1(z)$. Let J_z be the flux density of labeled particles across the plane in the positive z-direction, so that if n_1 is uniform then $J_z = 0$. Otherwise,

$$J_z = -D\frac{\partial n_1}{\partial z}, \tag{1-18}$$

known as Fick's law of diffusion (Fick, 1855), and D is the *coefficient of self diffusion*. When different species of particle are present we speak of mutual diffusion.

Finally, recall the similar but better-known linear relation of the type being discussed here known as Ohm's law,

$$j_z = \sigma_e \, E_z \,, \tag{1-19}$$

where σ_e is the electrical conductivity per unit volume. This parameter can also be written as

$$\sigma_e = nq^2 u \,, \tag{1-20}$$

where q is the charge on each particle and u is called the mobility. If the motion of the charges is due to a density gradient, then the Nernst-Einstein relation reads

$$u = \frac{D}{\kappa T} \,, \tag{1-21}$$

where κ is Boltzmann's constant. That is, the mobility is the steady velocity owing to unit force (Nernst, 1888; Einstein, 1905). We shall return to a deeper interpretation of this relation subsequently.

Equations (1–16)-(1–21) are known as constitutive relations and provide the necessary link with the material properties of the specific medium under study. One must next consider how these parameters appear in the stress tensor, knowledge of which will then complete the phenomenological theory. As an example, combination of Eq.(1–15a) with the definition of a fluid given by Eq.(1–3) yields

$$m\frac{\partial J^i}{\partial t} = -\delta^{ij} \, P_{,j} + p^{ik}{}_{,k} + mnF^i \,. \tag{1-22}$$

A Newtonian fluid, to be defined more precisely presently, is characterized by the shear tensor

$$p_{ij} = \eta \left[\frac{\partial v_i}{\partial x_j} + \frac{\partial v_j}{\partial x_i} - \frac{2}{3}\delta_{ij}\frac{\partial v_k}{\partial x_k} \right] + \varsigma\delta_{ij}\frac{\partial v_k}{\partial x_k} \,, \tag{1-23}$$

where η is again the shear viscosity, and ς is called the *bulk* viscosity. The quantity $\nu \equiv \eta/mn$ is referred to as the *kinematic viscosity*. Bulk viscosity is a measure of the energy transferred from the translational mode to other energy modes, so that the result of Problem 1.2 implies that $\varsigma = 0$ in a monatomic gas, in which mechanical and thermodynamic pressures are identical.

Let us now proceed to a more precise classification of bulk matter in terms of the characteristic structure of its stress tensor. We first consider *purely viscous* substances, which are exemplified by fluids and which flow under any shearing stress, no matter how small. Work done on these substances is converted completely to heat and they are characterized by the phenomenological equations

$$p^{ij} = p^{ij}(\gamma^{mn}), \qquad p^{ij}(0) = 0 \,. \tag{1-24}$$

A useful theoretical model is that called a Stokesian fluid, which is an isotropic purely viscous fluid. Its description is greatly simplified by observing that there are no isotropic vectors, so that the only second-rank isotropic tensor is a constant multiple of δ_{ij} (or g_{ij} in a general coordinate system). In this model, then, the shear tensor must have the structure

$$p^{ij} = \alpha_0 \delta^{ij} + \alpha_1 \gamma^{ij} + \alpha_2 \gamma^{im} \gamma_m^j \,, \tag{1-25}$$

where the α_i are polynomials in the principal invariants of Eq.(1–1). When $\gamma^{ij} = 0$, then $\alpha_0 = 0$.

An *incompressible fluid* is distinguished by a vanishing velocity divergence,

$$v^k{}_{,k} = 0 \,, \tag{1-26}$$

in which case P becomes a dynamical variable. (For compressible fluids P is always the thermodynamic pressure). One can then incorporate α_0 into P and we have

$$S^{ij} = P \delta^{ij} - \alpha_1 \gamma^{ij} - \alpha_2 \gamma^{im} \gamma_m^j \,. \tag{1-27}$$

This is the equation of state of an incompressible Stokesian fluid, in which $\frac{1}{2}\alpha_1$ is called the apparent viscosity and α_2 the cross viscosity. Apparently no Stokesian fluids exist experimentally, except in that they are Newtonian, but the idea is often a useful theoretical construct. When the condition of incompressibility (1–26) is combined with Eqs.(1–22) and (1–23) they are together called the *Navier-Stokes equations*. For an inviscid (nonviscous) fluid the set of equations is called the Euler equations, whereas linearizing them yields Stokes' equations. These expressions provide the basic description of classical hydrodynamics.

A *Newtonian fluid* is characterized by the following parameter values:

$$\alpha_0 = \mu \lambda_1 \,, \quad \alpha_1 = 2\eta \,, \quad \alpha_2 = 0 \,, \tag{1-28}$$

where μ and η are constants. Equation (1–25) then yields

$$p^{ij} = \mu \lambda_1 \delta^{ij} + 2\eta \gamma^{ij} \,. \tag{1-29}$$

It is customary to define an additional constant ς by writing

$$\mu \equiv \varsigma - \tfrac{2}{3}\eta \,, \tag{1-30}$$

so that we regain Eq.(1–23) and identify the viscosity coefficients. For an incompressible Newtonian fluid we must set $\lambda_1 = 0$, and then Eq.(1–29) yields Newton's law of friction:

$$p^{ij} = 2\eta \gamma^{ij} \,. \tag{1-31}$$

This expression was used long ago, for example, to describe steady flow of a simple fluid through a long slender pipe (Hagen, 1839; Poiseuille, 1841).

But it is really necessary to consider compressible fluids— so as to include the elasticity, or stiffness, needed for propagation of sound waves, say. Then P is just

the thermodynamic pressure and for small variations can be expanded through first
order in the density:

$$p - p_0 = \left(\frac{\partial p}{\partial n}\right)_{T, n_0} (n - n_0) + \cdots,$$ (1–32)

where p_0 and n_0 refer to the equilibrium state. The derivative is just the bulk
modulus, or inverse isothermal compressibility κ_T^{-1}, and is thus related to the longitudinal speed of sound in the medium. A compressible Newtonian fluid is therefore
described by

$$S^{ij} = p_0 \delta^{ij} + \kappa_T^{-1} \left(\frac{n - n_0}{n_0}\right) \delta^{ij} + \varsigma \lambda_1 \delta^{ij} + 2\eta \left(\gamma^{ij} - \tfrac{1}{3}\lambda_1 \delta^{ij}\right).$$ (1–33)

The primary use of the general Stokesian model is to estimate the effects in real
fluids of departures from Newtonian behavior.

Non-Newtonian behavior emerges when the constitutive equations no longer
take the simple form of Eq.(1–31), for example. If even a simple fluid is subject to
shear, say, then one finds that $\eta = \eta(\gamma)$ and the system can exhibit shear thinning:
the viscosity decreases with strain rate. Or, the density can decrease when the
system is subjected to shear at constant temperature and pressure, a phenomenon
known as shear dilatancy. Computer simulations of these effects have been reviewed
by Evans, *et al* (1984). One also finds non-Newtonian behavior in certain exotic
fluids such as polymeric liquids, which have only been studied intensely during
the second half of the twentieth century, as the applications of plastics developed.
Numerous experiments on such liquids reveal properties *qualitatively* different from
those of a Newtonian fluid, a number of which are reviewed by Bird and Curtiss
(1984). For example, unlike elastic substances, polymeric fluids might not return
all the way to their initial state after a small perturbation. Rather, they seem to
possess a 'fading memory', a phenomenon discussed in some detail by Coleman and
Noll (1961).

In contrast to purely viscous substances are the *purely elastic* materials, which
convert work into recoverable deformations. We shall consider for the moment only
incompressible isotropic solids, and it is then customary to utilize the tensor σ^{ij}
[see Eq.(1–13)] and introduce another tensor Σ^{ij} in place of p^{ij}. The constitutive
relations are now given by the Cauchy-Hooke equations (Cauchy, 1823; Hooke,
1676):

$$\Sigma^{ij} = 2\mu_E \sigma^{ij} + \lambda_E \sigma^m{}_m \delta^{ij},$$ (1–34)

where λ_E and μ_E are called moduli of elasticity. Equation (1–34) applies primarily
in the case of small displacements from an equilibrium configuration, and therefore
represents a linear theory of elasticity. Rather than the above parameters, one
often encounters instead the *shear modulus*, G, and the *bulk modulus*, K (the high-
frequency version of that defined for fluids):

$$G \equiv \mu_E, \qquad K \equiv \lambda_E + \tfrac{2}{3}G.$$ (1–35)

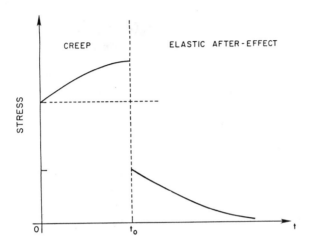

Fig. 1–1. A schematic illustration of memory retention, relaxation, and creep in anelastic solids.

In an ideal solid Cauchy's identity is valid,

$$K = \tfrac{5}{3}G, \qquad (1\text{-}36)$$

only a single elastic parameter is needed to describe the medium, and the familiar form of Hooke's law is found.

Pure fluids and pure elastics are two extremes between which there exists an enormous range of perhaps more interesting materials. These are the *viscoelastic* substances which exhibit both viscosity and elasticity simultaneously, and in which one can observe memory-retaining phenomena. In addition, they can support shear waves, and can exhibit thixotropy. When the elastic character dominates, viscoelasticity is manifested, for example, in materials exhibiting *creep*: a strain gradually decreases at a decreasing rate, so that it requires infinite time to recover an initial strain attained over a finite time interval. In anelastic solids the additional features of memory retention and relaxation are found, as illustrated in Figure 1–1.

Of major interest here is the case when viscous effects are dominant and we consider the *viscoelastic fluid*. Perhaps the simplest approach to these fluids is the dual representation essentially formulated by Maxwell (1866, 1867), in which the stress tensor of the medium is taken to be a combination of the quantities given by Eqs.(1–23) and (1–34). That is,

$$
\begin{aligned}
S_{ij} &= P\,\delta_{ij} - \Sigma_{ij} - p_{ij} \\
&= P\,\delta_{ij} - G\big(\sigma_{i,j} + \sigma_{j,i}\big) - \big(K - \tfrac{2}{3}G\big)\delta_{ij}\,\nabla\cdot\boldsymbol{\sigma} \\
&\quad + \eta\big(v_{i,j} + v_{j,i}\big) + \big(\varsigma - \tfrac{2}{3}\eta\big)\delta_{ij}\,\nabla\cdot\mathbf{v}\,.
\end{aligned}
\tag{1-37}
$$

If desired, G can be related to the transverse sound velocity and, because it is proportional to the longitudinal sound velocity, K can be incorporated into P. The above expression describes, of course, only a linear phenomenological theory of viscoelasticity, which is adequate for most purposes. A comprehensive review of this subject has been provided by Truesdell (1952).

There are numerous other rheological phenomena which can be studied in a similar manner. For example, elasticoviscous solids exhibit no steady state of shear flow, but only creep. Plastics manifest viscosity and elasticity simultaneously, but in different portions of the substance. The plasticity of ordinary paint is common: while not highly viscous, it nevertheless does not run off a wall when applied. Though interesting, the phenomenological theories of these substances have not been completely developed unambiguously, and so will not be discussed further here. Nor shall we discuss the coupling of elastic characteristics of certain media with their electric and magnetic properties, known as piezoelectricity, piezomagnetism, electrostriction and magnetostriction.

B. The Conventional Microscopic Connection

Unquestionably, the overwhelming amount of work directed toward a microscopic understanding of hydrodynamic and transport phenomena during the past 100 years has been based on kinetic theory and the Boltzmann equation. This equation was discussed in some detail in Volume I, for example, to which we shall have occasion to refer. It is useful to omit consideration of external forces for present purposes, so that one writes the Boltzmann transport equation as

$$
\frac{\partial\rho}{\partial t} + (\mathbf{v}\cdot\nabla_{\mathbf{r}})\rho = \left[\frac{\partial\rho}{\partial t}\right]_{\text{collisions}} \equiv J(\rho,\rho)\,.
\tag{1-38}
$$

We recall that $\rho(\mathbf{r},\mathbf{v},t)$ is Boltzmann's distribution function in μ-space, and that J is the so-called collision term characterizing the effects of particle collisions on ρ.

The function ρ is rather singular, unfortunately, because it is a sum of δ-functions. If one goes to the larger phase space and Liouville's equation, however, we see that the ensemble average of ρ is proportional to the single-particle phase-space distribution $f_1(\mathbf{x},\mathbf{p},t)$. In fact, it is just this function which satisfies Eq.(1-38) in the zero-density limit (e.g., Lanford, 1975), so that it will be convenient to replace ρ in Eq.(1-38) with the generic single-particle phase-space distribution function f and presume that any ambiguities regarding the meaning of 'averages' have been resolved by means of appropriate Γ-space considerations.

In Volume I we have discussed the limits of validity of the Boltzmann equation, noting that it can be valid at most in the very-low-density limit. Underlying this restriction is the *Stoßzahlansatz* of Boltzmann, part of which asserts that the two-particle distribution factors: $f_{12} = f_1 f_2$. That is, there are no correlations among the constituent particles of the system—in most cases only binary collisions are considered. (The undoubted failure of this presumption has been discussed in detail by Blatt and Opie, 1975). Thus, in what follows one should have in mind a dilute gas of atoms or molecules, although we shall find it necessary to discuss extensions to higher densities before long.

It was noted in the preceding section that classical hydrodynamics envisions the existence of a local density $n(\mathbf{x}, t)$, and a local fluid velocity $\mathbf{v}(\mathbf{x}, t)$, which refer to a small volume of the gas centered on some spatial point at time t. These are taken as velocity-space averages in the sense discussed above, so that we write

$$n(\mathbf{x}, t) \equiv \int f(\mathbf{x}, \mathbf{v}, t)\, d^3v\,, \tag{1-39a}$$

$$n(\mathbf{x}, t) v_i(\mathbf{x}, t) \equiv \int v_i\, f(\mathbf{x}, \mathbf{v}, t)\, d^3v\,. \tag{1-39b}$$

We shall find it convenient to introduce the notation $V(\mathbf{x}, t) \equiv \mathbf{v} - \mathbf{v}(\mathbf{x}, t)$, sometimes called the *peculiar velocity*, and to define the hydrodynamic derivative

$$\frac{D}{Dt} \equiv \frac{\partial}{\partial t} + v_i \frac{\partial}{\partial x_i}\,, \tag{1-40}$$

sums over repeated indices being understood. Then, for the elementary volume, we make the further local definitions of absolute temperature

$$\frac{3}{2} n(\mathbf{x}, t) \kappa T(\mathbf{x}, t) \equiv \int f(\mathbf{x}, \mathbf{v}, t) \tfrac{1}{2} m V^2(\mathbf{x}, t)\, d^3v\,, \tag{1-41a}$$

pressure tensor

$$P_{ij}(\mathbf{x}, t) \equiv \int m f(\mathbf{x}, \mathbf{v}, t) V_i V_j\, d^3v\,, \tag{1-41b}$$

and heat-flux vector

$$q_i(\mathbf{x}, t) \equiv \int f(\mathbf{x}, \mathbf{v}, t) \tfrac{1}{2} m V^2 V_i\, d^3v\,. \tag{1-41c}$$

In these expressions m is the mass of a constituent particle, and κ is Boltzmann's constant.

It is obvious that the assertion that the quantities of Eqs. (1–39) and (1–41) describe the system adequately amounts to a presumption of some kind of 'local thermodynamic equilibrium'. Even the particle interactions take place only within the fundamental volume element, so that tacitly only short-range forces are to be included.

With this identification of the basic entities of the theory, as well as their meanings, the next step is to obtain the equations of motion relating them. These are just the fundamental conservation laws in the system and are readily obtained by taking velocity moments of the Boltzmann equation (1–38), an exercise left to the problems. The results are

$$\frac{Dn}{Dt} = -n\nabla \cdot \mathbf{v}\,, \tag{1-42a}$$

$$nm\frac{Dv_i}{Dt} = -\frac{\partial P_{ij}}{\partial x_j}\,, \tag{1-42b}$$

$$\frac{3}{2}n\kappa\frac{DT}{Dt} = -P_{ij}\gamma^{ij} - \frac{\partial q_i}{\partial x_i}\,, \tag{1-42c}$$

which are effectively Eqs.(1–14) and (1–15). The rate-of-strain tensor γ_{ij} was defined in Eq.(1–10).

Finally, one must solve the Boltzmann equation in some order of approximation so as to obtain definite expressions for the local functions describing the gas, and thus close the system of equations. We shall outline briefly the procedure for carrying out these calculations, but for complete details the reader is referred to the classic works of Chapman and Cowling (1960), and Hirschfelder, *et al* (1954).

The most widely used approach to determining f is Enskog's perturbation solution of the Boltzmann equation (Enskog, 1917), in which one presumes the series expansion

$$f = f_0 + f_1 + f_2 + \cdots . \tag{1-43}$$

[A similar approach was also developed at about the same time by Chapman (1916-17)]. One takes f_0 to be a normalized Maxwell distribution about $\mathbf{v}(\mathbf{x}, t)$, thereby obtaining immediately the zero-order results by straightforward integration. From Eqs.(1–41) we find the approximations

$$P_{ij} \simeq P_{ij}^{(0)} = P\delta_{ij} = n\kappa T\delta_{ij}\,, \tag{1-44a}$$

$$q_i \simeq q_i^{(0)} = 0\,, \tag{1-44b}$$

where P is the pressure in the coordinate system moving with the stream velocity. Substitution into Eqs.(1–42) yields the Euler equations of hydrodynamics describing an inviscid fluid. Note that the effect of Eq.(1–42c) in this approximation is to maintain $(n^{-2/3}\kappa T)$ constant in the moving coordinate system, which is just the adiabatic expansion law for an ideal monatomic gas.

In the next approximation one introduces a perturbation function $\phi(\mathbf{x}, \mathbf{v}, t)$ and writes $f_1 = f_0\phi$, so that to this order $f \simeq f_0(1+\phi)$. Substitution into the Boltzmann equation then yields an integral equation for ϕ, the solution to which defines an enormous project. Moreover, the presence of the collision term in Eq.(1–38) now requires adoption of definite particle interaction models. The standard procedure— now known as the Chapman-Enskog method—is to include only binary collisions, expand everything in terms of Sonine polynomials, and then make approximations by retaining only a few terms in the expansions (e.g., Chapman and Cowling, 1960).

At this point, however, it is only important to note that the result of this exercise allows one to write the next approximations after Eqs.(1–44) as

$$P_{ij} \simeq P_{ij}^{(0)} + P_{ij}^{(1)} = P\delta_{ij} - 2\eta\left(\gamma_{ij} - \tfrac{1}{3}\gamma^k{}_k\delta_{ij}\right), \qquad (1\text{–}45a)$$

$$q_i \simeq q_i^{(0)} + q_i^{(1)} = -\lambda\frac{\partial T}{\partial x^i}. \qquad (1\text{–}45b)$$

That is, the first-order results are proportional to gradients, and the proportionality factors are the expected transport coefficients. The bonus here is that one can now calculate explicitly the shear viscosity η and the thermal conductivity λ by evaluation of the complicated integrals involving Sonine polynomials. It is with these expressions that the microscopic theory makes contact with experiment.

Substitution of Eqs.(1–45) into Eqs.(1–42) yields the Navier-Stokes equations (when $\nabla \cdot \mathbf{v} = 0$), in which we note that no bulk-viscosity term appears for a low-density monatomic gas. In next approximation, when the term f_2 is retained in Eq.(1–43), Burnett (1935) carried out the calculations leading to terms $P_{ij}^{(2)}$ and $q_i^{(2)}$ in Eqs.(1–45). These terms are proportional to squares of gradients, effectively, involving higher-order transport coefficients. Hence, although the theory is fundamentally restricted to low densities, the expansion (1–42) appears to be in powers of gradients, and each term describes a further departure of the system from equilibrium.

TRANSPORT COEFFICIENTS

The Chapman-Enskog method yields expressions for dilute-gas transport coefficients which, in the Navier-Stokes approximation, can be written as functions of certain *collision integrals* containing all the effects of two-body interactions. We can write these functions in the following form (e.g., Hirschfelder, *et al*, 1954; p.525):

$$\Omega^{(\ell,s)}(T) \equiv \left(\frac{\pi\kappa T}{2\mu}\right)^{1/2}\beta^{s+2}\int_0^\infty e^{-\beta E}E^{s+1}\,dE\int_0^\infty(1-\cos^\ell X)b\,db, \qquad (1\text{–}46)$$

where μ is the reduced mass of two colliding particles, $\beta \equiv (\kappa T)^{-1}$, and X is the scattering angle for a binary collision with relative total energy E and impact parameter b:

$$X(E,b) \equiv \pi - 2b\int_{r_m}^\infty \frac{r^{-2}\,dr}{\left[1 - \frac{b^2}{r^2} - \frac{\phi(r)}{E}\right]^{1/2}}. \qquad (1\text{–}47)$$

The quantity r_m is the largest zero of the denominator in the integral of Eq.(1–47), and $\phi(r)$ is the spherically-symmetric two-body potential. Because the approximation constitutes only a small departure from thermal equilibrium, T can be given the well-defined meaning of Kelvin temperature of the equilibrium state.

It is customary to normalize the collision integrals by the values for pure hard-sphere interactions, the latter being relatively easy to obtain. Thus, for hard spheres of *diameter a* we define the reduced collision integrals as

$$\Omega^{(\ell,s)*} \equiv \frac{\Omega^{(\ell,s)}}{\Omega_a^{(\ell,s)}}, \tag{1-48a}$$

and in Problem 1.4 it is found that

$$\Omega_a^{(\ell,s)} = \left(\frac{\kappa T}{2\pi\mu}\right)^{1/2} \frac{(s+1)!}{2} \left[1 - \frac{1+(-1)^\ell}{2(\ell+1)}\right] \pi a^2. \tag{1-48b}$$

Then, for a monatomic gas with no internal states, the shear viscosity and thermal conductivity are given respectively by

$$\eta = \frac{5}{16} \frac{(m\kappa T/\pi)^{1/2}}{a^2 \Omega^{(2,2)*}}, \tag{1-49}$$

$$\lambda = \frac{75}{64} \frac{\kappa}{m} \frac{(m\kappa T/\pi)^{1/2}}{a^2 \Omega^{(2,2)*}}, \tag{1-50}$$

and m is now the atomic mass. Note that these expressions are independent of particle density.

A major difficulty in comparing theory and experiment for transport coefficients is that the results depend on both the approximations of the kinetic theory, and on the choice of interaction potential. When X is a smooth function of E and b, as is the case for purely repulsive potentials, the calculation is simply a straightforward numerical integration. But for potentials with attractive wells X possesses singularities, and more careful techniques are required. Programs are now available which extend considerably the class of potentials for which collision integrals can be evaluated (e.g., Rainwater, *et al*, 1982).

Quite generally, the kinetic theory of transport coefficients obtained through the Boltzmann equation for a very dilute gas is felt to be rather satisfactory, particularly in view of its limitations. Comparison of some viscosity data for argon with several potential models is illustrated in Figure 1–2, and the agreement is rather good over a broad range of temperatures. One must ask, however, whether or not this is an artifact of the very-low-density approximation, for η and λ are here independent of density and it may be that any reasonable potential can be adjusted to fit the data over a substantial temperature range. The need to extend the theory somehow to higher densities is clear, despite the fact that the Boltzmann equation itself is fundamentally limited to low densities.

Although Enskog (1922) had carried out an extension of kinetic theory to higher densities which could provide transport coefficients for hard spheres, the first systematic development of kinetic theory toward higher densities is due to Bogoliubov (1946). Beginning with the Liouville equation, Bogoliubov was able to derive the Boltzmann equation and to provide a prescription for obtaining all higher-order density corrections. Just as the Boltzmann equation is restricted to two-body interactions, he showed that succeeding nth order generalizations depend on n-body

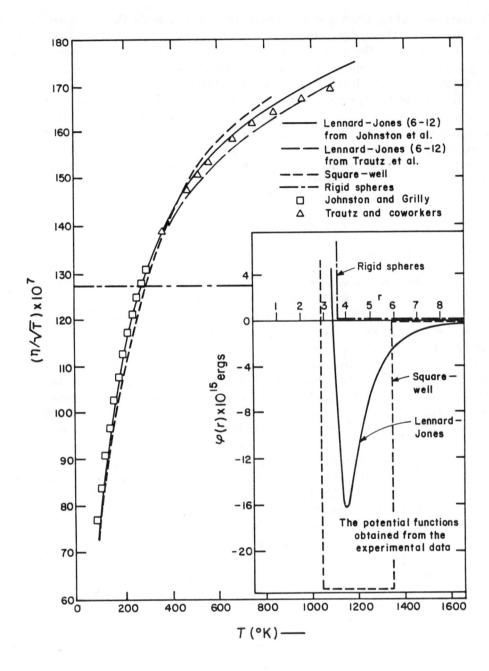

Fig. 1–2. The viscosity coefficient for argon as calculated for several potential models. Data are from H.L. Johnston and E.R. Grilly, *J. Phys. Chem.* **46**, 938(1942); M. Trautz and H.E. Binkele, *Ann. Phys.* **5**, 561(1930); M. Trautz, A. Melster, and R. Zink, *Ann. Phys.* **7**, 409(1930). [*Reproduced with permission from Hirschfelder, et al, 1954.*]

interactions. Thus, Choh and Uhlenbeck were able to study the first correction to the Boltzmann theory by including three-particle interactions (Choh, 1958). Later it was realized that the underlying features of this method were just those of cluster expansions in equilibrium statistical mechanics, an observation exploited in detail by Green and Piccirelli (1963), and Cohen (1962, 1963).

In all these developments there is to be found an analogue of the old *Stoßzahlansatz* which brings into serious question the rigor of the various derivations, but the essential thrust of these efforts was to establish the notion of density, or virial expansions of the transport coefficients. Any such series would doubtless be restricted to moderate densities by practical matters, but the improvement in understanding the phenomena should nevertheless be considerable. If C represents an arbitrary transport coefficient, then we can write

$$C = C_0 + C_1 n + C_2 n^2 + \cdots, \qquad (1\text{--}51)$$

where C_0 comes from the Boltzmann equation and C_k depends on the dynamics of $(k + 2)$ particles.

Unfortunately, it was soon discovered that all the coefficients C_k for $k \geq 2$ were infinite! That is, their analytic expressions were divergent. Parallel to the above generalizations of the Boltzmann equation there also emerged a theory of transport coefficients in terms of time-correlation functions which in principle was valid at any density. But when the so-called Green-Kubo expressions were expanded in a virial series the resulting coefficients were found to be identical with the C_k above, and so the divergences remained. One expects that this equivalence is related to the fact that both approaches rely in a fundamental way on 'local-equilibrium' presumptions. We shall have more to say about the correlation-function formulas in later chapters, where similar expressions will be derived in a rigorous and systematic manner.

The origin of the divergences was located in the four-body and higher interaction terms, and ultimately one began to understand the physical origin as a collective effect. That is, there exist both long-range and persistent correlations in the nonequilibrium fluid that preclude naive density expansions. This phenomenon is familiar from equilibrium systems in which long-range forces are involved, so that the virial expansion is non-analytic. Moreover, the solution is also the same— namely, one must sum the basic divergent terms to all orders so as to yield a finite infinite-order perturbation theory. [See, e.g., Chapter 8 in Volume I.] This procedure was first carried out in the present context by Kawasaki and Oppenheim (1964), resulting in a logarithmic term in the density (as might have been expected from the Coulomb problem). Thus, the conventional belief is that the expansion (1–51) must be rewritten as

$$C = C_0 + C_1 n + \overline{C}_2 n^2 \ln n + C_2 n^2 + \cdots, \qquad (1\text{--}52)$$

in which one can no longer be sure that only $(k + 2)$-particle interactions contribute to C_k. The density expansion of the viscosity has been compared with the data for krypton (van Den Berg and Trappeniers, 1978) and carbon dioxide (Kestin, *et al*, 1980), and in both cases the data are at least consistent with a term in $n^2 \ln n$.

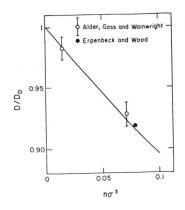

Fig. 1–3. The diffusion coefficient ratio D/D_0 for hard spheres compared with some results of computer simulations. [*Reproduced with permission from Kamgar-Parsi and Sengers (1983).*]

A comprehensive and definitive theory of the coefficients C_k does not yet exist, but some progress has been booked. The general situation regarding calculation of C_1 has been reviewed by Rainwater (1981), and numerous authors have investigated the coefficients for the hard-sphere system. Most of the latter work is referenced by Kamgar-Parsi and Sengers (1983), in which results are given for the coefficients in Eq.(1–52) corresponding to self-diffusion, viscosity, and thermal conductivity. Although experimental data are hard to come by for comparison with this model, Figure 1–3 indicates reasonably good agreement with the results of computer simulations. We shall return to an analysis of the virial expansions of transport coefficients subsequently, but the preceding discussion is sufficient to convey some feeling for the present state of conventional theory.

C. Thermodynamics

Traditionally, the thermodynamic description of nonequilibrium phenomena has evolved by mimicking the equilibrium theory, rather than emerging from a more fundamental foundation. One presumes that elementary subsystems, or volume elements can be chosen such that variation of state variables throughout dV is negligible. The values assigned to the element are taken to be some kind of average over dV, so that once more there is a presumption of local equilibrium. A macroscopic theory of irreversible processes is then based on the entropy function S. We provide only a brief outline here for perspective, and the reader is referred to standard references for further details (e.g., de Groot and Mazur, 1962).

Suppose S is a function of the relevant state variables $\{\alpha_i\}$, defined so that they vanish in a state of thermal equilibrium:

$$dS = \sum_i \frac{\partial S}{\partial \alpha_i}\, d\alpha_i \geq 0. \tag{1–53}$$

One also defines forces

$$X_i \equiv \frac{\partial S}{\partial \alpha_i},$$

(1–54)

and associated fluxes in the nonequilibrium state,

$$J_i \equiv \dot{\alpha}_i,$$

(1–55)

the superposed dot denoting a time derivative. Then the internal production of entropy can be described by

$$\sigma \equiv \frac{dS}{dt} = \sum_i \frac{\partial S}{\partial \alpha_i} \dot{\alpha}_i$$

$$= \sum_i J_i X_i \geq 0.$$

(1–56)

But how are we to define something called 'entropy' in a nonequilibrium state? Up to now this quantity has had an unambiguous meaning only for states of thermal equilibrium, in which the experimental entropy S_E is that of Clausius and $dS_E = dQ/T$. When equilibrium prevails it is also possible to construct a theoretical entropy from Gibbs' algorithm and equate it to S_E (see, e.g., Volume I), but there is not yet any justification for thinking the two are related—or even defined—outside of equilibrium. For this reason the macroscopic theory has generally been felt to be valid only for states close to that of equilibrium, in which case the conceptual problems *might* be neglected to first approximation.

The first systematic attempt at such a theory is due to Onsager (1931a, ˜b), in which he envisions a closed system characterized by the variables $\{\alpha_i\}$. In a neighborhood of equilibrium S can be expanded about the equilibrium value S_0 as follows:

$$S = S_0 - \tfrac{1}{2} G_{ij} \alpha_i \alpha_j + \cdots,$$

(1–57)

where $G = \tilde{G}$ is positive definite and the superposed tilde denotes the transpose. The scenario is that somehow the system is displaced slightly from equilibrium and then allowed to relax back to that state. Equation (1–54) for the forces (entropy gradients) can now be rewritten as

$$X_i = \frac{\partial S}{\partial \alpha_i} = -G_{ij} \alpha_j,$$

(1–58)

which are interpreted as driving the system back to equilibrium.

Further direction is provided by reference to the phenomenological relations discussed earlier, such as those of Fick and Fourier. These all provide linear relations between fluxes and gradients, and of course can be generalized to coupled processes. Thus, Onsager presumed that near equilibrium the following linear homogeneous relationships are valid:

$$J_i = L_{ik} X_k,$$

(1–59)

where the L_{ik} are called the *phenomenological coefficients*. Substitution of J_i into Eq.(1–56) reveals σ to be a positive semi-definite quadratic form, so that identities like $L_{11}L_{22} - L_{12}L_{21} \geq 0$ are valid, for example.

Just as the equilibrium state is the time-invariant ideal state for a closed system, the steady state is the time-invariant ideal state for an open system. Onsager was able to characterize this generalization by noting that for a stationary flow of heat, say, the rate of entropy production σ is a *maximum*. Actually, a special case of this assertion can already be found in a theorem involving Rayleigh's dissipation function (Rayleigh, 1873).

Unfortunately, it is not at all clear at this point that there is a general principle at work here regarding σ. Ziman (1956) discusses at some length a certain class of physical problems for which σ certainly is a maximum. But Keller (1970) formulates and studies some general variational principles of this type and finds mixed results. In particular, he finds some special situations for which σ is a minimum, and shows that in these cases the result follows from known physical laws. Prigogine (1945) had earlier noted that in the near-equilibrium case wherein *some* of the X_i are prescribed σ is a minimum if the remaining X_j are fixed such that their associated fluxes (1–59) vanish. From this he asserted the validity of a 'principle of minimum entropy production' for stationary states near equilibrium. But Landauer (1975) has pointed out an incredibly simple counterexample provided by a battery, linear resistance, and inductance in series: if the voltage is increased slightly both the current and the rate of dissipation *increase* continuously to the new steady state. Because the voltage increase can be made arbitrarily small, the process can be maintained arbitrarily close to equilibrium. Further counterexamples have been discussed by Jaynes (1980), who also observes that Onsager's viewpoint seems to have the right physical 'ring' to it. Rather than minimize σ, the system seems to *maximize* it in order to return to equilibrium as rapidly as possible under whatever constraints are present in opposition to this.

As a consequence of these observations we must conclude that the precise role of σ is neither clear nor rigorous at this time. Some further progress has been made by Tykodi (1967), but much remains to be clarified. We shall return to the problem in later chapters and attempt to establish a somewhat different level for the discussion.

Let us now return to Onsager's presumption (1–59) and continue the development from there. With $\dot{\alpha}_i = J_i$, the implication is that the solutions to Eq.(1–59) simply provide the trajectories along which the α_i relax to zero. Jaynes (1980) has provided a concise matrix description of this process which we follow here, the exercise being rather useful because it illustrates some points which will emerge in a much deeper context later. Thus, if Eq.(1–59) is rewritten as $\dot{\alpha} = -LG\alpha$, then integration yields the description

$$\alpha(t + \tau) = e^{-LG\tau}\alpha(t), \quad \tau > 0. \tag{1–60}$$

Onsager now presumes that the equilibrium distribution of the α_i at equal times has the Boltzmann form $f(\alpha) \propto \exp[\kappa^{-1}S(\alpha)]$, so that the same function providing the forces X_i also supplies the probability distribution of small thermal fluctuations of the α_i about equilibrium. By our previous convention for the α_i, their averages over this distribution must be $\langle \alpha_i \rangle = 0$, whereas their covariances will be denoted by $K_{ij} \equiv \langle \alpha_i \alpha_j \rangle$. One readily shows that K and G are mutually inverse: $KG = GK = \kappa I$, where I is the unit matrix. In order to ascertain how rapidly the α_i fluctuate about zero Onsager makes a third presumption: the average regression of fluctuations follows the law (1–60), so that if $\alpha(t)$ is averaged over many repetitions we have

$$\langle \alpha(t + \tau) \rangle = e^{-LG\tau} \alpha(t), \quad \tau > 0. \tag{1–61}$$

Note that this average is *not* an expectation value computed with the distribution $f(\alpha)$ introduced above. But if we account for both types of average in the symbol $\langle \cdots \rangle$, then we can define a time-dependent covariance matrix:

$$K_{ij}(\tau) \equiv \langle \alpha_i(t + \tau) \alpha_j(t) \rangle = K_{ji}(-\tau), \tag{1–62}$$

and K is actually independent of t.

With the aid of some elementary identities on functions of matrices, combination of Eqs.(1–61) and (1–62) shows that $K(\tau)$ also decays according to the law (1–60):

$$K(\tau) = e^{-LG\tau} K(0) = K(0)e^{-GL\tau}, \quad \tau > 0, \tag{1–63}$$

where $K(0) \equiv \kappa G^{-1}$. Because $K(-\tau) = \tilde{K}(\tau)$, we can also write

$$K(-\tau) = K(0)e^{-G\tilde{L}\tau}, \quad \tau > 0, \tag{1–64}$$

for both G and $K(0)$ are symmetric. This leads finally to Onsager's fourth presumption, that of 'microscopic reversibility': $K(-\tau) = K(\tau)$. Comparison of Eqs.(1–63) and (1–64) then yields the famous *Onsager reciprocity relations*:

$$L = \tilde{L}. \tag{1–65}$$

As Jaynes (1980) has pointed out, these relations have stood the test of time despite the tortuous path taken to obtain them. In the sequel we shall find that they are in fact rather trivial consequences of a more general development.

Applications of the linear theory are now straightforward. For example, if $L_{ij} = 0$ for $i \neq j$ the phenomena are uncoupled and

$$J_i = L_{ii} X_i. \tag{1–66}$$

Thus the fluxes are directly and linearly proportional to forces or gradients, as in the empirical discussion of Section A, so that one readily identifies transport coefficients, and from the work of Section B we obtain a microscopic connection to the macroscopic theory. Some caution must be exercised in identifying transport coefficients with the phenomenological coefficients, however, for consistent definitions for fluxes and forces are required. The dangers in this exercise are illustrated by an example from de Groot and Mazur (1962), where they assert that steady-state heat conduction is characterized by a minimum in the rate of entropy production (despite Onsager's earlier result!). The proof requires that L_{qq} in the relevant version of Eq.(1–66)—$J_q = L_{qq}\nabla(T^{-1})$—be independent of temperature. But if this expression is identified with $J_q = -\lambda\nabla T$, then the thermal conductivity must vary with temperature as T^{-2}. There is, of course, no known substance obeying such a law, leading us to conclude that perhaps general assertions are to be treated with some care in this macroscopic theory.

Coupled processes are described quite naturally by Eqs.(1–57) and (1–65), examples being thermal diffusion and the Dufour effect (involving both temperature and density gradients), and various galvanomagnetic and thermomagnetic effects. Callen (1948) demonstrated specifically how Onsager's macroscopic theory describes the entire body of thermoelectric phenomena (Seebeck, Peltier, and Thomson effects), which we summarize here. A small temperature difference ΔT across a wire gives rise to a heat current q, whereas a small potential difference $\Delta\mathcal{E}$ yields an electric current j. Hence, one has the coupled equations

$$\frac{q}{T} = L_{11}\frac{\Delta T}{T} + L_{12}\frac{\Delta\mathcal{E}}{T},$$
$$j = L_{21}\frac{\Delta T}{T} + L_{22}\frac{\Delta\mathcal{E}}{T}. \tag{1–67}$$

Invoke the reciprocity relations to define a parameter

$$\epsilon \equiv \frac{L_{12}}{L_{22}} = \frac{L_{21}}{L_{22}}, \tag{1–68}$$

called the *Seebeck coefficient*. Then, by setting ΔT and j to zero independently, we obtain two separate physical interpretations:

$$\epsilon = \begin{cases} \frac{1}{T}\frac{q}{j}, & \Delta T = 0 \\ -\frac{\Delta\mathcal{E}}{\Delta T}, & j = 0 \end{cases}. \tag{1–69}$$

All three effects are readily analyzed from this basis. In Chapter 8, however, we shall be able to arrive at these results directly from a comprehensive microscopic theory.

D. Critique

It has been recognized for many years, of course, that the Boltzmann equation is fundamentally limited to very low densities, although this does not seem to have inhibited its use at higher densities. There is also evidence that there exists a broad class of common problems for which the equation is not valid even at low densities (Jaynes, 1971). Bearing these difficulties in mind, one finds it all the more remarkable that reasonable agreement with the data can be obtained for dilute gases. Part of the reason for this success, surely, is that model potentials usually have a few parameters available which can be adjusted so as to obtain a fit. Given this, it is tempting to conclude that for these systems almost any sensible theory might do as long as it includes binary interactions qualitatively, and that the transport coefficients are fairly insensitive to other details. A similar phenomenon occurs for the second virial coefficient in the equilibrium theory.

Whatever the density, the conventional approaches to nonequilibrium phenomena are practicably limited to near-equilibrium scenarios, although the Enskog expansion is supposed to extrapolate away from the equilibrium state. There is no way at this point to judge the validity of the apparent gradient expansion, and subsequent discussions regarding the persistence of correlations in fluids will cast deep doubt on such a presumption. Phenomenologically, of course, a gradient *approximation* close to equilibrium is strongly supported by the facts.

Extensions to higher densities—either beginning with Bogoliubov's generalization, or based on time-correlation functions—are not rooted in entirely solid ground. Both contain presumptions analogous in spirit to, but different from the *Stoßzahlansatz*. In addition, they also contain the hypothesis of 'local equilibrium', which more and more seems to be untenable. This point has been emphasized recently by Evans, *et al*, (1984). On the one hand, when the transport coefficients are expanded straightforwardly in the density, the presence of divergences dictates a resummation of the perturbation theory in order to obtain finite, though non-analytic, results. On the other hand there is nothing particularly novel in this phenomenon, for there are many examples of similar behavior in equilibrium statistical mechanics (see, e.g., Table 8–1 in Volume I). Nevertheless, the naive use of perturbation theory has perhaps provided fair warning that familiar mathematical techniques are not necessarily easily extended to irreversible phenomena. These techniques will come under close scrutiny in what follows.

Nonequilibrium thermodynamics remains essentially a near-equilibrium phenomenology, with no convincing theoretical foundation—at least not one which is generally accepted. Despite shortcomings in its derivation discussed earlier, Onsager's linear theory has provided a useful matrix within which to describe coupled processes in particular, and one has the impression that there is much which is fundamentally correct about it. Indeed, in subsequent chapters we shall be able to put most of it on a much firmer footing.

The major goal of the following chapters will be to resolve many of these problems by providing a single sound foundation for their investigation. At the same time we shall be able to include a study of various other attempts in this direction and see just where they fit in a more general scheme.

With respect to the scenario of classical hydrodynamics, let us recapitulate just what should be expected of statistical mechanics in this regard. A comprehensive theory should (a) establish macroscopic constitutive relations quite naturally in terms of microscopic dynamics, (b) establish rigorously the macroscopic conservation laws as consequences of the well-understood microscopic laws, (c) provide a clear and unobjectionable derivation of the macroscopic equations of motion, and (d) illustrate precisely how a full nonlinear theory can be constructed. These questions, among a number of others, will be addressed in the following chapters.

Problems

1.1 The vorticity vector of Eq.(1–12) is often defined without the factor of $1/2$, in which case we write $\boldsymbol{\omega} \equiv \nabla \times \mathbf{v}$. Consider an incompressible fluid ($\nabla \cdot \mathbf{v} = 0$) of constant density n and constant viscosity η, so that the kinematic viscosity $\nu \equiv \eta/mn$ is also constant.

    (a) Show from Eqs.(1–42b) and (1–45a), say, that the Navier-Stokes equations are

$$\frac{\partial \mathbf{v}}{\partial t} + (\mathbf{v} \cdot \nabla)\mathbf{v} = -\nabla\left(\frac{p}{mn}\right) + \nu\nabla^2\mathbf{v}\,.$$

    (b) Use these equations to obtain the following equation of motion for $\boldsymbol{\omega}$:

$$\frac{\partial \boldsymbol{\omega}}{\partial t} + (\mathbf{v} \cdot \nabla)\boldsymbol{\omega} = (\boldsymbol{\omega} \cdot \nabla)\mathbf{v} + \nu\nabla^2\boldsymbol{\omega}\,.$$

1.2 Let $\overline{P}$ be the mechanical pressure in a medium, and P the thermodynamic pressure. By means of Eq.(1–23), and the fact that $-P = \frac{1}{3}(p_{11} + p_{22} + p_{33})$, show that

$$P - \overline{P} = \varsigma\frac{\partial v_k}{\partial x_k}\,,$$

summation convention implied.

1.3 By evaluating the first three (0,1,2) velocity moments of the Boltzmann equation, derive the conservation laws of Eqs.(1–42).

1.4 Verify Eq.(1–48b) for the hard-sphere collision integrals.

REFERENCES

Bird, R.B., and C.F. Curtiss: 1984, 'Fascinating Polymeric Liquids', *Physics Today*, January, p.36.

Bogoliubov, N.N.: 1946, 'Expansion into a Series of Powers of a Small Parameter in the Theory of Statistical Equilibrium; Kinetic Equations', *J. Phys. U.S.S.R.* **10**, 257, 265.

Burnett, D.: 1935, 'The Distribution of Molecular Velocities and the Mean Motion in a Non-Uniform Gas', *Proc. London Math. Soc.* **40**, 382.

Cauchy, A.-L.: 1823, 'Recherches sur l'équilibre et le mouvement interieur des corps solides ou fluides, élastiques ou non élastiques', *Bull. Soc. Philomath.* , 9.

Chapman, S.: 1916-17, 'The Kinetic Theory of Simple and Composite Monatomic Gases: Viscosity, Thermal Conduction, and Diffusion', *Proc. Roy. Soc. London* **93**, 1.

Chapman, S., and T.G. Cowling: 1952, *The Mathematical Theory of Non-Uniform Gases*, 2nd ed., Cambridge Univ. Press, Cambridge.

Choh, S.T.: 1958, 'The Kinetic Theory of Phenomena in Dense Gases', *Ph.D. thesis*, Univ. of Michigan (unpublished).

Cohen, E.G.D.: 1962, 'On the Generalization of the Boltzmann Equation to General Order in the Density', *Physica* **28**, 1025.

Cohen, E.G.D.: 1962, 'Cluster Expansions and the Hierarchy. I.Non-Equilibrium Distribution Functions', *Physica* **28**, 1045.

Cohen, E.G.D.: 1962, 'Cluster Expansions and the Hierarchy. II.Equilibrium Distribution Functions', *Physica* **28**, 1060.

Cohen, E.G.D.: 1963, 'On the Kinetic Theory of Dense Gases', *J. Math. Phys.* **4**, 183.

Coleman, B.D., and W. Noll: 1961, 'Foundations of Linear Viscoelasticity', *Rev. Mod. Phys.* **33**, 239.

de Groot, S.R., and P. Mazur: 1962, *Non-Equilibrium Thermodynamics*, North-Holland, Amsterdam.

Einstein, A.: 1905, 'Über die von der molekularkinetischen Theorie der Wärme geforderte Bewegung von in ruhenden Flüssigkeiten suspendierten Teilchen', *Ann. d. Phys.* **17**, 549.

Enskog, D.: 1917, *Kinetische Theorie der Vorgänge in Mäßig verdünten Gasen*, Almqvist and Wiksells, Uppsala.

Enskog, D.: 1922, 'Kinetische Theorie der Wärmeleitung, Reibung und Selbstdiffusion in gewißen verdichteten Gasen und Flüssigkeiten', *Kungl. Sr. Vetenskapsakad. Handl.* **63**, No.4.

Evans, D., H.J.M. Hanley, and S. Hess: 1984, 'Non-Newtonian Phenomena in Simple Fluids', *Physics Today*, January. p.26.

Fick, A.: 1855, *Pogg. Ann.* **94**, 59.

Fourier, J.: 1822, *Théorie Analytique de la Chaleur*, Paris.

Fredrickson, A.G.: 1964, *Principles and Applications of Rheology*, Prentice-Hall, Englewood Cliffs, NJ.

Green, M.S., and R.A. Piccirelli: 1963, 'Basis of the Functional Assumption in the Theory of the Boltzmann Equation', *Phys. Rev.* **132**, 1388.

Hagen, G.: 1839, *Pogg. Ann.* **46**, 423.

Hirschfelder, J.O., C.F. Curtiss, and R.B. Bird: 1954, *Molecular Theory of Gases and Liquids*, Wiley, New York.

Hooke, R.: 1676, *A Description of Helioscopes and Some Other Instruments*, London. [Reprinted in: R.T. Gunther, *Early Sciences in Oxford* **8**, 119(1931)].

Jaynes, E.T.: 1971, 'Violation of Boltzmann's H-Theorem in Real Gases', *Phys. Rev.* **A4**, 747.

Jaynes, E.T.: 1980, 'The Minimum Entropy Production Principle', *Ann. Rev. Phys. Chem.* **31**, 579.

Kamgar-Parsi, B., and J.V. Sengers: 1983, 'Logarithmic Density Dependence of the Transport Properties of Gases', *Phys. Rev. Letters* **51**, 2163.

Kawasaki, K., and I. Oppenheim: 1965, 'Logarithmic Term in the Density Expansion of Transport Coefficients', *Phys. Rev.* **139**, A1763.

Keller, J.B.: 1970, 'Extremum Principles for Irreversible Processes', *J. Math. Phys.* **11**, 2919.

Kestin, J., Ö. Korfali, and J.V. Sengers: 1980, 'Density Expansion of Carbon Dioxide Near the Critical Point', *Physica* **100A**, 335.

Landauer, R.: 1975, 'Inadequacy of Entropy and Entropy Derivatives in Characterizing the Steady State', *Phys. Rev. A* **12**, 639.

Lanford, O.E., III: 1975, 'Time Evolution of Large Classical Systems', in J. Moser (ed.), *Dynamical Systems, Theory and Applications (Lecture Notes in Physics, Vol.38)*, Springer-Verlag, Berlin.

Martin, P.C., O. Parodi, and P.S. Pershan: 1972, 'Hydrodynamic Theory for Crystals, Liquid Crystals, and Normal Fluids', *Phys. Rev. A* **6**, 2401.

Maxwell, J.C.: 1866, 'Viscosity or Internal Friction of Air and Other Gases', *Phil. Trans. Roy. Soc. (London)* **A156**, 249.

Maxwell, J.C.: 1867, 'On the Dynamical Theory of Gases', *Phil. Trans. Roy. Soc. (London)* **A157**, 49.

Nernst, W.: 1888, *Z. Physik Chem.* **2**, 613.

Onsager, L.: 1931a, 'Reciprocal Relations in Irreversible Processes.I', *Phys. Rev.* **37**, 405.

Onsager, L.: 1931b, 'Reciprocal Relations in Irreversible Processes.II', *Phys. Rev.* **38**, 2265.

Poiseuille, J.: 1841, 'Recherches expérimentales sur le mouvement des liquides dans les tubes de très petits diamètres', *Compt. Rend.* **12**, 112.

Prigogine, I.: 1945, *Bull. Acad. Belgium Class Sci.* **31**, 600.

Rainwater, J.C.: 1981, 'Softness Expansion of Gaseous Transport Properties. II.Moderately Dense Gases', *J. Chem. Phys.* **74**, 4130.

Rainwater, J.C., P.M. Holland, and L. Biolsi: 1982, 'Binary Collision Dynamics and Numerical Evaluation of Dilute Gas Transport Properties for Potentials with Multiple Extrema', *J. Chem. Phys.* **77**, 434.

Reiner, M.: 1958, 'Rheology', in S. Flügge (ed.), *Handbuch der Physik, Vol.VI*, Springer-Verlag, Berlin, p.434.

Truesdell, C.: 1952, 'The Mechanical Foundations of Elasticity and Fluid Dynamics', *J. Ratl. Mech. Anal.* **1**, 125.

Van Den Berg, H.R., and N.J. Trappeniers: 1978, 'Experimental Determination of the Logarithmic Term in the Density Expansion of the Viscosity Coefficient of Krypton at 25°C', *Chem. Phys. Letters* **58**, 12.

Ziman, J.M.: 1956, 'The General Variational Principle of Transport Theory', *Can. J. Phys.* **34**, 1256.

Chapter 2

General Theory of Nonequilibria

The modern era in nonequilibrium statistical mechanics begins about 40 years ago, essentially at mid-century. A major advance toward an understanding of irreversibility was the discovery by Kirkwood (1946), and Callen and Welton (1951), that a general theory of linear irreversible processes could be approached by means of fluctuations about the equilibrium state of the thermodynamic system. This suggestion is inherent in the Nyquist theorem and its explanation of Johnson noise, as is very well known (Nyquist, 1928; Johnson, 1928). The idea was subsequently confirmed for dynamical perturbations starting from equilibrium by Green (1952, 1954), Kubo (1957), and Mori (1958), and then extended to parts of the nonlinear domain by Bernard and Callen (1959). A complete nonlinear formalism was eventually developed to encompass arbitrary departures from equilibrium (Peterson, 1967).

Although these formulations are undoubtedly correct and of direct utility in studying *some* aspects of the nonequilibrium problem, they are still limited in generality. For example, they are used mainly to describe reasonably small departures from a known equilibrium state, and are restricted to nonequilibrium situations generated by the addition of an explicit driving force to the system Hamiltonian. That is, the perturbation of the equilibrium system is presumed to be a clearly specified dynamical interaction. These are certainly important processes, but a complete theory must surely encompass as well situations in which the initial state consists of an already-established temperature or density gradient, say, whose descriptions in terms of dynamical perturbations is at least somewhat unconvincing. Of course it is often possible to construct fictitious Hamiltonians which are presumed to describe such processes, and this is sometimes done in order to surmount the latter difficulty. Examples are provided by Kadanoff and Martin (1963), Luttinger (1964), and Puff and Gillis (1968). But these procedures are clearly artificial, and at best not completely unambiguous, as will be discussed in some detail in Chapter 5. This observation had already been made earlier by Mori (1956). Furthermore, the interesting aspects of irreversible processes beginning from an arbitrary nonequilibrium state, and not obviously describable in terms of a Hamiltonian, are often nonlinear and nonlocal, and a viable theory must automatically incorporate these features. Immediate examples are plentiful: astrophysical systems which, in the entire history of the universe, have never been close to equilibrium; ultrasonic attenuation; and, general viscoelastic media. It is now clear that such a theory can be based straightforwardly on Eqs.(P–2)-(P–4), to a detailed study of which we now turn.

A. Theory of the Partition Functional

As noted earlier, the derivation of Eqs.(P–2)-(P–4) is just an extension of that for the canonical equilibrium statistical operator based on the principle of maximum entropy (PME). Justification for the latter procedure and its relationship to the original ideas of Gibbs have been discussed at great length in Volume I, say, to which the reader is referred for further detail. Let us now for simplicity consider the single Heisenberg operator $\hat{F}(t)$, whose general time development is described by the unitary transformation

$$\hat{F}(t) = \hat{U}^{\dagger}(t, t_0)\hat{F}(t_0)\hat{U}(t, t_0). \tag{2-1}$$

The time-development operators are solutions of the equation of motion

$$i\hbar \frac{d\hat{U}(t, t_0)}{dt} = \hat{\mathsf{H}}(t)\hat{U}(t, t_0), \tag{2-2}$$

subject to the initial condition $\hat{U}(t_0, t_0) = \hat{1}$. Although the statistical operator, or density matrix, $\hat{\rho}(t_0)$, remains stationary in the Heisenberg picture, this is conventionally taken to coincide with the Schrödinger picture at $t = t_0$. In the latter $\hat{\rho}$ evolves in time according to the prescription

$$\hat{\rho}(t) = \hat{U}(t, t_0)\hat{\rho}(t_0)\hat{U}^{\dagger}(t, t_0). \tag{2-3}$$

That is, $\hat{U}$ maps the Hilbert space of possible states onto itself. As discussed in Volume I, say, where we reviewed the elements of quantum dynamics, $\hat{\rho}(t)$ is a solution of the equation of motion

$$i\hbar \frac{d\hat{\rho}(t)}{dt} = [\hat{\mathsf{H}}(t); \hat{\rho}(t)]. \tag{2-4}$$

Should $\hat{\mathsf{H}}$ not be explicitly dependent on the time, Eq.(2–2) has the solution

$$\hat{U}_0(t, t_0) = e^{i(t-t_0)\,\hat{\mathsf{H}}\,/\hbar}. \tag{2-5}$$

In general, however, it is extremely difficult to find an expression for $\hat{U}$ in closed form.

Suppose now that at $t = t_0$ we know the expectation value of $\hat{F}$, which is defined as in Eq.(P–5):

$$\langle \hat{F}(t_0) \rangle \equiv \text{Tr}(\hat{\rho}_0 \hat{F}_0), \tag{2-6}$$

and where $\hat{\rho}_0 \equiv \hat{\rho}(t_0)$, $\hat{F}_0 \equiv \hat{F}(t_0)$. This information, along with the normalization condition $\text{Tr}\,\hat{\rho}_0 = 1$, leads to the maximum-entropy estimate

$$\hat{\rho}_0 = \frac{1}{Z_0}e^{\lambda_0 \hat{F}_0}, \quad Z_0 = \text{Tr}\,e^{\lambda_0 \hat{F}_0}. \tag{2-7}$$

The Lagrange multiplier is determined by substitution of $\hat{\rho}_0$ into the equation of constraint (2–6).

At time $t = t_1 > t_0$ a new expectation value is obtained:

$$\langle \hat{F}(t_1) \rangle = \mathrm{Tr}(\hat{\rho}_1 \hat{F}_1) \neq \langle \hat{F}(t_0) \rangle \, . \tag{2–8}$$

The PME now utilizes both pieces of data, Eqs.(2–6) and (2–8), to give the new estimate

$$\hat{\rho}_1 = \frac{1}{Z_1} e^{\lambda_0 \hat{F}_0 + \lambda_1 \hat{F}_1} \, , \quad Z_1 = \mathrm{Tr}\, e^{\lambda_0 \hat{F}_0 + \lambda_1 \hat{F}_1} \, . \tag{2–9}$$

If the procedure is repeated n times, such that $t_{i+1} > t_i$, then at $t = t_n$

$$\hat{\rho}_n = \frac{1}{Z_n} \exp\left(\sum_{i=0}^{n} \lambda_i \hat{F}_i \right) \, , \quad Z_n = \mathrm{Tr} \exp\left(\sum_{i=0}^{n} \lambda_i \hat{F}_i \right) \, . \tag{2–10}$$

In the event that we have continuous information on the expectation value of $\hat{F}(t)$, we can set $t_n = \tau$, take all the time intervals equal and n very large, and in the Riemann limit obtain the replacement

$$\sum_{i=0}^{n} \lambda_i \hat{F}_i \longrightarrow \int_{t_0}^{\tau} \lambda(t) \hat{F}(t) \, dt \, , \tag{2–11}$$

where now $\lambda(t)$ is an undetermined Lagrange-multiplier function. The equation of constraint in the interval (t_0, τ) determines $\lambda(t)$ by functional differentiation, and $Z[\lambda(t)]$ has become the *partition functional*.

Note that this derivation in no way depends upon the interpretation of t as time, so that t can actually be taken as any parameter capable of being ordered monotonically. Thus the extension to Eqs.(P–2)-(P–5) is straightforward and the generalized Gibbs algorithm is established. These latter equations constitute, from our point of view, the most general prescription of statistical mechanics possible based on the given data. Because we shall want to refer to them frequently, let us restate them here. Given several pieces of data $\langle \hat{F}_k(\mathbf{x}, t) \rangle$ over space-time regions $R_k(\mathbf{x}, t)$, the initial statistical operator encompassing only this information and maximizing the entropy is

$$\hat{\rho} = \frac{1}{Z} \exp\left[\sum_k \int_{R_k} \lambda_k(\mathbf{x}, t) \hat{F}_k(\mathbf{x}, t) \, d^3 x \, dt \right] \, , \tag{2–12}$$

where

$$Z[\{\lambda_k\}] = \mathrm{Tr} \exp\left[\sum_k \int_{R_k} \lambda_k(\mathbf{x}, t) \hat{F}_k(\mathbf{x}, t) \, d^3 x \, dt \right] \, , \tag{2–13}$$

is the partition functional. The Lagrange-multiplier functions $\lambda_k(\mathbf{x}, t)$ are identified from the initial data by means of the coupled functional differential equations

$$\langle \hat{F}_k(\mathbf{x}, t) \rangle \equiv \mathrm{Tr}[\hat{\rho} \hat{F}_k(\mathbf{x}, t)] = \frac{\delta}{\delta \lambda_k(\mathbf{x}, t)} \ln Z \, , \quad (\mathbf{x}, t) \in R_k \, , \tag{2–14}$$

whereas the predicted expectation of any other Heisenberg operator $\hat{J}$ at $(\mathbf{x}, t)$ is

$$\langle \hat{J}(\mathbf{x}, t) \rangle = \mathrm{Tr}[\hat{\rho}\hat{J}(\mathbf{x}, t)] = \mathrm{Tr}[\hat{\rho}(t)\hat{J}(\mathbf{x})]. \tag{2--15}$$

The maximum entropy itself is a functional of the initial values $F_k(\mathbf{x}, t) \equiv \langle \hat{F}_k(\mathbf{x}, t) \rangle \in R_k$, and follows directly from Eq.(2--12):

$$S[\{F_k(\mathbf{x}, t)\}] = -\kappa \, \mathrm{Tr}(\hat{\rho} \ln \hat{\rho})$$
$$= \kappa \ln Z[\{\lambda_k(\mathbf{x}, t)\}] + \kappa \sum_k \int \lambda_k(\mathbf{x}, t) F_k(\mathbf{x}, t) \, d^3x \, dt. \tag{2--16}$$

As might be expected from such general considerations, the entropy functional plays a broader role here than it did in equilibrium. The most obvious point, of course, is that the presence of time-dependent data imparts a natural time dependence to S. This will be explored in more detail subsequently, and we shall see that its maximization in order to obtain an initial $\hat{\rho}_0$ is only one of many possible scenarios in which S plays a role. Note, though, that here S is a measure of all microscopic states consistent with the history of a system over the $R_k(\mathbf{x}, t)$ as defined by the measured values $F_k(\mathbf{x}, t)$. That is, one can visualize the evolution of a microstate as a path in 'phase space-time', so that S is the cross section of a tube formed by all paths by which the given history could have been realized. For this reason, Jaynes (1980) has referred to S for any particular history as the *caliber* of that history. Thus, one could also refer to a 'maximum caliber principle'.

In a real sense $S[\{F_k(\mathbf{x}, t)\}]$ as given by Eq.(2--16) governs the theory of irreversible processes in much the same way as the Lagrangian governs mechanical processes. The associated variational principle determines the states of stable equilibrium as a special case, but more generally we now see that it actually determines equations of motion for nonequilibrium processes. Observe carefully, though, that this theoretical construct may or may not have anything to do with some experimental quantity called 'entropy'—such a quantity has generally only been defined for equilibrium states.

Prior to proceeding with further developments, it is useful for the historical record to point out that fragments of the general idea contained in Eq.(2--12) had been noticed earlier. For example, in 1951 Bergmann attempted to construct a relativistically invariant statistical mechanics by extending the canonical ensemble of Gibbs (Bergmann, 1951). His modification in terms of relativistic invariants lacked any further generality, however, despite subsequent developments in other directions (Bergmann and Lebowitz, 1955; Lebowitz and and Bergmann, 1957). Several years later McLennan (1959) developed an ensemble described by a phase-space distribution function that is a functional of the nonconservative forces in the system. Although his method of derivation is somewhat restrictive and difficult to appreciate, it does lead to time-correlation expressions for the dissipative coefficients of linear transport processes. But, again, the technique lacks complete generality and seems to contain a number of arbitrary presumptions. This and other approaches to time-correlation descriptions have been reviewed extensively by Zwanzig (1965).

More recently, Zubarev (1974) and his collaborators have discovered an expression for the statistical operator that bears a resemblance to that of Eq.(2–12), but has neither the content nor the full generality of the latter expression. He constructs an ensemble somewhat in the manner of Bergmann, but employing conserved operators of the system having generalized thermodynamic coefficients. The resulting $\hat{\rho}(t)$ satisfies the equation of motion (2–4) and can be shown to maximize the entropy subject to constraints imposed at either the infinitely remote past or future. Although this approach contains some elements of the philosophy advocated here, only in a very limited sense does it address the problem of stage (i) of the Prologue. To the extent that it does, it is a rather special case of the present theory. Indeed, one gains the impression that the $\hat{\rho}(t)$ constructed by Zubarev is intended to adequately describe a nonequilibrium system for all time, and that now seems an unrealistic goal. We return to this point presently.

It is rather important to distinguish the theory being discussed here from another class which bears a superficial similarity to it—namely, the so-called *local equilibrium theories*. Basically, one envisions data $F_k(\mathbf{x})$ available at one particular time t, and then constructs a local-equilibrium statistical operator $\hat{\rho}_t$ based on this one-time information. Because S_t provides a measure of all microscopic states compatible with the given data at a single time t, it is independent of any one history for the macroscopic state. In fact, S_t accounts for all possible histories, irrespective of constraints provided by the data! Thus, all memory of the past is discarded and the theory can not possibly encompass things such as fading memory effects.

One strange consequence of this local-equilibrium formulation is the existence of a 'time delay' before currents settle down to some steady, or 'plateau' values. Having reached such values, though, one can now calculate near-equilibrium transport coefficients, and the modifications appropriate to encompass some kind of memory terms were developed by Robertson (1966, 1978). This work was based on projection-operator techniques, employed by Zwanzig (1961) and Mori (1965a, b), as well. Grabert (1982) has provided a more recent review of these methods.

Let us return to Eqs.(2–12) and (2–13), which describe the initial state of the system based on the data obtained in regions R_k, and which characterize the system only in those regions of space and time. This statistical operator is *not* a function of space and time, in general—it merely provides an initial condition which, when coupled with the equations of motion, determines the evolution of $\hat{\rho}$ outside the regions R_k (but see Chapter 7 for a generalization). Lack of additional information outside these regions—in the future, say—renders $\hat{\rho}$ less and less reliable, but it is still the best estimate possible based on the available data. Without new information about the present there is no way that $\hat{\rho}(t)$ can be thought to give an exact description of the 'actual' situation at present, without new information. As we shall see later, $\hat{\rho}$ *can* be made a function of time if considerable care is exercised.

For a closed system in thermal equilibrium the characteristic operators are constants of the motion, and continuous measurement yields the same data, as is well known. Thus the full integral in Eq.(2–12) contains redundant information and is equivalent to $\lambda \hat{F}$, where $\hat{F}$ is the total system operator and λ is just an undetermined constant. More generally, in this case

$$\int_{R_k} \lambda_k(\mathbf{x}, t) \hat{F}_k(\mathbf{x}, t) \, d^3x \, dt \quad \longrightarrow \quad \lambda_k \hat{F}_k \,. \tag{2-17}$$

The most familiar example of the resulting equilibrium theory occurs when there is only one operator of interest, the Hamiltonian $\hat{H} \neq \hat{H}(t)$. The Lagrange multiplier is then well known to be $\lambda \equiv \beta = (\kappa T)^{-1}$, where κ is Boltzmann's constant and T is the absolute temperature. Then,

$$\hat{\rho}_0 = \frac{1}{Z_0(\beta)} e^{-\beta \hat{H}} \,, \qquad Z_0(\beta) = \text{Tr} \, e^{-\beta \hat{H}} \,, \tag{2-18a}$$

and the expectation value of any other operator is given by

$$\langle \hat{J} \rangle \equiv \text{Tr}(\hat{\rho}_0 \hat{J}) \,. \tag{2-18b}$$

Of course, the equilibrium theory is much richer than just this simple example would indicate. By incorporating initial data on particle number into the calculation, one obtains the familiar equations describing the grand canonical ensemble, for example. Not as familiar is Gibbs' demonstration of how to incorporate angular momentum into the ensemble (Gibbs, 1902; p.39), a problem considered earlier by Maxwell (1878). Thus, if the given data are the total energy of the system, $E = \langle \hat{H} \rangle_0$, and a component of total angular momentum, $M_i \equiv \langle \hat{J}_i \rangle_0$, then the *rotational ensemble* is described by

$$\hat{\rho}_r = \frac{1}{Z_r} e^{-\beta(\hat{H} - \omega_i \hat{J}_i)} \qquad Z_r = \text{Tr} \, e^{-\beta(\hat{H} - \omega_i \hat{J}_i)} \,, \tag{2-19}$$

where ω_i is the ith component of the uniform and constant angular velocity ω.

Perhaps the simplest step away from equilibrium theory is to irreversible processes which result from a dynamical perturbation of what would otherwise be an equilibrium situation. That is, the departure from equilibrium is caused entirely by explicitly known terms of the form $\hat{V}(t)$ in the Hamiltonian. In the most common of theses circumstances the system is described by the constant time-independent Hamiltonian $\hat{H}_0$ for all times $t < t_0$. At time t_0 one or more conservative forces $F_j(t)$ are applied to the system, so that the Hamiltonian becomes

$$\hat{H}_t = \hat{H}_0 - \sum_j \hat{A}_j F_j(t) \,, \qquad t > t_0 \,, \tag{2-20}$$

where the $\hat{A}_j$ are linear operators coupling the external forces to the system.

Reflection indicates that the only information available in such problems is that supplied by Eq.(2–20), as well as the knowledge that the system has been in equilibrium in the (possibly remote) past. Consequently, the initial statistical operator can only be the equilibrium quantity described by Eq.(2–18a), or the corresponding operators for other equilibrium ensembles for which we can define effective Hamiltonians. This is also the view adopted by Bernard and Callen (1959). The system evolves in time, however, in a well-defined way described by Eqs.(2–1) and (2–2), so that explicit knowledge of the Hamiltonian (2–20) allows predictions of expectation values of other operators for times $t > t_0$. The expectation values

$$\langle \hat{J}(t) \rangle = \text{Tr}[\hat{\rho}_0 \hat{J}(t)] \tag{2–21}$$

are no longer constant, because $\hat{U}$ no longer commutes with $\hat{\rho}_0$. A perturbation of the time-evolution operators then yields a series of approximations corresponding to the various orders of what is conventionally called dynamic response theory. This aspect of the formalism serves to re-emphasize that nothing more can be learned about a system than what is gained from experiment in the form of additional measurements.

The generalized Gibbs algorithm described by Eqs.(2–12)-(2–15) is extraordinarily rich, as well as rather subtle, and the remainder of this volume is devoted to extracting from it a number of these riches. Basically, we shall focus on four major scenarios. The first of these is the general, or *standard scenario* described as is by Eqs.(2–12)-(2–15), based on data from some space-time region which can characterize an arbitrary initial nonequilibrium state. Zwanzig (1983) has written that "there seems to be a considerable insensitivity to details of the initial state. How can we see that this is so from purely theoretical arguments?" Whether or not this is so, we submit that Eq.(2–12) provides just that theoretical expression needed to study this question. Fundamentally this scenario is one of relaxation, both logical and physical. Pending the acquisition of further information, the predictions of expectation values must become less reliable as time passes.

Some of those additional data, of course, could specify a steady or equilibrium state, in which case the reliability of our predictions increases enormously. When defined carefully, the *steady-state scenario* provides a time-invariant picture of an open system and is often realizable experimentally.

Much more interesting are driven processes, for which part of our knowledge is that the system is being driven. The *thermal-driving scenario* is perhaps the richest of all and will be studied in detail in Chapter 7, and subsequently. It is what describes the process of heating water in a kettle discussed earlier, but in principle refers to arbitrary driving sources. *Dynamic driving* has already been formulated, in Eq.(2–20). A sub-scenario of some interest is that in which a system is driven from a known equilibrium state, either thermally or dynamically, and allowed to relax back to equilibrium. We shall discuss both of these processes in later chapters. Now, however, we take up a more detailed study of the dynamic response of a physical system to a known external stimulus, for this is the easiest of these scenarios to investigate.

REFERENCES

Bergmann, P.G.: 1951, 'Generalized Statistical Mechanics', *Phys. Rev.* **84**, 1026.

Bergmann, P.G., and J.L. Lebowitz: 1955, 'New Approach to Nonequilibrium Processes', *Phys. Rev.* **99**, 578.

Bernard, W., and H.B. Callen: 1959, 'Irreversible Thermodynamics of Nonlinear Processes and Noise in Driven Systems', *Rev. Mod. Phys.* **31**, 1017.

Callen, H.B., and T.A. Welton: 1951, 'Irreversibility and Generalized Noise', *Phys. Rev.* **83**, 34.

Gibbs, J.W.: 1902, *Elementary Principles in Statistical Mechanics*, Yale Univ. Press, New Haven.

Grabert, H.: 1982, *Projection Operator Techniques in Nonequilibrium Statistical Mechanics*, Springer-Verlag, Berlin.

Green, M.S.: 1952, 'Markoff Random Processes and the Statistical Mechanics of Time-Dependent Phenomena', *J. Chem. Phys.* **20**, 1281.

Green, M.S.: 1954, 'Markoff Random Processes and the Statistical Mechanics of Time-Dependent Phenomena. II.Irreversible Processes in Fluids', *J. Chem. Phys.* **22**, 398.

Jaynes, E.T.: 1980, 'The Minimum Entropy Production Principle', *Ann. Rev. Phys. Chem.* **31**, 579.

Johnson, J.B.: 1928, 'Thermal Agitation of Electricity in Conductors', *Phys. Rev.* **32**, 97.

Kadanoff, L.P., and P.C. Martin: 1963, 'Hydrodynamic Equations and Correlation Functions', *Ann. Phys. (N.Y.)* **24**, 419.

Kirkwood, J.G.: 1946, 'The Statistical Mechanical Theory of Transport Processes', *J. Chem. Phys.* **14**, 180.

Kubo, R.: 1957, 'Statistical Mechanical Theory of Irreversible Processes.I', *J. Phys. Soc. Japan* **12**, 570.

Lebowitz, J.L., and P.G. Bergmann: 1957, 'Irreversible Gibbsian Ensembles', *Ann. Phys. (N.Y.)* **1**, 1.

Luttinger, J.M.: 1964, 'Theory of Thermal Transport Coefficients', *Phys. Rev.* **135**, A1505.

Maxwell, J.C.: 1878, 'On Boltzmann's Theorem on the Average Distribution of Energy in a System of Material Points', *Camb. Phil. Trans.* **12**, 547.

McLennan, J.A., Jr.: 1959, 'Statistical Mechanics of the Steady State', *Phys. Rev.* **115**, 1405.

Mori, H.: 1956, 'A Quantum-Statistical Theory of Transport Processes', *J. Phys. Soc. Japan* **11**, 1029.

Mori, H.: 1958, 'Statistical-Mechanical Theory of Transport in Fluids', *Phys. Rev.* **112**, 1829.

Mori, H.: 1965a, 'Transport, Collective Motion, and Brownian Motion', *Prog. Theor. Phys.* **33**, 423.

Mori, H.: 1965b, 'A Continued-Fraction Representation of the Time-Correlation Functions', *Prog. Theor. Phys.* **34**, 399.

Nyquist, H.: 1928, 'Thermal Agitation of Electric Charge in Conductors', *Phys. Rev.* **32**, 110.

Peterson, R.L.: 1967, 'Formal Theory of Nonlinear Response', *Rev. Mod. Phys.* **39**, 69.

Puff, R.D., and N.S. Gillis: 1968, 'Fluctuations and Transport Properties of Many-Particle Systems', *Ann. Phys. (N.Y.)* **46**, 364.

Robertson, B.: 1966, 'Equations of Motion in Nonequilibrium Statistical Mechanics', *Phys. Rev.* **144**, 151.

Robertson, B.: 1978, 'Application of Maximum Entropy to Nonequilibrium Statistical Mechanics', in R.D. Levine and M. Tribus (eds.), *The Maximum Entropy Formalism*, M.I.T. Press, Cambridge, p.289.

Zubarev, D.N.: 1974, *Nonequilibrium Statistical Thermodynamics*, Consultants Bureau, New York.

Zwanzig, R.: 1961, 'Statistical Mechanics of Irreversibility', in W.E. Brittin, B.W. Downs, and J. Downs (eds.), *Lectures in Theoretical Physics, Vol.III*, Interscience, New York.

Zwanzig, R.: 1965, 'Time-Correlation Functions and Transport Coefficients in Statistical Mechanics', *Ann. Rev. Phys. Chem.* **16**, 67.

Zwanzig, R.: 1983, 'Where Do We Go from Here?', in H.J. Raveché (ed.), *Perspectives in Statistical Physics*, North-Holland, Amsterdam, p.124.

Chapter 3

Theory of Dynamical Response

In developing the nonequilibrium theory we shall first examine some of the more familiar aspects in which a well-defined dynamical perturbation drives the system from equilibrium. This serves not only as a review of a part of the formalism that has been worked out in great detail in recent years, but also provides a guide for exploring those features of the general theory which are less well known.

If a system is perturbed dynamically by means of an explicit addition to the Hamiltonian, then the full Hamiltonian takes the form of Eq.(2–20):

$$\hat{H}_t = \hat{H}_0 - \sum_j \hat{A}_j F_j(t), \quad t > t_0, \tag{3-1}$$

where the $\hat{A}_j$ are linear operators through which the external forces couple to the system. As noted in the preceding chapter, if the system was in thermal equilibrium in the remote past the statistical operator can only be taken as that describing the equilibrium system: $\hat{\rho}_0$. Time development, however, is now governed by the full Hamiltonian of Eq.(3–1). This implies that expectation values are no longer constant, for the time-development operators $\hat{U}$ of Eq.(2–1) no longer commute with $\hat{\rho}_0$.

It is sometimes convenient to suppose that the external forces are applied continuously and forever. This is occasionally accomplished by introducing a mathematical device known as 'adiabatic switching', which brings the system smoothly from the equilibrium state to the perturbed state. A factor $\exp(-\epsilon|t|)$ is inserted everywhere and the positive quantity ϵ is set equal to zero at the end of the calculation. This is often necessary to ensure convergence of time integrals with infinite limits, and has the additional merit of ensuring that an exact eigenstate of $\hat{H}_t$ is generated from each eigenstate of $\hat{H}_0$, as is well known (Gell-Mann and Low, 1951). We shall rarely include this factor in what follows, but the procedure is available if needed. At time $t > t_0$, then, one is interested in the expectation value of the Heisenberg operator $\hat{B}_i$,

$$\langle \hat{B}_i(t) \rangle = \text{Tr}[\hat{\rho}_0 \hat{B}_i(t)] = \text{Tr}[\hat{\rho}(t)\hat{B}_i]. \tag{3-2}$$

A. Response Functions

We define the total response of the system at time t to the external forces as the difference

$$\Delta B_i(t) \equiv \langle \hat{B}_i(t) \rangle - \langle \hat{B}_i \rangle_0, \tag{3-3}$$

where the subscript zero on expectation values refers to the equilibrium expectations in which $\hat{U} = \hat{U}_0$ [see Eq.(2–5)]. Now, owing to the freedom of choice implied by Eq.(3–2), we can shift temporarily to the Schrödinger picture in a formal way and note that

$$\hat{\rho}(t) = \hat{U}(t,t_0)\hat{\rho}_0\hat{U}^\dagger(t,t_0)$$

is simply a solution of the equation of motion

$$i\hbar\frac{d\hat{\rho}}{dt} = [\hat{H}_t, \hat{\rho}].$$

One verifies by direct substitution and use of the similar equation $i\hbar d\hat{U}/dt = \hat{H}\,\hat{U}$ that

$$\hat{\rho}(t) = \hat{\rho}_0 + \frac{i}{\hbar}\int_{t_0}^{t} \hat{U}(t,t')\Big[\sum_j \hat{A}_j F_j(t'), \hat{\rho}_0\Big]\hat{U}^\dagger(t,t')\,dt' \tag{3-4}$$

is also a solution to the equation of motion.

Combination of Eqs.(3–2) and (3–4) gives the total response of the system at time t to the perturbation (3–1):

$$\langle \hat{B}_i(t)\rangle - \langle \hat{B}_i\rangle_0 = \sum_j \int_{t_0}^{t} \Phi_{ij}(t,t';t_0)F_j(t')\,dt', \tag{3-5}$$

where we have defined the *response functionals*

$$\Phi_{ij}(t,t';t_0) \equiv \frac{1}{i\hbar}\langle [\hat{A}_j(t_0), \hat{B}_i(t,t')]\rangle_0. \tag{3-6}$$

Note that no approximations have been made and that Φ_{ij} is a real nonlinear functional of the applied forces. That is, the time-evolution operators contain the full Hamiltonian of Eq.(3–1). From here on we shall tend to omit explicit reference to t_0 in these functions—if $t_0 = -\infty$ it disappears anyway.

The importance of these nonlinear expressions is that they can be used to calculate derivatives of the response with respect to the amplitudes of the applied forces. One can then understand in principle the complete behavior of the system in terms of the response and its derivatives evaluated at the actual values of arbitrarily large forces. Some useful examples are provided by Peterson (1967).

Despite its generality and theoretical importance, the full nonlinear response theory is of limited practical value, because it is mathematically intractable. Eventually it is necessary to make approximations, and the conventional procedure is to carry out the well-known perturbation expansion of the time-evolution operators through first order in applied forces. This calculation is presented in Appendix A, along with a number of other mathematical results of use in this and succeeding chapters. Thus,

$$\hat{U}(t,t') \simeq \hat{U}_0(t,t')\left\{\hat{1} + \frac{i}{\hbar}\int_{t'}^{t} dt_1\, F_j(t_1)\hat{U}_0^\dagger(t_1,t')\hat{A}_j\hat{U}_0(t_1,t')\right\}, \tag{3-7}$$

where $\hat{U}_0$ is given by Eq.(2–5) with Hamiltonian $\hat{H}_0$. The response functionals of Eq.(3–6) become in this approximation

$$
\begin{aligned}
\Phi_{ij}(t,t') &= \frac{1}{i\hbar} \langle [\hat{A}_j, \hat{U}^\dagger(t,t') \hat{B}_i \hat{U}(t,t')] \rangle_0 \\
&\simeq \frac{1}{i\hbar} \langle [\hat{A}_j, \hat{B}_i(t-t')] \rangle_0 \\
&\quad + \left(\frac{i}{\hbar}\right)^2 \int_{t'}^{t} dt_1 \, \langle [\hat{A}_j, [\hat{A}_k(t_1-t'), \hat{B}_i(t-t')]] \rangle_0 F_k(t_1) \, dt_1 \\
&\quad + \cdots .
\end{aligned}
\tag{3–8}
$$

Substitution into Eq.(3–5), along with some straightforward algebra, yields the approximation

$$
\begin{aligned}
\Delta B_i(t) &\simeq \int_{t_0}^{t} \phi_{ij}(t-t') F_j(t') \, dt' \\
&\quad + \int_{t_0}^{t} dt_1 \int_{t_0}^{t} dt_2 \, \phi_{ijk}(t-t_2, t_1-t_2) F_k(t_1) F_j(t_2),
\end{aligned}
\tag{3–9}
$$

where

$$
\phi_{ij}(t-t') \equiv \frac{1}{i\hbar} \langle [\hat{A}_j, \hat{B}_i(t-t')] \rangle_0 ,
\tag{3–10}
$$

$$
\phi_{ijk}(t-t_2, t_1-t_2) \equiv \frac{1}{(i\hbar)^2} \langle [\hat{A}_j, [\hat{A}_k(t_1-t_2), \hat{B}_i(t-t_2)]] \rangle_0 ,
\tag{3–11}
$$

are called the first- and second-order *response functions*, respectively. In the last four equations we have employed the summation convention over repeated indices, and the time dependence of all operators is that of Eq.(3–15) below with Hamiltonian $\hat{H}_0$. Ordinary linear-response theory utilizes only the expression in Eq.(3–10), and is thus seen to be the simplest possible approximation to the full theory (Kubo, 1957). Note that Eqs.(3–9)-(3–11) could have been obtained directly by making the expansion (3–7) in Eq.(3–2).

For convenience, and without significant loss of generality, we consider from here on only a single external force, so that the perturbed Hamiltonian takes the form

$$
\hat{H}_t = \hat{H}_0 - \hat{A}F(t), \quad t > t_0 ,
\tag{3–12}
$$

where $\hat{H}_0$ describes the equilibrium system. Then the first-order linear response of the system to the dynamical perturbation applied at time $t = t_0$, in terms of the operator $\hat{B}$, is

$$
\langle \hat{B}(t) \rangle - \langle \hat{B} \rangle_0 = \int_{t_0}^{t} \phi_{BA}(t-t') F(t') \, dt' ,
\tag{3–13}
$$

where

$$
\phi_{BA}(t-t') \equiv \frac{1}{i\hbar} \langle [\hat{A}, \hat{B}(t-t')] \rangle_0 ,
\tag{3–14}
$$

and

$$\hat{B}(t - t') = e^{i(t-t')\hat{H}_0/\hbar} \hat{B} e^{-i(t-t')\hat{H}_0/\hbar} \, . \tag{3-15}$$

If we set $t_0 = -\infty$ and change integration variables in Eq.(3–13), we obtain

$$\langle \hat{B}(t) \rangle - \langle \hat{B} \rangle_0 = \int_0^\infty \phi_{BA}(t') F(t - t') \, dt' \, . \tag{3-16}$$

With the specific choice $F(t - t') = \delta(t - t')$, it is seen that the linear-response function is just the response of the system to a unit impulse. Also, because $\hat{\rho}_0$ and $\hat{U}_0$ commute, cyclic invariance of the trace immediately yields the reciprocity relation

$$\phi_{BA}(-t) = -\phi_{AB}(t) \, , \tag{3-17}$$

with the tacit agreement that any axial-vector fields contained in ϕ_{BA} must also change their signs. We must hasten to emphasize that the properties reflected in Eqs.(3–15)-(3–17) are not necessarily valid in the nonlinear theory.

Another useful interpretation is generated from Eq.(3–13) in the case that $t_0 = -\infty$ if we suppose that the external force is constant and then removed at $t = 0$. A change of integration variable yields

$$\langle \hat{B}(t) \rangle - \langle \hat{B} \rangle_0 = F \, R_{BA}(t) \, , \tag{3-18}$$

where

$$R_{BA}(t) \equiv \int_t^\infty \phi_{BA}(t') \, dt' \tag{3-19}$$

is called the *relaxation function*. It is clear that in this form the theory has the potential for describing relaxation of the system, although we shall return to a broader discussion of these processes in the sequel. For the moment we merely note that if ϕ_{BA} indeed vanishes as $t \to \infty$, then

$$\phi_{BA}(t) = -\frac{\partial}{\partial t} R_{BA}(t) \, . \tag{3-20}$$

That is, $R_{BA}(t)$ contains much more information than the response function.

Generally the $\hat{A}_j$ are not explicitly time dependent, so that the Heisenberg equation of motion reads

$$i\hbar \dot{\hat{A}}_j = [\hat{A}_j(t), \hat{H}_t] \, , \tag{3-21}$$

where the superposed 'dot' denotes a total time derivative. Cyclic invariance of the trace in Eq.(3–10) allows that equation to be rewritten as

$$\begin{aligned}
\phi_{ij}(t - t') &= \frac{1}{i\hbar} \mathrm{Tr}\{\hat{\rho}_0[\hat{A}_j, \hat{B}_i(t - t')]\} \\
&= \frac{1}{i\hbar} \mathrm{Tr}\{[\hat{\rho}_0, \hat{A}_j]\hat{B}_i(t - t')\} \, .
\end{aligned} \tag{3-22}$$

These two expressions lead us to an alternative form for the linear response function if we also note from Appendix A the identity

$$[e^{\hat{A}}, \hat{B}] = e^{\hat{A}}[\hat{A}, \overline{\hat{B}}] \tag{3-23}$$

where we define the *Kubo transform* of $\hat{B}$ with respect to $\hat{A}$:

$$\overline{\hat{B}} \equiv \int_0^1 e^{-x\hat{A}} \hat{B} e^{x\hat{A}} \, dx. \tag{3-24}$$

If the time-evolution operator is that of Eq.(2–5) with time-independent $\hat{H}_0$, then with $\hat{A} = -\beta \hat{H}_0$ we can also write

$$\overline{\hat{B}} = \frac{1}{\beta} \int_0^\beta \hat{B}(t - i\hbar u) \, du, \tag{3-25}$$

in the notation of Eq.(3–15).

We now combine Eqs.(3–21)-(3–24), and note that in linear approximation $\hat{H}_t$ in Eq.(3–21) can be replaced $\hat{H}_0$. Then,

$$\phi_{ij}(t) = \beta \langle \overline{\dot{\hat{A}}_j} \hat{B}_i(t) \rangle_0$$
$$= -\beta \langle \overline{\hat{A}_j} \dot{\hat{B}}_i(t) \rangle_0. \tag{3-26}$$

Thus, the linear response function can also be expressed as a *time-correlation function*.

B. Linear Response and Fluctuations

Experimental implications of the theory are most readily obtained through the spectral representation of the response. Throughout we shall adopt the following conventions for space-time Fourier transformations:

$$f(\mathbf{x}, t) = \int \frac{d^3k}{(2\pi)^3} e^{i\mathbf{k}\cdot\mathbf{x}} \int_{-\infty}^\infty \frac{d\omega}{2\pi} e^{-i\omega t} f(\mathbf{k}, \omega), \tag{3-27a}$$

$$f(\mathbf{k}, \omega) = \int d^3x \, e^{-i\mathbf{k}\cdot\mathbf{x}} \int_{-\infty}^\infty dt \, e^{i\omega t} f(\mathbf{x}, t). \tag{3-27b}$$

Although it is certainly possible to attribute spatial dependence to the operators $\hat{A}$ and $\hat{B}$, and thus to the response function, we shall not need to consider space variables explicitly in this section. The necessary modifications are trivial at any rate.

Equation (3–13) has the appearance of a convolution, owing to the time independence of $\hat{H}_0$, so that the equation is readily analyzed by means of the faltung theorem for Fourier transforms. Introduction of the step-function $\theta(t)$, which is unity for $t > 0$ and vanishes otherwise, generates the relation

$$\Delta B(\omega) = \chi_{BA}(\omega) F(\omega)\,, \tag{3-28}$$

where

$$F(\omega) = \int_{t_0}^{\infty} e^{i\omega t}\, F(t)\, dt\,, \tag{3-29}$$

$$\chi_{BA}(\omega) \equiv \int_{0}^{\infty} e^{i\omega t}\, \phi_{BA}(t)\, dt\,. \tag{3-30}$$

Although the specific nature of $F(\omega)$ depends on the choice of t_0, as well as on inclusion of a possible adiabatic-switching factor, the form of χ_{BA} is dictated by the context in which ϕ_{BA} arises. That is, χ_{BA} is the Fourier transform of a causal function $\theta(t)\phi_{BA}(t)$, and is therefore a causal transform. This observation suggests a possibly deeper analysis in the complex plane and, because we shall require similar analyses in the sequel, it is worth carrying out in some detail.

Because $\chi_{BA}(\omega)$ is in general complex, we can write

$$\chi_{BA}(\omega) = \chi'_{BA}(\omega) + i\chi''_{BA}(\omega)\,, \tag{3-31}$$

where χ'_{BA} and χ''_{BA} are real functions of ω. With $\phi_{BA}(t)$ real it follows that χ'_{BA} is an even function of ω and χ''_{BA} is odd. If $\chi_{BA}(\omega)$ exists, it defines an analytic function of z which is regular, continuous, and bounded in the upper half-plane, and which vanishes as $\omega = \mathrm{Re}\,z$ tends to infinity there. A well-known theorem on causal functions of this kind (e.g., Titchmarsh, 1948) implies that the real and imaginary parts of the Fourier transform constitute a pair of Hilbert transforms. That is, as demonstrated in Problem 3.2,

$$\chi'(\omega) = -\frac{1}{\pi} \mathrm{P} \int_{-\infty}^{\infty} \frac{\chi''(u)}{\omega - u}\, du\,, \tag{3-32a}$$

$$\chi''(\omega) = \frac{1}{\pi} \mathrm{P} \int_{-\infty}^{\infty} \frac{\chi'(u)}{\omega - u}\, du\,, \tag{3-32b}$$

where the integrals are Cauchy principal values. These are often referred to as Kramers-Kronig relations, and their limiting values are known as sum rules:

$$\chi'(0) = \frac{1}{\pi} \int_{-\infty}^{\infty} \chi''(u) \frac{du}{u}\,, \tag{3-33a}$$

$$\lim_{\omega \to \infty} \omega \chi''(\omega) = \frac{1}{\pi} \int_{-\infty}^{\infty} \chi'(u)\, du\,. \tag{3-33b}$$

(Operator subscripts will be omitted when irrelevant and no confusion can occur.) These and other sum rules are often useful in analyzing spectra, and detailed discussions have been given elsewhere (e.g., Kubo and Ichimura, 1972).

Now consider the full Fourier transform of the response function itself. Although $\phi(t)$ appears causally in the above context, it is not necessarily causal as defined in Eq.(3–14) as a general function of t. Thus,

$$\phi(\omega) \equiv \int_{-\infty}^{\infty} e^{i\omega t} \phi(t) \, dt = \int_{0}^{\infty} e^{i\omega t} \phi(t) \, dt + \int_{-\infty}^{0} e^{i\omega t} \phi(t) \, dt \equiv \phi_+(\omega) + \phi_-(\omega) \, .$$

(3–34)

This expression defines two functions which are boundary values on the real axis of an analytic function $\tilde{\phi}(z)$ that is regular in the entire complex plane, with the exception of the real axis. Note that $\phi_+(\omega)$ is a causal transform. The function $\tilde{\phi}(z)$ has the following integral representations (e.g., Roos, 1969):

$$\tilde{\phi}(z) = \int_{0}^{\infty} e^{izt} \phi(t) \, dt \, , \quad \mathrm{Im}\, z > 0$$

$$= \int_{-\infty}^{0} e^{izt} \phi(t) \, dt \, , \quad \mathrm{Im}\, z < 0$$

$$= \int_{-\infty}^{\infty} \frac{d\omega}{2\pi i} \frac{\phi(\omega)}{\omega - z} \, , \quad \mathrm{Im}\, z \neq 0 \, .$$

(3–35)

That is, there is a branch cut along the entire real axis.

The physical response is the limit of the sectionally analytic function $\tilde{\phi}(z)$ as z approaches the real axis from above, from the 'physical' sheet. In order to proceed further it is necessary to examine more closely the operators $\hat{A}$ and $\hat{B}$ appearing in the definition (3–14) of $\phi_{BA}(t)$. It is generally presumed that the linear operators relevant to many-body theories possess definite properties of parity, P, and time reversal, T. Although we shall analyze these ideas in detail later for correlation functions, here it suffices to note that most operators of interest are even under PT, though there are exceptions. If this is the case, or if the equally reasonable presumption that both operators have the same signature under time reversal is made, then $\phi(\omega)$ is imaginary.

With this observation, and use of the identity

$$\lim_{\gamma \to 0^+} \frac{1}{x \pm i\gamma} = \mathrm{P}\left(\frac{1}{x}\right) \mp i\pi\delta(x) \, ,$$

(3–36)

the limit of $\tilde{\phi}(z)$ in Eq.(3–35) yields

$$\lim_{\gamma \to 0^+} \tilde{\phi}(\omega + i\gamma) = X(\omega) = X'(\omega) + \tfrac{1}{2}\phi(\omega) \, ,$$

(3–37)

where

$$X'(\omega) = -\mathrm{P} \int_{-\infty}^{\infty} \frac{du}{2\pi i} \frac{\phi(u)}{\omega - u} \, .$$

(3–38)

Thus, the dissipation in the system owing to the external force, and which is described by the imaginary part of the susceptibility X, is actually given by the Fourier transform of the response function, $\phi(\omega)$. At high frequencies, though, the sum rule of Eq.(3–33b) brings into play the reactive part $\chi'(\omega)$.

Further insight into the linear theory is obtained by introducing the space-time correlation function between operators $\hat{A}(\mathbf{r}, t)$ and $\hat{B}(\mathbf{r}, t)$:

$$S_{AB}(\mathbf{r}, t; \mathbf{r}', t') \equiv \langle \delta\hat{A}(\mathbf{r}, t)\, \delta\hat{B}(\mathbf{r}', t')\rangle_0 \,, \qquad (3\text{--}39)$$

in terms of deviations $\delta\hat{A} \equiv \hat{A} - \langle\hat{A}\rangle_0\hat{1}$. Denote the Fourier-transformed quantity by $S_{AB}(\mathbf{k}, \omega)$. If one employs the representation independence of the trace and writes out both sides of the following expression in a representation in which $\hat{H}_0$ is diagonal, then for arbitrary operators $\hat{X}(t)$, $\hat{Y}(t)$ it is straightforward to verify the identity

$$\int_{-\infty}^{\infty} \langle\hat{X}(0)\hat{Y}(t)\rangle_0\, e^{i\omega t}\, dt = e^{-\beta\hbar\omega} \int_{-\infty}^{\infty} \langle\hat{Y}(t)\hat{X}(0)\rangle_0\, e^{i\omega t}\, dt \,. \qquad (3\text{--}40)$$

It then follows from the definitions that

$$\chi''_{AB}(\mathbf{k}, \omega) = \frac{1}{2\hbar}\left(1 - e^{-\beta\hbar\omega}\right)S_{AB}(\mathbf{k}, \omega)\,, \qquad (3\text{--}41)$$

often called a *fluctuation-dissipation theorem*.

This relation between the equilibrium fluctuations in the system and the dissipative response induced by external forces has, as noted earlier, provided the major impetus for the development of many modern discussion of irreversible processes. But the notion of fluctuation-dissipation relations is much older, and perhaps the first such theorem is the Nernst-Einstein relation of Eq.(1–21) for electrical mobility and the diffusion coefficient.

In practice it is often more useful to describe the fluctuations in a slightly different way, and to this end it will suffice to continue omitting the explicit space dependence of the operators. Thus, we define a *time-correlation function* for two arbitrary operators as

$$\begin{aligned} C_{AB}(t, t') &\equiv \tfrac{1}{2}\langle\hat{A}(t)\hat{B}(t') + \hat{B}(t')\hat{A}(t)\rangle_0 \\ &= \langle\{\hat{A}(t), \hat{B}(t')\}\rangle_0 \\ &= C_{AB}(t, t')\,, \end{aligned} \qquad (3\text{--}42)$$

where the curly brackets will denote one-half an anticommutator. The symmetrized product serves not only to avoid ambiguity in the event of non-commutativity, but also assures that the sum is Hermitian if $\hat{A}$ and $\hat{B}$ are so. Hence, C_{AB} is real when $\hat{A}$ and $\hat{B}$ are Hermitian. Note that the definition (3–42) contains only the Hamiltonian $\hat{H}_0$, which is time independent, so that $C_{AB}(t, t') = C_{AB}(t - t') \equiv C_{AB}(\tau)$.

Problem 3.4 asks one to derive the identity (3–40), as well as the following:

$$\int_{-\infty}^{\infty} \langle[\hat{X}(0), \hat{Y}(0)]\rangle_0 e^{i\omega\tau}\, d\tau = -2\tanh\left(\tfrac{1}{2}\beta\hbar\omega\right) \int_{-\infty}^{\infty} \langle\{\hat{X}(0), \hat{Y}(\tau)\}\rangle_0 e^{i\omega\tau}\, d\tau\,. \quad (3\text{--}43)$$

This provides an effective relationship between the linear response and the spontaneous equilibrium fluctuations, a relationship we now seek to make more explicit. Note that the numerical factor multiplying the integral on the right-hand side of this last expression is essentially the reciprocal of the average energy of an harmonic oscillator.

By taking inverse transforms of the results of Problem 3.5 we obtain the forms

$$\chi'(\omega) = \frac{1}{2} \int_{-\infty}^{\infty} \phi(\tau) e^{i\omega\tau} \, d\tau \,, \quad \phi(\tau) \text{ even} \,, \tag{3-44a}$$

$$\chi''(\omega) = -\frac{i}{2} \int_{-\infty}^{\infty} \phi(\tau) e^{i\omega\tau} \, d\tau \,, \quad \phi(\tau) \text{ odd} \,. \tag{3-44b}$$

Now substitute from Eq.(3–10) for the linear response function and then employ Eq.(3–43) to obtain the desired result for the fluctuations—which must be done with some care.

This last *caveat* is best illustrated by solving the equation $xf(x) = g(x)$ for $f(x)$, in an interval including the origin. The solution is $f(x) = g(x)/x$ only up to a constant multiple of $\delta(x)$, and the constant can not be determined without additional information. In the present context the implication is that we must include the zero-frequency components explicitly. The point appears implicitly in the work of Bernard and Callen (1959), and is discussed at length by Case (1972).

By means of the above procedure, then, we find that

$$C_{BA}(\tau) = \frac{\hbar}{i} P \int_{-\infty}^{\infty} e^{-i\omega\tau} \chi'_{BA}(\omega) \coth(\tfrac{1}{2}\beta\hbar\omega) \frac{d\omega}{2\pi}$$

$$+ \frac{\hbar}{i} C \chi'_{BA}(0) \,, \quad \phi_{BA} \text{ even} \,, \tag{3-45a}$$

$$= \hbar P \int_{-\infty}^{\infty} e^{-i\omega\tau} \chi''_{BA}(\omega) \coth(\tfrac{1}{2}\beta\hbar\omega) \frac{d\omega}{2\pi}$$

$$+ \hbar C \chi''_{BA}(0) \,, \quad \phi_{BA} \text{ odd} \,, \tag{3-45b}$$

where C is an arbitrary constant. These, too, are called fluctuation-dissipation theorems.

As an example let $\hat{A}$ be driven from $t_0 = -\infty$, and suppose it is also $\hat{A}$ whose expectation value we wish to predict at time t. Take the driving force to be

$$F(t) = \tfrac{1}{2}\left(F_0 e^{-i\omega t} + F_0^* e^{i\omega t}\right) \tag{3-46}$$

so that Eqs.(3–13) and (3–30) yield immediately

$$\langle \hat{A}(t) \rangle - \langle \hat{A} \rangle_0 = \tfrac{1}{2} F_0 \chi_{AA}(\omega) e^{-i\omega t} + \tfrac{1}{2} F_0^* \chi_{AA}^*(\omega) e^{i\omega t} \,. \tag{3-47}$$

From Eq.(3–12) we calculate the time rate-of-change of energy in the system:

$$Q \equiv \frac{dE}{dt} = \left\langle \frac{\partial \hat{H}_t}{\partial t} \right\rangle = -\langle \hat{A}(t) \rangle \frac{dF}{dt} \,, \tag{3-48a}$$

where

$$\frac{dF}{dt} = -i\frac{\omega}{2}\left(F_0 e^{-i\omega t} - F_0^* e^{i\omega t}\right). \tag{3-48b}$$

For convenience we presume $\langle \hat{A} \rangle_0 = 0$, so that Eqs.(3-47) and (3-48) yield

$$Q(\omega) = \tfrac{1}{2}|F_0|^2 \,\omega \chi''_{AA}(\omega), \tag{3-49}$$

plus oscillatory terms that do not contribute to the discussion. Experimentally one could carry out a time average at any rate. Therefore, the rate at which the system absorbs energy from the external source is governed completely by the imaginary part of the susceptibility. Again, though, at very high frequencies, the sum rule of Eq.(3-33b) brings into play the real part of χ_{AA}.

We know that $\phi_{AA}(\tau)$ is an odd function of τ. Hence, from Eq.(3-45b), we find at $t = 0$

$$\langle \hat{A}(0)\hat{A}(0) \rangle = \frac{\hbar}{\pi} \int_0^\infty \chi''(\omega) \coth(\tfrac{1}{2}\beta\hbar\omega)\,d\omega \;+ 2\hbar C\chi''(0). \tag{3-50}$$

That is, the mean-square equilibrium fluctuations of $\hat{A}$ are directly related to the average rate of energy dissipation when $\hat{A}$ is driven by the external force.

C. The Scope of Linear Response Theory

Because it is practically very difficult to utilize the full nonlinear formalism, little has been done in that area except in the expanded form of the response. When departures from linearity are not too large, it is useful to study the second-order linear-response function of Eq.(3-11). Although the function ϕ_{ijk} has no obvious symmetry properties in general, if the measured observable is that actually driven by the external forces one can extract a number of special symmetries, as well as some interesting relations involving higher-order driven moments and correlation functions (Bernard and Callen, 1959). We shall explore some of these in the next chapter. Nevertheless, the nonlinear theory remains essentially undeveloped, and the most useful form at the moment remains the linear approximation.

And just when is the linear approximation valid and useful? To say simply that 'something' must be 'small' will not suffice, as is amply demonstrated by a number of strong criticisms of linear response theory (van Kampen, 1965, 1971; van Vliet, 1978, 1979; van Velsen, 1978; Lippert, *et al*, 1984). Rather, one must attempt to delineate the kind of physical situation to which the theory *should* apply, and we begin by re-stating our viewpoint in regard to this scenario.

When an external 'force'—usually an electric or magnetic field—is imposed on a macroscopic system—usually in thermal equilibrium—this results in the addition of an explicit and well-defined term $\hat{H}'(t)$ to the system Hamiltonian. Nothing new has been learned about the system itself, other than that the internal dynamics have been changed owing to a change in the well-understood total Hamiltonian. As a consequence, the statistical operator must remain unchanged as that describing the equilibrium system prior to application of the external perturbation: $\hat{\rho}_0$. If one believes the probability for heads in flipping an honest coin to be an intrinsic physical property of the coin—rather than something we assign owing to symmetry—then the above comments may present a difficulty for that person. But if one appreciates that probabilities can only be assigned based on available evidence —as expounded at great length in Volume I—then the statistical operator must remain $\hat{\rho}_0$.

What does change is the time-development operator: $\hat{U}_0 \rightarrow \hat{U}(\hat{H}_t = \hat{H}_0 + \hat{H}')$. Because this operator no longer commutes with $\hat{\rho}_0$, it follows that the expectation of any Heisenberg operator $\hat{B}$ will now be time dependent. Note, however, that $\hat{U}$ remains unitary (see Appendix A), so that the maximum entropy of the equilibrium system remains constant:

$$S_I = -\kappa \operatorname{Tr}[\hat{\rho}_0 \ln \hat{\rho}_0] = -\kappa \operatorname{Tr}[\hat{\rho}(t) \ln \hat{\rho}(t)]. \qquad (3\text{-}51)$$

This will no doubt be unsettling to some, but it is precisely what is to be expected. The point is that in the equilibrium system prior to $t = t_0$ we were able to equate S_I with the experimental entropy S_E, *and the latter is defined only for equilibrium states.* For many problems of interest the system is certainly out of equilibrium in the presence of the external force, but neither S_I nor S_E are any longer descriptive of the system with complete accuracy. If the perturbation is removed and the system allowed to return to equilibrium, then based on that information S_I can be re-calculated and S_E measured anew—and both will undoubtedly be different from the previous values, owing to dissipation of energy.

This last scenario in fact forms the basis for deriving the weak second law of thermodynamics, as discussed in detail in Volume I, and by Jaynes (1965). In the sequel we shall return to the matter of *predicting* this behavior, but that must await the development of more powerful tools. For the present we merely note that no approximations have been made, and that the rigorous expressions for (nonlinear) dynamical response can only stand or fall with quantum mechanics itself.

As noted earlier, however, the complete theory is mathematically intractable at this point in our computational development, so that the first and most useful step is a linear approximation. By this we mean a linearization of the *response*— $\Delta B(t) \equiv \langle \hat{B}(t) \rangle - \langle \hat{B} \rangle_0$— and effectively that means linearization of $\langle \hat{B}(t) \rangle$. It must be emphasized that what we are linearizing is the departure from equilibrium, and this will appear in a much clearer light in the more general discussion of Chapter 5. In turn, the only mathematical means for doing this is to expand the time-development operators, which are defined by their action on the *system* states. (Recall that $\hat{H}_t$ is an N-body Hamiltonian.) But even if one chooses to view the scenario differently from that advocated here, remarkably the result is the same no matter how one chooses to linearize! It is unique.

Following construction of a linearized theory, though, the questions as to when and to what it applies are entirely different matters, and the scope of applicability is eventually to be decided by experiment. Despite some modern writings to the contrary, no theory is of 'everything'. The response of some systems to some fields may indeed be a first-order effect, whereas other situations are possibly described accurately only in higher-order perturbation theory. If, on the one hand, particle-particle collisions are significant and the system is characterized by short relaxation times, then the effects of the external perturbation are rapidly absorbed and the collective response remains small. It appears that at each instant the external force acts on an essentially equilibrium system. On the other hand a very dilute system, say, can rapidly be driven far from equilibrium by small perturbations and the linear approximation may be very poor. (We shall return to a further discussion of this very-dilute-gas problem subsequently.) To assert that the system is undergoing a steady-state process, for example, would surely be at variance with the facts. It is not only the amplitude of the applied force that is important in addressing the question of utility, but also the intrinsic 'rigidity' of the system being studied. This theory, as with any other, can certainly lead to difficulties if applied indiscriminately. Indeed, independent of linearization, one expects long-time application of any external field to generate difficulties owing to Joule heating, and the current in an isolated conductor can cease because charge collects on the surfaces.

As will become amply clear in the sequel, response and correlation functions are not easily calculated from microscopic considerations without thoughtful approximations. But as a matter of principle, all the physics of the many-body system is to be found in these functions: dissipation, relaxation, correlation and relaxation times, etc. Calculations are indeed difficult, but that has little to do with the *logic* of the theory. Incidentally, these comments apply also to the full nonlinear theory, independent of linearization.

A fair reading of some of the objections raised to linear response theory seems to sort the detractors into one of two camps. There is either an unfortunate over-reliance on the kinetic-theory/ergodicity view of many-body physics; or, the theory is found wanting because it was applied to a situation in which it was not meant to be applied, or the full information about the system was not employed. Kubo, *et al* (1985; Sec. 4.7.1) have addressed the question of kinetic methods versus linear response theory. As has been suggested with respect to the *H*-theorem (Jaynes, 1971), and as noted in Chapter 1 regarding minimum entropy production, perhaps it would be more reasonable in such cases to attempt a classification of those situations in which the proposed principle might be *expected* to apply. In the same spirit, it would be most useful to develop criteria for determining when and for what systems linear response theory should provide a useful description. Although one has the feeling that an essential requirement would amount to the perturbing frequency being a good deal less than the collision frequency, we are far from having such rigorous criteria at hand.

Problems

3.1 Consider a set of external forces to be turned on in step-function fashion at $t = t_0$, such that for $t > t_0$ their amplitudes are constants $\{F_j\}$.

(a) Find the formal expression for the response, indicating the time dependence of operators explicitly.

(b) If, in addition, measurement indicates that the system has reached equilibrium even though the static forces are still present, obtain an explicit expression for $\langle \hat{B}_i \rangle$, and for the 'susceptibility'

$$\chi_{ij}(F_1, \ldots, F_n) \equiv \frac{\partial \langle \hat{B}_i \rangle}{\partial F_j} .$$

(The derivatives are often referred to as *response coefficients*, and one should refer to Appendix A for certain operator identities needed for their calculation.)

3.2 By employing the identity

$$\theta(t) = \lim_{\gamma \to 0^+} \frac{i}{2\pi} \int_{-\infty}^{\infty} \frac{dx}{x + i\gamma} e^{-ixt}$$

for the step-function, as well as Eq.(3–36), prove the relations (3–32).

3.3 Verify in detail the expansion in Eq.(3–8).

3.4 Given the appropriate convergence conditions on the functions involved, verify the Fourier-transform identities of Eqs.(3–40) and (3–43).

3.5 Again, presume the appropriate convergence conditions on the functions involved.

(a) Develop the inverse Fourier transforms of the susceptibility functions $\chi'(\omega)$, and $\chi''(\omega)$:

$$\chi'(t) = \begin{cases} \phi(t), & t > 0 \\ \phi(-t), & t < 0 \end{cases} ,$$

$$\chi''(t) = -i\chi'(t)\,\mathrm{sgn}(t) ,$$

where $\mathrm{sgn}(t) \equiv 2\theta(t) - 1$. Hence, $\chi'(t) + i\chi''(t) = \theta(t)\phi(t)$.

(b) Conversely, show that

$$\phi(t) = 2 \begin{cases} \int_{-\infty}^{\infty} \frac{d\omega}{2\pi} e^{-i\omega t} \chi'(\omega), & \phi(t) \text{ even} \\ i \int_{-\infty}^{\infty} \frac{d\omega}{2\pi} e^{-i\omega t} \chi''(\omega), & \phi(t) \text{ odd} \end{cases} .$$

3.6 Envision a simple RC-circuit such that the complex impedance is

$$Z(\omega) = R + \frac{1}{i\omega C} ,$$

from which the complex admittance $Y(\omega) = Y'(\omega) + i\omega Y''(\omega)$ can be constructed. By making the identification $X(\omega) \equiv \omega Y''(\omega)$ and employing the result of Eq.(3–50), show that for $\beta\hbar\omega \ll 1$ and observations made through a very narrow band-pass filter,

$$\langle I^2 \rangle_0 \simeq \frac{2}{\pi} \kappa T\, Y''(\omega)\, \Delta\omega .$$

This is essentially Nyquist's result.

REFERENCES

Bernard, W., and H.B. Callen: 1959, 'Irreversible Thermodynamics of Nonlinear Processes and Noise in Driven Systems', *Rev. Mod. Phys.* **31**, 1017.

Case, K.M.: 1972, 'On Fluctuation-Dissipation Theorems', *Transport Th. and Stat. Phys.* **2**, 129.

Gell-Mann, M., and F. Low: 1951, 'Bound States in Quantum Field Theory', *Phys. Rev.* **84**, 350.

Jaynes, E.T.: 1965, 'Gibbs vs Boltzmann Entropies', *Am. J. Phys.* **33**, 391.

Jaynes, E.T.: 1971, 'Violation of Boltzmann's *H*-Theorem in Real Gases', *Phys. Rev.* **A4**, 747.

Kubo, R.: 1957, 'Statistical-Mechanical Theory of Irreversible Processes. I', *J. Phys. Soc. Japan* **12**, 570.

Kubo, R., and M. Ichimura: 1972, 'Kramers-Kronig Relations and Sum Rules', *J. Math. Phys.* **13**, 1454.

Kubo, R., M. Toda, and N. Hashitsume: 1985, *Statistical Physics II*, Springer-Verlag, Berlin.

Lippert, E., C.A. Chatzidimitriou-Dreismann, and K.-H. Naumann: 1984, 'Structure, Dynamics, and Dissipation in Hard-Core Molecular Liquids', *Adv. Chem. Phys.* **57**, 311.

Peterson, R.L.: 1967, 'Formal Theory of Nonlinear Response', *Rev. Mod. Phys.* **39**, 69.

Roos, B.W.: 1969, *Analytic Functions and Distributions in Physics and Engineering*, Wiley, New York.

Titchmarsh, E.C.: 1948, *Introduction to the Theory of Fourier Integrals*, 2nd ed., Clarendon Press, Oxford.

van Kampen, N.G.: 1965, 'Fluctuations in Nonlinear Systems', in R.E. Burgess (ed.), *Fluctuation Phenomena in Solids*, Academic Press, New York.

van Kampen, N.G.: 1971, 'The Case Against Linear Response Theory', *Phys. Norvegica* **5**, 279.

van Velsen, J.F.C.: 1978, 'On Linear Response Theory and Area Preserving Mappings', *Phys. Repts.* **41C**, 135.

van Vliet, K.M.: 1978, 'Linear Response Theory Revisited. I.The Many-Body van Hove Limit', *J. Math. Phys.* **19**, 1345.

van Vliet, K.M.: 1979, 'Linear Response Theory Revisited. II.The Master Equation Approach', *J. Math. Phys.* **20**, 2573.

Chapter 4

Dynamical Response
of
Real Systems

This chapter is almost a digression from the general chain of theoretical development, but it is to be hoped the reader will nevertheless find it useful. The aim here
is to illustrate just how one goes about using the theory of dynamical response.
Because the most obvious external perturbations of this kind consist of electric and
magnetic fields, the ensuing discussion is devoted to development of some of the
specific formal expressions for the response associated with these forces. Further
applications—e.g., to resonance absorption, and dielectric relaxation—are treated
by Kubo, *et al*, (1985).

Among other things, statistical mechanics should provide macroscopic expressions for the material parameters of a system in terms of the microscopic constituents, so that one expects the formalism of the preceding chapter to do exactly
that for situations appropriate to that scenario. But any theory requires considerable interpretation when applied to real systems, implying that some care must be
exercised in identifying mathematical quantities with their physical counterparts .
Toward that end, we begin by reviewing the rather elementary phenomenological
equations of continuous media, a necessary exercise nevertheless.

Let us first discuss dielectric media and their response to external electric fields.
In such a medium the total polarization density is defined as

$$P_i \equiv P_i - \partial_j Q_{ji} + \cdots ,\tag{4-1}$$

where P_i is a component of the total dipole density, Q_{ji} refers to the total quadrupole density, etc. It usually suffices, as we shall almost always presume, to consider
only the dipole contributions and write $P_i \simeq P_i$. The total electric dipole moment of
the medium then has components $M_i \equiv V P_i$, where V is the volume of the dielectric.
In general, $\mathbf{M}$ is the vector sum of all the individual dipoles in the dielectric, and is
defined instantaneously as the instantaneous sum

$$\boldsymbol{P} = \sum_{\ell=1}^{N} \mathbf{m}(\ell) .\tag{4-2}$$

53

Expressions for the dipole moments of the individual particles in the medium usually take the form

$$m_i(\ell) = \mu_i(\ell) + \overline{\alpha}_{ij} E_j'(\ell) + \overline{\beta}_{ijk} E_j'(\ell) E_k'(\ell) + \cdots , \qquad (4\text{--}3)$$

where $\mu(\ell)$ is the permanent dipole moment of the ℓth particle, $\mathbf{E}'(\ell)$ is the total local electric field at the position of the particle, $\overline{\alpha}_{ij}$ is called the atomic polariz-ability tensor, and $\overline{\beta}_{ijk}$, etc., are known as hyperpolarizabilities. (The reason for the overbars will emerge presently.) This particular coupling scheme may also be affected by inhomogeneities. Note that $\mathbf{E}'(\ell)$ is the complete field external to the particle, so that it is a functional of any external fields, as well as of the microscopic fields produced by all other particles. Normally we shall presume that the medium is not an electret—that is, there is no spontaneous polarization and all the $\mu(\ell)$ are zero. The nonlinear terms can also usually be presumed negligible. Thus, with our continued use of the summation convention over repeated indices, the *induced dipole moment* of the particle is defined as

$$p_i(\ell) \equiv \overline{\alpha}_{ij} E_j'(\ell) , \qquad (4\text{--}4)$$

which, owing to the microscopic fields of other particles, may be nonzero even in the absence of external fields. In linear approximation, then,

$$\mathbf{m}(\ell) = \boldsymbol{\mu}(\ell) + \mathbf{p}(\ell) , \qquad (4\text{--}5)$$

and μ vanishes for non-electrets.

A familiar elementary example of the macroscopic manifestation of these points is found in the parallel-plate capacitor shown in Figure 4–1. The space between the plates is completely filled with dielectric and end effects are neglected. When the uniform external field $\mathbf{E}_0$ is applied, the elementary dipoles are aligned in the same direction to give an overall polarization density $\mathbf{P}$, and this latter is equivalent to an induced electric field of opposite sign opposing the applied field. The total field within the dielectric is then

$$\mathbf{E} = \mathbf{E}_0 + \mathbf{E}' = \mathbf{E}_0 - 4\pi\mathbf{P} . \qquad (4\text{--}6)$$

In the general dielectric it is found convenient to relate $\mathbf{P}$ to the total field $\mathbf{E}$ by means of a material parameter characteristic of the medium. We write, per-unit volume,

$$\mathbf{P} = \chi_e \mathbf{E}, \qquad (4\text{--}7)$$

where χ_e is called the *electric susceptibility* of the medium. In fact, when this relation is valid it is taken as the definition of a dielectric. More generally, the relationship is actually a tensor equation and is nonlinear:

$$P_i = \chi_{ij} E_j + \chi_{ijk} E_j E_k + \chi_{ijk\ell} E_j E_k E_\ell + \cdots . \qquad (4\text{--}8)$$

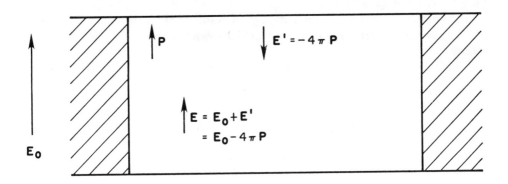

Fig. 4–1. Illustration of the electric and polarization fields within a simple parallel-plate capacitor.

Many substances are isotropic, or have a center of inversion symmetry (e.g., calcite), in which case even powers of $\mathbf{E}$ do not appear and $\mathbf{P}(\mathbf{E}) = -\mathbf{P}(-\mathbf{E})$. Piezoelectric crystals, on the other hand, do not exhibit this type of symmetry.

Another material parameter of the system found quite useful is the *dielectric coefficient*:

$$\varepsilon_{ij} \equiv \delta_{ij} + 4\pi X_{ij}\,, \qquad (4\text{–}9)$$

or in isotropic substances,

$$\varepsilon = 1 + 4\pi X_e\,, \qquad (4\text{–}10)$$

which is often a constant. Combination of Eqs.(4–9) and (4–8) with the definition of $\mathbf{M}$ yields, for a linear substance,

$$\left(\varepsilon_{ij} - \delta_{ij}\right) E_j = \frac{4\pi}{V} M_i\,, \qquad (4\text{–}11)$$

where it is again emphasized that $\mathbf{E}$ is the total field within the material. Although this last equation is most often written for static fields only, it is also valid for time-varying fields if ε_{ij} is interpreted as a complex function.

Equation (4–4) suggests it may be useful to relate $\mathbf{P}$ or $\mathbf{M}$ to the external field $\mathbf{E}_0$, similar to the relations (4–7) and (4–8) for $\mathbf{E}$. Hence, we write

$$M_i \equiv \alpha_{ij}\left[\mathbf{E}_0\right]_j\,, \qquad (4\text{–}12)$$

defining the *macroscopic polarization tensor*. There are, of course, higher-order nonlinear terms defining macroscopic hyperpolarizabilities, but for now we consider only linear effects. One often considers the medium, as well as the external field, to be homogeneous. Whether or not this is actually the case, there is no implication that the total field within the dielectric is homogeneous, and the presence of field gradients can not be excluded.

A completely general phenomenological result is the expression $\nabla \cdot \mathbf{P} = -\rho_b$, where ρ_b is the charge density bound in place within the dielectric. Because the relation

$$\nabla \cdot \mathbf{E} = 4\pi \rho_t, \qquad \rho_t \equiv \rho_b + \rho_f \tag{4-13}$$

is always true, it follows that

$$\nabla \cdot (\mathbf{E} + 4\pi \mathbf{P}) = 4\pi \rho_f, \tag{4-14}$$

where ρ_f is the density of free charges in the medium. This last statement is true independent of any relation between $\mathbf{E}$ and $\mathbf{P}$. Thus, it has been found useful to define an electric displacement vector

$$\mathbf{D} \equiv \mathbf{E} + 4\pi \mathbf{P}, \tag{4-15}$$

and rewrite Eq.(4–14) as

$$\nabla \cdot \mathbf{D} = 4\pi \rho_f. \tag{4-16}$$

In the linear isotropic situation, substitution for $\mathbf{P}$ from Eq.(4–7) and reference to Eq.(4–10) yields

$$\mathbf{D} = \varepsilon \mathbf{E}. \tag{4-17}$$

Introduction of this new vector is usually thought not to be of great physical significance, and is conventionally taken to be nothing more than an artifice.

Now consider time-varying fields and note that a real current owing to a time-varying polarization is given by $d\mathbf{P}/dt$. Also, Maxwell's equations describing microscopic phenomena in a vacuum have the form

$$\nabla \cdot \mathbf{E} = 4\pi \rho, \qquad \nabla \cdot \mathbf{B} = 0, \tag{4-18a}$$

$$\nabla \times \mathbf{E} = -\frac{1}{c}\frac{\partial \mathbf{B}}{\partial t}, \qquad \nabla \times \mathbf{B} = \frac{1}{c}\frac{\partial \mathbf{E}}{\partial t} + \frac{4\pi}{c}\mathbf{J}, \tag{4-18b}$$

where $\mathbf{B}$ is the magnetic field strength, $\mathbf{J}$ is the charge current, and c is the speed of light in vacuum. In a macroscopic substance the first equation in Eq.(4–18a) is altered as indicated in Eq.(4–13). In a dielectric the equation for $\nabla \times \mathbf{B}$ must also be changed to read

$$\nabla \times \mathbf{B} = \frac{1}{c}\left[\frac{\partial \mathbf{E}}{\partial t} + 4\pi \frac{\partial \mathbf{P}}{\partial t} + 4\pi \mathbf{J}\right]$$

$$= \frac{1}{c}\left[\frac{\partial \mathbf{D}}{\partial t} + 4\pi \mathbf{J}\right], \tag{4-19}$$

and only the term $\partial \mathbf{E}/\partial t$ does *not* represent charge in motion. Although microscopic motion of all charges determines the total conduction current, measurements alone can not distinguish between the free and bound charge current densities, $\mathbf{J}$ and $\partial \mathbf{P}/\partial t$, respectively.

Another material coefficient characterizing the medium is obtained by noting that the induced current density owing to free charges will be a function of the total electric field within the substance. In linear isotropic media this relationship is written

$$\mathbf{J} = \sigma \mathbf{E}, \tag{4-20}$$

defining the *electric conductivity* σ. Again, this is in general a tensor equation and nonlinear for some media. Moreover, although $\mathbf{J}$ may be the total current induced in the medium by an external field, it is important to realize that $\mathbf{E}$ is actually the *total* field within the substance.

MAGNETIC FIELDS

All magnetic fields are caused by charge in motion, which also creates magnetic dipoles. Therefore, some speculations of elementary-particle physics notwithstanding, one always presumes magnetic monopoles not to exist, and the second equation in Eq.(4–18a) is always true in any medium. In a macroscopic substance we define the magnetization $\mathbf{M}$ as the total magnetic moment per unit volume, and note that the magnetic field produced by $\mathbf{M}$ can equally well be produced by the current density

$$\mathbf{J}_b \equiv c \nabla \times \mathbf{M}. \tag{4-21}$$

The macroscopic magnetic field $\mathbf{B}$ is a volume average of the microscopic fields, and in a macroscopic material which is purely magnetic Eq.(4–19) becomes

$$\nabla \times \mathbf{B} = \frac{4\pi}{c} \left[\mathbf{J}_b + \mathbf{J}_f \right] = \frac{4\pi}{c} \mathbf{J}_t. \tag{4-22}$$

But one can also write this equation as

$$\nabla \times (\mathbf{B} - 4\pi \mathbf{M}) = \frac{4\pi}{c} \mathbf{J}_f, \tag{4-23}$$

thereby suggesting it may be useful to define a new vector

$$\mathbf{H} \equiv \mathbf{B} - 4\pi \mathbf{M}, \tag{4-24}$$

called the magnetic field intensity. Equation (4–23) is then rewritten as

$$\nabla \times \mathbf{H} = \frac{4\pi}{c} \mathbf{J}_f. \tag{4-25}$$

Contrary to the case with $\mathbf{D}$, the quantity $\mathbf{H}$ is very useful, because it is related to the free current with which we work in experiments. The macroscopic field $\mathbf{H}$ arises in media because $\mathbf{B}$ is originally defined in terms of the atomic and molecular currents only, but the particle magnetic moments $\boldsymbol{\mu}$ also contribute to the total field within the medium. Although it is true that $\mathbf{B} = \nabla \times \mathbf{A}$ and $\mathbf{A}$ is a proper average of microscopic vector potentials, the actual macroscopic field in matter depends additionally on the magnetization.

We exclude here any general discussion of materials possessing a permanent magnetization, such as ferromagnets. As we know, induced magnetization falls into two broad classes. Diamagnetic materials have $\mathbf{M}$ antiparallel to $\mathbf{H}$, manifesting Lenz's law: currents induced by an external magnetic field will in turn set up fields opposing the original field. Paramagnetic substances have $\mathbf{M}$ parallel to $\mathbf{H}$, and $\mathbf{M}$ arises from orientation of the molecular magnetic moments in the external field. In reality systems are either diamagnetic or paramagnetic depending on the outcome of competition between the two effects.

Constitutive relations connect the molecular characteristics of matter to the macroscopic fields and, except near a possible phase transition to a magnetically ordered state, $\mathbf{M}$ can be expanded as follows:

$$\mathbf{M} = \mathbf{M}(0) + \left(\frac{\partial M}{\partial H}\right)_{H=0} \mathbf{H} + O(\mathbf{H}^2). \tag{4-26}$$

It is conventional to consider $\mathbf{M}$ as a function of $\mathbf{H}$, rather than of $\mathbf{B}$. Again with the exception of permanent magnets, the relation is generally found to be linear, so that one writes

$$\mathbf{M} = \chi_m \mathbf{H}, \qquad \chi_m \equiv \left(\frac{\partial M}{\partial H}\right)_{H=0}, \tag{4-27}$$

defining the *magnetic susceptibility* per unit volume. Of course χ_m is in general a tensor, and in permanent magnets $\chi_m = \chi_m(\mathbf{H})$.

It is often convenient to define another material parameter by writing

$$\begin{aligned} \mathbf{B} &= \mathbf{H} + 4\pi \mathbf{M} \\ &= (1 + 4\pi \chi_m)\mathbf{H} \\ &= \mu \mathbf{H}, \end{aligned} \tag{4-28}$$

where $\mu \equiv (1 + 4\pi \chi_m)$ is the *magnetic permeability*, and in Gaussian units $\mu = 1$ in vacuum. Because χ_m is usually of order 10^{-5}, $\mathbf{B} \simeq \mathbf{H}$ is often a good approximation under appropriate circumstances.

As an example of the above relationships, consider a permeable sphere in the presence of an external field $\mathbf{H}_0 = \mathbf{B}_0$, as illustrated in Figure 4–2. The internal fields are related by $\mathbf{B}_{\text{in}} = \mu \mathbf{H}_{\text{in}}$, so that

$$\mathbf{M} = \frac{3}{4\pi}\left(\frac{\mu-1}{\mu+2}\right)\mathbf{H}_0 = \frac{\mu-1}{4\pi}\mathbf{H}_{\text{in}} \equiv \chi_m \mathbf{H}_{\text{in}}, \tag{4-29}$$

where we have noted the relation

$$\mathbf{H}_{\text{in}}\frac{3}{\mu+2}\mathbf{H}_0. \tag{4-30}$$

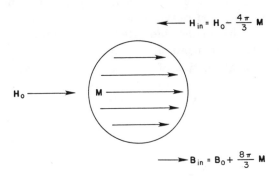

Fig. 4–2. Magnetization and field relations for a permeable sphere.

These expressions emphasize the importance of recognizing that $\mathbf{H}$ in Eqs.(4–27) and (4–28) is the total internal field $\mathbf{H}_{\text{in}}$. For large μ the use of $\mathbf{H}_0$ would lead to considerable error. [It is interesting to note that these same equations hold for a dielectric sphere if we make the replacements $\mathbf{B} \to \mathbf{D}$, $\mathbf{H} \to \mathbf{E}$, $\mu \to \varepsilon$, $\mathbf{M} \to \mathbf{P}$, and note that $\mathbf{E}$ and $\mathbf{D}$ are parallel.]

In summary, these considerations for a general medium show that the Maxwell equation (4–19) should be replaced by

$$\nabla \times \mathbf{H} = \frac{1}{c}\frac{\partial D}{\partial t} + \frac{4\pi}{c}\mathbf{J}_{\text{f}} \tag{4–31}$$

for an arbitrary time-varying electromagnetic field. Moreover, the constitutive relations of Eqs.(4–26) and (4–27) are generally tensor relations and can be nonlinear, as in Eq.(4–8).

Finally, many substances are not only anisotropic, but also inhomogeneous. Thus, the constitutive relations we have discussed here can occasionally be nonlocal, as well as memory retaining. This means that, in the linear approximation, for example, we often must consider expressions of the form

$$P_i(\mathbf{r},t) = \int d^3r' \int_{-\infty}^{t} \chi_{ij}^{e}(\mathbf{r},\mathbf{r}';t,t')E_j(\mathbf{r}',t')\,dt'\,, \tag{4–32a}$$

$$M_i(\mathbf{r},t) = \int d^3r' \int_{-\infty}^{t} \chi_{ij}^{m}(\mathbf{r},\mathbf{r}';t,t')H_j(\mathbf{r}',t')\,dt'\,, \tag{4–32b}$$

$$J_i(\mathbf{r},t) = \int d^3r' \int_{-\infty}^{t} \sigma_{ij}(\mathbf{r},\mathbf{r}';t,t')E_j(\mathbf{r}',t')\,dt'\,, \tag{4–32c}$$

again employing the summation convention.

A. Magnetic Polarization

As a first example of the utility of linear response theory we consider a dynamical perturbation in the form of an external magnetic field, and only systems with zero spontaneous magnetization will be examined. Perhaps the simplest application of this type concerns the so-called isothermal susceptibilities, which are calculated as response coefficients in the manner indicated in Problem 3.1. The external forces F_j are taken to be the components of a uniform static magnetic field $\mathbf{H}$, so that the Hamiltonian is now a function of this external parameter: $\hat{\mathsf{H}} = \hat{\mathsf{H}}(\mathbf{H})$. Such external parameters generate 'generalized forces' within the system in the form of derivatives with respect to these parameters. Hence, in the present case the observable of interest is described by the total magnetization operator, defined as

$$\hat{M}_i \equiv -\frac{\partial \hat{\mathsf{H}}}{\partial H_i}, \tag{4-33}$$

and therefore the isothermal susceptibility is found to be

$$\chi_{ij}(\mathbf{H}) \equiv \frac{\partial \langle \hat{M}_i \rangle}{\partial H_j} = \frac{\beta}{V} \left[\langle \overline{\hat{M}_j}\hat{M}_i \rangle - \langle \hat{M}_j \rangle \langle \hat{M}_i \rangle \right]. \tag{4-34}$$

In arriving at this expression we referred to Appendix A for the identity

$$\frac{\partial}{\partial H_j} e^{-\beta \hat{\mathsf{H}}} = \beta e^{-\beta \hat{\mathsf{H}}} \overline{\hat{M}_j}. \tag{4-35}$$

This is a completely nonlinear expression, in that the expectation values and Kubo transform, though calculated with a canonical statistical operator, utilize the full Hamiltonian. The context in which Eq.(4-34) applies, it will be recalled, is that the relevant information includes the fact that the system remains in thermal equilibrium following application of the external forces.

Although Eq.(4-34) is nonlinear, it is nonlinear in the *external* field, and so is not the material parameter defined in Eq.(4-27). The latter involves the total *internal* field and, owing to the long-range character of the dipolar fields, this will be strongly dependent on the geometry of the material sample. We can, however, define the actual material parameter as

$$\chi_{ij} = \frac{\partial H_i}{\partial H_k^0} \frac{\partial \langle \hat{M}_k \rangle}{\partial H_j}, \tag{4-36}$$

under the supposition that the external field $\mathbf{H}^0$ can be determined as a function of the internal field in any specific experiment. Thus, we shall agree that it is actually the second factor on the right-hand side of Eq.(4-36) that is being discussed in the present section.

It is the linearized version of Eq.(4-34) that is usually encountered in the literature, in which one sets $\mathbf{H} = 0$. Owing to the specification of zero spontaneous magnetization, Eq.(4-34) in this limit becomes

$$\chi_{ij} = \frac{\beta}{V} \langle \overline{\hat{M}_j}\hat{M}_i \rangle_0, \tag{4-37}$$

Let us now consider an arbitrary time-varying magnetic field, in which case the Hamiltonian is

$$\hat{H}_t = \hat{H}_0 - \sum_j \hat{M}_j H_j(t) . \qquad (4\text{-}38)$$

From our earlier work we recall that the total magnetic moment of the system has the linear form

$$\langle \hat{M}_i(t) \rangle = \int_{t_0}^{t} \phi_{ij}(t - t') H_j(t') \, dt' , \qquad (4\text{-}39)$$

with

$$\phi_{ij}(t - t') = \frac{1}{\hbar} \langle [\hat{M}_j, \hat{M}_i(t - t')] \rangle_0 . \qquad (4\text{-}40)$$

The interpretation here is that the field is turned on at time t_0, and t is the time at which the measurement is made. A more useful form of this result is obtained by again recalling the relation of ϕ_{ij} to the time-correlation function and writing

$$\langle \hat{M}_i(t) \rangle = \beta \int_{t_0}^{t} \frac{d}{dt'} \langle \overline{\hat{M}_j(t')} \hat{M}_j(t) \rangle_0 \, H_j(t') \, dt'$$

$$= \beta \langle \overline{\hat{M}_j} \hat{M}_i \rangle_0 \, H_j(t) - \beta \int_{t_0}^{t} dt' \, \langle \overline{\hat{M}_j} \hat{M}_i(t - t') \rangle_0 \frac{dH_j}{dt'} , \qquad (4\text{-}41)$$

after an integration by parts. With a change of integration variables we can rewrite this as

$$\langle \hat{M}_i(t) \rangle = \beta \langle \overline{\hat{M}_j} \hat{M}_i \rangle_0 \, H_j(t) + \beta \int_{0}^{t - t_0} dt' \, \langle \overline{\hat{M}_j} \hat{M}_i(t') \rangle_0 \frac{dH_j(t - t')}{dt'} . \qquad (4\text{-}42)$$

Now, in standard applications it is customary to let $t_0 = -\infty$ and write the magnetic field explicitly as

$$\mathbf{H}(t) = \mathbf{H}e^{i\omega t} . \qquad (4\text{-}43)$$

Then, Eq.(4-42) becomes

$$\langle \hat{M}_i(t) \rangle = \beta \langle \overline{\hat{M}_j} \hat{M}_i \rangle_0 \, H_j e^{i\omega t} - i\omega\beta H_j e^{i\omega t} \int_{0}^{\infty} dt' \, e^{-i\omega t'} \langle \overline{\hat{M}_j} \hat{M}_i(t') \rangle_0 , \qquad (4\text{-}44)$$

and one immediately identifies the complex frequency-dependent magnetic susceptibility by dividing by the volume:

$$\chi_{ij}(\omega) = \frac{\beta}{V} \langle \overline{\hat{M}_j} \hat{M}_i \rangle_0 - \frac{i\omega\beta}{V} \int_{0}^{\infty} e^{-i\omega t'} \langle \overline{\hat{M}_j} \hat{M}_i(t') \rangle_0 \, dt' . \qquad (4\text{-}45)$$

This expression was first obtained by Kubo and Tomita (1954), and illustrates how the susceptibility should relax to the zero-frequency result of Eq.(4-37). These authors, as well as Argyres and Kelley (1964), and Oppenheim, *et al* (1964), have based theories of nuclear magnetic relaxation on expressions such as Eq.(4–45). We shall examine relaxation processes in more detail in later chapters.

B. Electric Polarization

A problem somewhat analogous to that of the preceding section is the case of an external electric field applied to a macroscopic dielectric. As we know, calculation of the material parameters is complicated by their intrinsic relation to the field internal to the medium, rather than to the external field. In order to clarify this in the present context, we first define the total electric polarization operator as

$$\hat{M}_i \equiv -\left(\frac{\partial \hat{H}}{\partial E_i^0}\right)_{\mathbf{E}^0=0}, \tag{4–46}$$

where $\mathbf{E}^0$ is the external field. This is analogous to the definition (4–33) for magnetic fields, except that here we shall linearize from the beginning.

Now refer to Eq.(4–11), in which $\mathbf{E}$ is the total internal field, and rewrite that expression as

$$\varepsilon_{ij} - \delta_{ij} = \frac{4\pi}{V}\frac{\partial}{\partial E_j}\langle \hat{M}_i\rangle, \tag{4–47}$$

thereby connecting the phenomenological and statistical-mechanical theories. Because the derivative is with respect to the total internal field in the dielectric, it is rather difficult to calculate. Instead, we consider $\mathbf{E}$ as a function of the external field, $\mathbf{E}_0$, and rewrite Eq.(4–47) as

$$\varepsilon_{ij} - \delta_{ij} = \frac{4\pi}{V}\left(\frac{\partial E_k^0}{\partial E_j}\right)_{\mathbf{E}=0}\left(\frac{\partial \langle \hat{M}_i\rangle}{\partial E_k^0}\right)_{\mathbf{E}^0=0}, \tag{4–48}$$

where again the summation convention is implied. It is well to point out that we have here made a further linearity approximation, in that V may vary with $\mathbf{E}$ if the field is inhomogeneous. Clearly, such effects are of higher order.

The dependence of $\mathbf{E}_0$ on $\mathbf{E}$ is by no means simple in a general dielectric, because the dipolar interactions are of long range and cause this relationship to depend crucially on the geometry. For example, in an ideal parallel-plate capacitor completely filled with dielectric,

$$\mathbf{E}^0 = \mathbf{D} = \varepsilon\mathbf{E}, \tag{4–49}$$

whereas for a sphere in vacuum and a uniform external field,

$$\mathbf{E}^0 = \frac{\varepsilon+2}{3}\mathbf{E}. \tag{4–50}$$

In this latter example the isotropic form of Eq.(4–48) yields

$$\frac{\varepsilon-1}{\varepsilon+2} = \frac{4\pi}{V}\left(\frac{\partial \langle \hat{M}\rangle}{\partial E^0}\right)_{E^0=0}. \tag{4–51}$$

For many substances the derivative is just $N\alpha$, where N is the total particle number and α the scalar electric polarizability, and this becomes the familiar Clausius-Mossotti relation. In general, when this relation is valid it takes the form

$$\frac{\varepsilon - 1}{\varepsilon + 2} = \frac{4\pi}{3} \left(\frac{N}{V}\right) \alpha[1 + O(\alpha^2)], \qquad (4\text{-}52)$$

and holds for numerous geometries other than a sphere. Therefore, a very detailed examination of the derivatives in Eq.(4-48) is required before Eq.(4-52) can be derived in general, under the appropriate approximations (e.g., Böttcher, 1973; Fröhlich, 1949).

As a consequence of these observations, it seems reasonable to postpone the problem of calculating the derivatives of $\mathbf{E}^0$ until specific applications are made, and focus on what can be obtained directly from the present theory. Equations (4-12) and (4-48) then lead us to define the macroscopic electric polarizability tensor as

$$\alpha_{ij} \equiv \left(\frac{\partial \langle \hat{M}_i \rangle}{\partial E_j^0}\right)_{\mathbf{E}^0 = 0}. \qquad (4\text{-}53)$$

In the remainder of this section we shall omit the superscript '0' and interpret $\mathbf{E}$ as the external electric field, for there can be no confusion in this context.

Take the total Hamiltonian of the system to be

$$\hat{H}_t = \hat{H}_0 - \sum_j \hat{M}_j E_j(t), \qquad (4\text{-}54)$$

where the external field is turned on at time $t = t_0$. Just as in the case of magnetic fields, response theory yields for the expectation value of the polarization at time t the nonlinear expression

$$\langle \hat{M}_i(t) \rangle = \frac{1}{i\hbar} \int_{t_0}^{t} \langle [\hat{M}_j(t_0), \hat{M}_i(t, t')] \rangle_0 E_j(t') \, dt', \qquad (4\text{-}55)$$

where the time-development operators depend on the full Hamiltonian of Eq.(4-54). The linear expression is readily obtained, with the external field being written in the form

$$\mathbf{E}(t) = \mathbf{E}e^{i\omega t}. \qquad (4\text{-}56)$$

Equation (4-55) then becomes

$$\langle \hat{M}_i(t) \rangle = \beta \langle \overline{\hat{M}_j \hat{M}_i} \rangle_0 E_j e^{i\omega t} - i\omega\beta E_j e^{i\omega t} \int_0^\infty e^{-i\omega t'} \langle \overline{\hat{M}_j \hat{M}_i(t')} \rangle_0 \, dt', \qquad (4\text{-}57)$$

when there is no spontaneous polarization. Thus, we identify the polarizability tensor as

$$\alpha_{ij}(\omega) = \beta \langle \overline{\hat{M}_j \hat{M}_i} \rangle_0 - i\omega\beta \int_0^\infty e^{-i\omega t'} \langle \overline{\hat{M}_j \hat{M}_i(t')} \rangle_0 \, dt', \qquad (4\text{-}58)$$

and for an isotropic medium this reduces to

$$\alpha(\omega) = \tfrac{1}{3}\beta\langle\overline{\hat{\mathbf{M}}}\cdot\hat{\mathbf{M}}\rangle_0 - \frac{i\omega\beta}{3}\int_0^\infty e^{-i\omega t'}\langle\overline{\hat{\mathbf{M}}}\cdot\hat{\mathbf{M}}(t')\rangle_0\,dt'. \tag{4-59}$$

Further approximations suggest themselves at this point, the first of which is found by rewriting the expectation values in the form

$$\langle\overline{\hat{B}}\hat{B}\rangle_0 = \langle\hat{B}^2\rangle_0 + \int_0^1 dx\,\big\langle\big[e^{x\beta\,\hat{H}_0},\hat{B}\big]e^{-x\beta\,\hat{H}_0}\,\hat{B}\big\rangle_0. \tag{4-60}$$

But that part of $\hat{H}_0$ which fails to commute with $\hat{\mathbf{M}}$ is generally a differential operator, so that the commutator is proportional to higher-order multipoles. We have neglected quadrupoles and higher from the beginning, so it is reasonable to omit such terms here and approximate Eq.(4-59) by

$$\alpha(\omega) \simeq \tfrac{1}{3}\beta\langle\hat{\mathbf{M}}^2\rangle_0 - \frac{i\omega\beta}{3}\int_0^\infty e^{-i\omega t'}\langle\hat{\mathbf{M}}\cdot\hat{\mathbf{M}}(t')\rangle_0\,dt'. \tag{4-61}$$

This type of expression was apparently first used by Glarum (1960) to study relaxation in dielectrics. In Eqs.(4-58), (4-59), and (4-61) we can see how the polarizability relaxes to the isothermal result as the time-correlation decays. Thus, there was no need to calculate this quantity separately.

As a final approximation, consider a very dilute gas in which we neglect contributions from the intermolecular interactions. Then $\mathbf{E} = \mathbf{E}_0$ and $\langle M^2\rangle_0 = N\langle\hat{\mu}^2\rangle_0$, so that the time-correlation function is just $N\alpha$. In the case of non-permanent dipoles $\langle\hat{\mu}^2\rangle_0 = 0$, and Eq.(4-61) leads to the well-known Debye equation (Debye, 1949).

C. Electrical Conductivity

The prototype calculation in response theory has generally involved the electrical conductivity. Because this parameter relates the induced current in the medium to the total internal field, the corresponding response function has given rise to some difficulties in its interpretation. We shall discuss this problem further presently, after first reviewing the standard results.

Let us consider an external static electric field applied to the medium in the x-direction. The Hamiltonian for the system is

$$\hat{H} = \hat{H}_0 - e\hat{x}E, \tag{4-62}$$

where $E \equiv E_x$, e is the electric charge on each particle, and $\hat{x}$ is the sum of the displacement operators for all the particles. We measure the induced current density in the x-direction, and note that $\langle\hat{j}_x\rangle_0 = 0$, because it is presumed that there are no currents in the equilibrium system. This is a static-field problem of the kind considered in Problem 3.7, so that

$$\langle\hat{j}_x(t)\rangle = \frac{1}{i\hbar}E\int_0^{t-t_0}\langle[e\hat{x},\hat{j}_x(\tau)]\rangle_0\,d\tau, \tag{4-63}$$

which is completely nonlinear—that is, the time-evolution operators contain the full Hamiltonian of Eq.(4–62).

As already abundantly demonstrated, this response functional can also be written as a time-correlation functional. Thus, with the additional observation that the time derivative of the dipole-moment operator is just the current, $e\dot{\hat{x}} = V\hat{j}_x$, we can rewrite Eq.(4–63) as

$$\langle \hat{j}_x(t) \rangle = \beta E V \int_0^{t-t_0} \langle \overline{\hat{j}_x} \hat{j}_x(\tau) \rangle_0 \, d\tau , \qquad (4\text{–}64)$$

and then identify the field-dependent conductivity tensor per unit volume:

$$\sigma_{xx}(t) \equiv \beta \int_0^{t-t_0} \langle \overline{\hat{j}_x} \hat{j}_x(\tau) \rangle_0 \, d\tau . \qquad (4\text{–}65)$$

If one allows $t_0 \to -\infty$, this is just the nonlinear result used by Miyake and Kubo (1962) to discuss the 'kink' effect observed by Esaki (1962). The linearized version of Eq.(4–65) is what is most often found in the literature.

The more general case of a time-varying electric field applied to a system which was in thermal equilibrium in the remote past will be discussed only in terms of the linear response, in the form of the induced current. We generalize the Hamiltonian of Eq.(4–62) to

$$\hat{H}_t = \hat{H}_0 - e\hat{x}_i E_i(t) , \qquad (4\text{–}66)$$

employing the summation convention over repeated indices. Then, the linear response of the system is

$$\langle \hat{j}_k(t) \rangle = \int_{-\infty}^{t} E_i(t') \phi_{ik}(t - t') \, dt' , \qquad (4\text{–}67a)$$

with

$$\phi_{ik}(t - t') = \frac{1}{i\hbar} \langle [e\hat{x}_i(t'), \hat{j}_k(t)] \rangle_0 . \qquad (4\text{–}67b)$$

Once again introduce the time-correlation function and, after changing integration variables, we obtain the alternative expression

$$\langle \hat{j}_k(t) \rangle = \beta V \int_0^{\infty} E_i(t - t') \langle \overline{\hat{j}_i} \, \hat{j}_k(t') \rangle_0 \, dt' . \qquad (4\text{–}68)$$

When the external field has the explicit form

$$\mathbf{E}(t) = \mathbf{E}^0 e^{i\omega t} , \qquad (4\text{–}69)$$

Eq.(4–68) can be rewritten as

$$\langle \hat{j}_k(t) \rangle = \beta V E_i^0 e^{i\omega t} \int_0^{\infty} e^{-i\omega t'} \langle \overline{\hat{j}_i} \, \hat{j}_k(t') \rangle_0 \, dt' . \qquad (4\text{–}70)$$

Hence, the frequency-dependent conductivity tensor per unit volume is

$$\sigma_{ik}(\omega) = \beta \int_0^\infty e^{-i\omega t} \langle \hat{\jmath}_i \, \hat{\jmath}_k(t) \rangle_0 \, dt \, . \tag{4–71}$$

This type of expression for the electrical conductivity has served as a starting point for various applications to models of real systems. For example, Langer (1960, 1961, 1962a,b), and others, have used it to study the electrical conductivity of metals, and a similar application has been carried out by Friedman (1964a, b). As remarked above, Eq.(4–70) relates the induced current to the *external* field, so that $\sigma_{ik}(\omega)$ is not necessarily the material parameter usually thought of as the conductivity. It is useful, therefore, to sketch a detailed calculation in which this problem is exhibited explicitly.

We consider a fully-ionized gas at low density and high temperature, and investigate its linear response to a weak external electromagnetic field. The unperturbed system of density n is characterized by the equilibrium canonical statistical operator $\hat{\rho}_0$, and the time-independent Hamiltonian $\hat{H}$. In the equilibrium system the charge-density operator, vector potential, and scalar potential, respectively, are written $\hat{\gamma}(\mathbf{r})$, $\mathbf{A}(\mathbf{r})$, $\phi(\mathbf{r})$. An external perturbation is described by the full Hamiltonian

$$\hat{H}'(t) = \hat{H} + \hat{H}^0(t) \, , \tag{4–72}$$

where $\hat{H}^0(t)$ is a functional of the external potentials $\mathbf{A}^0(\mathbf{r},t)$, $\phi^0(\mathbf{r},t)$, the external charge density $\hat{\gamma}^0(\mathbf{r},t)$, and the external current density $\hat{\jmath}^0(\mathbf{r},t)$. The external fields are then determined from the prescriptions

$$\mathbf{E}^0(\mathbf{r},t) = -\frac{1}{c}\frac{\partial}{\partial t}\mathbf{A}^0(\mathbf{r},t) - \nabla\phi^0(\mathbf{r},t) \, , \tag{4–73a}$$

$$\mathbf{B}^0(\mathbf{r},t) = \nabla \times \mathbf{A}^0(\mathbf{r},t) \, , \tag{4–73b}$$

which are in turn related by Maxwell's equations.

The applied fields perturb the equilibrium system by inducing local currents, as well as local fields $\langle \mathbf{E}'' \rangle$ and $\langle \mathbf{B}'' \rangle$. The total local fields within the medium can then be written as

$$\mathbf{E}'(\mathbf{r},t) \equiv \mathbf{E}^0(\mathbf{r},t) + \langle \mathbf{E}''(\mathbf{r},t) \rangle \, , \tag{4–74a}$$

$$\mathbf{B}'(\mathbf{r},t) \equiv \mathbf{B}^0(\mathbf{r},t) + \langle \mathbf{B}''(\mathbf{r},t) \rangle \, , \tag{4–74b}$$

owing to the linearity of Maxwell's equations.

It is most convenient to work in a Fourier-transformed representation, in which case the Maxwell equations for the external fields become

$$\begin{aligned}
i\mathbf{k} \cdot \mathbf{E}^0(\mathbf{k},\omega) &= 4\pi\gamma^0(\mathbf{k},\omega) \\
ic\mathbf{k} \times \mathbf{E}^0(\mathbf{k},\omega) &= i\omega\mathbf{B}^0(\mathbf{k},\omega) \\
i\mathbf{k} \cdot \mathbf{B}^0(\mathbf{k},\omega) &= 0 \\
ic\mathbf{k} \times \mathbf{B}^0(\mathbf{k},\omega) &= -i\omega\mathbf{E}^0(\mathbf{k},\omega) + 4\pi\mathbf{j}^0(\mathbf{k},\omega) \, .
\end{aligned} \tag{4–75}$$

Similarly, the total local fields in the perturbed system are

$$ik \cdot \mathbf{E}'(\mathbf{k}, \omega) = 4\pi[\gamma^0(\mathbf{k}, \omega) + \langle \hat{\gamma}''(\mathbf{k}, \omega) \rangle]$$
$$ick \times \mathbf{E}'(\mathbf{k}, \omega) = i\omega \mathbf{B}'(\mathbf{k}, \omega)$$
$$ik \cdot \mathbf{B}'(\mathbf{k}, \omega) = 0$$
$$ick \times \mathbf{B}'(\mathbf{k}, \omega) = -i\omega \mathbf{E}'(\mathbf{k}, \omega) + 4\pi[\mathbf{j}^0(\mathbf{k}, \omega) + \langle \hat{\mathbf{j}}''(\mathbf{k}, \omega) \rangle], \qquad (4\text{-}76)$$

where $\langle \hat{\gamma}'' \rangle$ and $\langle \hat{\mathbf{j}}'' \rangle$ are the induced local charge and current densities, respectively. In order to identify the calculable electromagnetic quantities of interest, it is useful to eliminate $\mathbf{B}^0$ from Eqs.(4-75) and $\mathbf{B}'$ from Eqs.(4-76), and then eliminate the external charge and current densities between them. This yields the following two relations:

$$ik \cdot (\mathbf{E}' - \mathbf{E}^0) = 4\pi \langle \hat{\gamma}'' \rangle, \qquad (4\text{-}77a)$$

$$c^2[\mathbf{k} \times (\mathbf{k} \times \mathbf{E}') - \mathbf{k} \times (\mathbf{k} \times \mathbf{E}^0)] = -\omega^2(\mathbf{E}' - \mathbf{E}^0) - i\omega 4\pi \langle \hat{\mathbf{j}}'' \rangle. \qquad (4\text{-}77b)$$

We next define the *external conductivity tensor* σ_{ij}^0 by means of the linear relation

$$\langle \hat{j}_i''(\mathbf{k}, \omega) \rangle = -i\sigma_{ij}^0(\mathbf{k}, \omega) E_j^0(\mathbf{k}, \omega), \qquad (4\text{-}78)$$

relating the induced current in the medium to the external electric field. As usual, we use the summation convention over repeated indices. Because the unperturbed system is presumed uniform and isotropic, the vector $\mathbf{k}$ defines the only preferred direction and σ_{ij}^0 can be decomposed as follows:

$$\sigma_{ij}^0(\mathbf{k}, \omega) = \sigma_L^0(\mathbf{k}, \omega)\frac{k_i k_j}{k^2} + \sigma_T^0(\mathbf{k}, \omega)\left(\delta_{ij} - \frac{k_i k_j}{k^2}\right), \qquad (4\text{-}79)$$

which defines the longitudinal and transverse conductivities. If one presumes the continuity equation holds in the perturbed system, then in Fourier-transformed notation

$$\mathbf{k} \cdot \langle \hat{\mathbf{j}}'' \rangle = \omega \langle \hat{\gamma}'' \rangle. \qquad (4\text{-}80)$$

From Eqs.(4-77)-(4-80) some straightforward algebra yields the longitudinal and transverse components of the total local fields in the perturbed system:

$$E_L'(\mathbf{k}, \omega) = E_L^0(\mathbf{k}, \omega)\left[1 - \frac{4\pi}{\omega}\sigma_L^0(\mathbf{k}, \omega)\right], \qquad (4\text{-}81a)$$

$$E_T'(\mathbf{k}, \omega) = E_T^0(\mathbf{k}, \omega)\left[1 - 4\pi\frac{\omega \sigma_T^0(\mathbf{k}, \omega)}{\omega^2 - c^2 k^2}\right]. \qquad (4\text{-}81b)$$

In the same manner one can introduce the *microscopic conductivity tensor* σ_{ij}' by the linear relation

$$\langle \hat{j}_i''(\mathbf{k}, \omega) \rangle = -i\sigma_{ij}'(\mathbf{k}, \omega) E_j'(\mathbf{k}, \omega), \qquad (4\text{-}82)$$

relating the induced current to the total local field. Again we can make the decomposition into longitudinal and transverse parts,

$$\sigma'_{ij}(\mathbf{k},\omega) = \sigma'_L(\mathbf{k},\omega)\frac{k_i k_j}{k^2} + \sigma'_T(\mathbf{k},\omega)\left(\delta_{ij} - \frac{k_i k_j}{k^2}\right), \qquad (4\text{-}83)$$

and so Eqs.(4-81) can be rewritten as

$$E'_L(\mathbf{k},\omega) = E^0_L(\mathbf{k},\omega) - E'_L(\mathbf{k},\omega)\frac{4\pi\sigma'_L(\mathbf{k},\omega)}{\omega}, \qquad (4\text{-}84a)$$

$$E'_T(\mathbf{k},\omega) = E^0_T(\mathbf{k},\omega) - E'_T(\mathbf{k},\omega)\frac{4\pi\omega\sigma'_T(\mathbf{k},\omega)}{\omega^2 - c^2 k^2}. \qquad (4\text{-}84b)$$

The preceding discussion allows us to define longitudinal and transverse dielectric functions as the following ratios:

$$\varepsilon_L(\mathbf{k},\omega) \equiv \frac{\sigma'_L(\mathbf{k},\omega)}{\sigma^0_L(\mathbf{k},\omega)} = 1 + \frac{4\pi}{\omega}\sigma'_L(\mathbf{k},\omega), \qquad (4\text{-}85a)$$

$$\varepsilon_T(\mathbf{k},\omega) \equiv \frac{\sigma'_T(\mathbf{k},\omega)}{\sigma^0_T(\mathbf{k},\omega)} = 1 + \frac{4\pi\omega'_T(\mathbf{k},\omega)}{\omega^2 - c^2 k^2}. \qquad (4\text{-}85b)$$

A significant result of these definitions resides in the fact that wave propagation in the unperturbed isotropic system is described by the expressions

$$\varepsilon_L(\mathbf{k},\omega) = 0, \qquad \varepsilon_T(\mathbf{k},\omega) = 0, \qquad (4\text{-}86)$$

which provide dispersion relations for longitudinal and transverse waves, respectively. That is, these equations delineate just those regions of the electromagnetic spectrum where the external fields are *not* absorbed by screening, so that propagation can occur. In addition, Eqs.(4-85) are readily inverted to provide desired expressions for the internal conductivities:

$$\sigma'_L(\mathbf{k},\omega) = \frac{\sigma^0_L(\mathbf{k},\omega)}{1 - \frac{4\pi}{\omega}\sigma^0_L(\mathbf{k},\omega)}, \qquad (4\text{-}87a)$$

$$\sigma'_T(\mathbf{k},\omega) = \frac{\sigma^0_T(\mathbf{k},\omega)}{1 - 4\pi\sigma^0_T(\mathbf{k},\omega)\frac{\omega}{\omega^2 - c^2 k^2}}. \qquad (4\text{-}87b)$$

A similar discussion of this problem has been given previously by Izuyama (1961).

Propagation of transverse waves is governed by the second of the conditions (4-86), from which we obtain

$$\text{Re}\,\varepsilon_T(\mathbf{k},\omega) = \left[1 - \frac{4\pi\sigma^0_T(\mathbf{k},\omega)}{\omega^2 - c^2 k^2}\omega\right]^{-1} = 0, \qquad (4\text{-}88)$$

and large values of $\text{Re}\,\sigma^0_T(\mathbf{k},\omega)$ correspond to the propagating waves. The appropriate dispersion relation has been obtained by DuBois, *et al* (1963), and it turns out that the only propagating mode for transverse waves in a high-temperature, low-density, nonrelativistic ionized gas corresponds to

$$\omega \simeq [c^2 k^2 + \omega_p^2]^{1/2}, \tag{4-89}$$

where $\omega_p = (4\pi e^2 n Z^2/m)^{1/2}$ is a representative plasma frequency for the system, and m is the particle mass.

D. Gyromagnetic Phenomena

As a further useful application of dynamical response theory we shall carry out a systematic study of gyromagnetic phenomena. More than any of the applications discussed to this point, such a study serves very well to elucidate the dynamic response theory as a special case of the more general theory derived in Chapter 2, and also exhibits the transition from the former to the latter.

The major physical feature underlying all the gyromagnetic effects is that nuclear and electronic spins, as well as orbital angular momenta, generate a magnetic moment parallel to the angular momentum with a magnitude fixed through a characteristic constant of proportionality called the *gyromagnetic ratio*. Coupled with the appropriate equations of motion, this observation leads to

Larmor's Theorem. *Provided that the angular momentum of the system in the direction of the field is a constant of the motion, the effect of a uniform magnetic field on a system of spins or particles can be transformed away by introducing a rotating coordinate system.*

This theorem can be derived by first defining a rotation operator

$$\hat{R}(t) \equiv e^{i\hat{\mathbf{J}}\cdot\boldsymbol{\omega}\, t/\hbar}, \tag{4-90}$$

where $\hat{\mathbf{J}}$ is the total angular-momentum operator of the system and $\boldsymbol{\omega}$ the angular velocity of the coordinate system about the axis defined by the vector $\boldsymbol{\omega}$. In the absence of a magnetic field the system wavefunction evolves in time according to

$$\psi(\mathbf{x},t) = e^{-i\hat{\mathrm{H}}_0\, t/\hbar}\,\psi(\mathbf{x}), \tag{4-91}$$

where $\hat{\mathrm{H}}_0$ is time independent. In the presence of a constant magnetic field $\mathbf{H}$, the wavefunction evolves according to the prescription

$$\psi_f(\mathbf{x},t) = e^{-i(\hat{\mathrm{H}}_0 + \hat{\mathrm{H}}_1)t/\hbar}\,\psi(\mathbf{x}), \tag{4-92}$$

where $\hat{\mathrm{H}}_1 = \hat{\mathrm{H}}_1(\mathbf{H}) \neq \hat{\mathrm{H}}_1(t)$. Suppose now that the zero-field system undergoes a rotation of the form (4-90), such that $[\hat{\mathbf{J}}\cdot\boldsymbol{\omega}, \hat{\mathrm{H}}_0] = 0$. Then the description of Eq.(4-91) becomes

$$\psi_r(\mathbf{x},t) = e^{-i(\hat{\mathrm{H}}_0 - \hat{\mathbf{J}}\cdot\boldsymbol{\omega})t/\hbar}\,\psi(\mathbf{x},t). \tag{4-93}$$

The Larmor theorem results from a comparison of ψ_f with ψ_r.

Consider a system of spins all having the same gyromagnetic ratio γ in a uniform magnetic field $\mathbf{H}$, with spin operator $\hat{\mathbf{S}}$. Then $\hat{H}_1 = -\gamma\hat{\mathbf{S}}\cdot\mathbf{H}$, and $\psi_f = \psi_r$ if $\mathbf{H} = \boldsymbol{\omega}\gamma$. Should the particles have different gyromagnetic ratios γ_k, then $\hat{H}_1 = -\sum_k \gamma_k \hat{\mathbf{S}}_k \cdot \mathbf{H}$, and Larmor's theorem holds if $\hat{\mathbf{J}}\cdot\boldsymbol{\omega} = \sum_k \boldsymbol{\omega}_k \cdot \hat{\mathbf{S}}_k$ and $\boldsymbol{\omega}_k \equiv \gamma_k\mathbf{H}$. Although the theorem holds exactly for spin systems, it can be extended to free electrons, say, in a uniform field $\mathbf{H}$. In this case,

$$\hat{H}_1 = \frac{e}{2mc}\mathbf{H}\cdot(\hat{\mathbf{L}}+2\hat{\mathbf{S}}) + \frac{e^2}{8mc^2}|\hat{\mathbf{r}}\times\mathbf{H}|^2 . \tag{4-94}$$

The quadratic term is very small for laboratory magnetic fields and can generally be neglected. Thus, Larmor's theorem is approximately true here, also, with space- and spin-rotational frequencies

$$\boldsymbol{\omega}_\ell = -\frac{e}{2mc}\mathbf{H}, \qquad \boldsymbol{\omega}_s = 2\boldsymbol{\omega}_\ell , \tag{4-95}$$

respectively. Finally, the theorem holds approximately in many cases for time-dependent fields and infinitesimal rotations, with $\mathbf{H}$ and $\boldsymbol{\omega}$ replaced by $\mathbf{H}(t)$ and $\boldsymbol{\omega}(t)$.

Despite Larmor's theorem, however, polarization in the direction of the field can actually take place in a real system, and field effects are not transformed away completely by a rotation. The effect persists because of the dipole-dipole coupling between spins, as an example, for then $\hat{H}_0$ does not commute with $\hat{\mathbf{S}}\cdot\boldsymbol{\omega}$ and this permits the necessary exchange of angular momentum to produce polarization. With a field present the spins are polarized. In the absence of a field, but with the system being physically rotated, the dipolar coupling produces a polarization of the same magnitude as that generated by a field producing a Larmor frequency equal to the rotation frequency. The magnetic field due to rotation alone represents the Barnett effect (Barnett, 1915), whereas the angular momentum due to the field alone is called the Einstein-de Haas effect (Einstein and de Haas, 1915). The following discussion of the theory of these effects follows that of Heims and Jaynes (1962).

The rotational ensemble of Gibbs derives from the general maximum-entropy formalism in an almost trivial way by presuming information available on the total energy and a component of total angular momentum of the system. One obtains from Eq.(2–19) the statistical operator

$$\hat{\rho} = \frac{1}{Z}e^{-\beta(\hat{H}_0 -\omega_i\hat{J}_i)}, \quad Z = \mathrm{Tr}\left[e^{-\beta(\hat{H}_0 -\omega_i\hat{J}_i)}\right], \tag{4-96}$$

with *no* summation convention implied here. If $\omega_i\hat{J}_i$ is a constant of the motion, then the Hamiltonian in the rotating coordinate frame is $\hat{H} = \hat{H}_0 -\omega_i\hat{J}_i$. We say that the system is in thermal equilibrium if all expectation values are constant in a coordinate frame rotating with the system. Then all other vector operators will have time-independent expectation values in the equilibrium ensemble described by Eq.(4–96).

For the remainder of this section we shall presume Boltzmann statistics to be adequate, so that the expectation value of the magnetic moment for a single atom is representative of that for N atoms. This is a reasonable premise if the spin-spin interaction is very weak—which excludes ferromagnetic systems, say, in what follows. In this region of temperature and density it is also valid to assert that $\beta\mu_0 H \ll 1$, where $\mu_0 = e\hbar/2mc$ is the Bohr magneton.

Now consider the electrons of an atom or ion exposed to a uniform magnetic field in the z-direction. The Hamiltonian is then written

$$\hat{H} = \hat{H}_0 + \hat{H}_1 H + \hat{H}_2 H^2 , \qquad (4\text{-}97)$$

where

$$\hat{H}_1 \equiv -\frac{e}{2mc}\left(\hat{L}_z + 2\hat{S}_z\right), \qquad (4\text{-}98a)$$

$$\hat{H}_2 \equiv \frac{e^2}{8mc^2}\sum_k \left(\hat{x}_k^2 + \hat{y}_k^2\right), \qquad (4\text{-}98b)$$

and $\mathbf{H} = \nabla \times \mathbf{A}$ is derived from

$$\mathbf{A} = \tfrac{1}{2}(-yH, xH, 0), \quad \nabla \cdot \mathbf{A} = 0, \quad H \equiv H_z . \qquad (4\text{-}99)$$

The case of a static external field with no rotation has been discussed previously, but it is instructive to review the results in the present notation. From Eqs.(4-97) and (4-33), the expectation of the magnetic-moment operator in the z-direction is

$$\langle \hat{M}_z \rangle = -\langle \hat{H}_1 \rangle - 2H\langle \hat{H}_2 \rangle . \qquad (4\text{-}100)$$

Presuming no spontaneous magnetization, and that the unperturbed system is isotropic, we proceed as in an earlier section to obtain through leading order in the field

$$\langle \hat{M}_z \rangle \simeq \beta H \langle \overline{\hat{H}_1 \hat{H}_1} \rangle_0 - 2H\langle \hat{H}_2 \rangle_0 . \qquad (4\text{-}101)$$

The magnetic susceptibility is then

$$\begin{aligned}
\chi &= \frac{\partial}{\partial H}\left(\frac{\langle \hat{M}_z \rangle}{V}\right) \\
&= \frac{\beta}{V}\langle \overline{\hat{H}_1 \hat{H}_1} \rangle_0 - \frac{2}{V}\langle \hat{H}_2 \rangle_0 .
\end{aligned} \qquad (4\text{-}102)$$

One immediately identifies the diamagnetic contribution as

$$\begin{aligned}
\chi_d &= -\frac{2}{V}\langle \hat{H}_2 \rangle_0 \\
&= -\frac{e^2}{4mc^2 V}\sum_k \langle (\hat{x}_k^2 + \hat{y}_k^2) \rangle_0 \\
&= -\frac{1}{V}\frac{e^2}{6mc^2}\sum_k \langle \hat{r}_k^2 \rangle_0 ,
\end{aligned} \qquad (4\text{-}103)$$

where the third line is valid if the system exhibits cubic or isotropic symmetry. The paramagnetic contribution is

$$\chi_p = \frac{\beta}{V}\langle \overline{\hat{H}_1\,\hat{H}_1}\rangle_0 , \tag{4-104}$$

which is identical with the isothermal result of Eq.(4–37). One can also make a further approximation with respect to the Kubo transform, as discussed following Eq.(4–60), so that the familiar result

$$\chi_p \simeq \frac{\beta\mu_0^2}{\hbar^2 V}\langle(\hat{L}_z + 2\hat{S}_z)^2\rangle_0 \tag{4-105}$$

is an excellent approximation. Of course, this must be multiplied by N to obtain the correct system quantity.

The entire point of this section is to generalize the preceding calculation so as to combine the effects of an external magnetic field with a rotation of the system. Let the stationary system in zero field be described by the Hamiltonian $\hat{H}_0$. Then a static uniform field $\mathbf{H}$ is applied and the system is physically rotated at angular frequency $\boldsymbol{\omega}$. Asterisks will be used to denote quantities evaluated in the rotating coordinate system, which is fixed with respect to the physically rotating crystal, and we shall retain only terms linear in the relevant parameters. The time-dependent Hamiltonian for the system is then

$$\hat{H}^*(t) = \hat{H}_0^* - \frac{e}{2mc}(\hat{\mathbf{L}} + 2\hat{\mathbf{S}})^* \cdot \mathbf{H}^*(t) + \frac{e^2}{8mc^2}\sum_k |\hat{\mathbf{r}}_k \times \mathbf{H}_k^*(t)|^2 - \hat{\mathbf{J}}\cdot\boldsymbol{\omega}. \tag{4-106}$$

Although $\hat{H}_0$ might be expected to be time dependent, $\hat{H}_0^*$ can quite generally be taken independent of the time.

With the summation convention over repeated indices and the presumption that the unperturbed system is unpolarized, we find in the usual way that

$$\langle \hat{M}_i^*\rangle = \chi_{ij}H_j^*(t) + \Theta_{ij}\omega_j , \tag{4-107a}$$

$$\langle \hat{J}_i^*\rangle = \Theta'_{ij}H_j^*(t) + \eta_{ij}\omega_j , \tag{4-107b}$$

where we have defined the following tensors:

$$\chi_{ij} \equiv \beta\left(\frac{e}{2mc}\right)^2 \langle\overline{(\hat{L}_j^* + 2\hat{S}_j^*)(\hat{L}_i^* + 2\hat{S}_i^*)}\rangle_0$$

$$+ \frac{e^2}{4mc^2}\sum_k \langle(\hat{\mathbf{r}}_k^*)_i(\hat{\mathbf{r}}_k^*)_j - \delta_{ij}(\hat{\mathbf{r}}_k^*)^2\rangle_0 , \tag{4-108}$$

$$\Theta_{ij} \equiv \beta\left(\frac{e}{2mc}\right)\langle\overline{\hat{J}_j^*(\hat{L}_i^* + 2\hat{S}_i^*)}\rangle_0 , \tag{4-109}$$

$$\Theta'_{ij} \equiv \beta\left(\frac{e}{2mc}\right)\langle\overline{(\hat{L}_j^* + 2\hat{S}_j^*)\hat{J}_i^*}\rangle_0 , \tag{4-110}$$

$$\eta_{ij} \equiv \langle\overline{\hat{J}_j^*\hat{J}_i^*}\rangle_0 . \tag{4-111}$$

Owing to the general symmetries of the correlation functions, these tensors have the following symmetry properties:

$$\Theta_{ij} = \Theta'_{ji}, \quad \chi_{ij} = \chi_{ji}, \quad \eta_{ij} = \eta_{ji}. \tag{4-112}$$

At this point it is convenient to specialize to a simplified experimental arrangement. This is partially motivated by the observation that the above expressions were derived under the supposition that the system is in thermal equilibrium. But this presumption is not generally valid, because $\mathbf{H}^*(t)$ is time dependent. Consequently, we shall choose the field parallel to the axis of rotation, and then only the transverse components have a time dependence and the gyromagnetic effects are associated with thermal equilibrium. One could, of course, make alternative statements comparing the rate of rotation with relaxation times in the system, but this seems to be more than we are entitled to assert in the present general problem.

Thus, we envision the field to be along the 3-axis, about which the system is rotating. A vector in the rotating system is related to its components in the laboratory frame by the relation

$$\begin{pmatrix} A_1^* \\ A_2^* \\ A_3^* \end{pmatrix} = \begin{pmatrix} \cos\omega t & \sin\omega t & 0 \\ -\sin\omega t & \cos\omega t & 0 \\ 0 & 0 & 1 \end{pmatrix} \begin{pmatrix} A_1 \\ A_2 \\ A_2 \end{pmatrix}, \tag{4-113}$$

and if $\sin\omega t$, $\cos\omega t$ are treated as c-numbers the commutation relations for vector operators are the same in both frames.

We shall also presume that there is sufficient crystal symmetry—such as reflection symmetry across a plane perpendicular to the 3-axis—so that all off-diagonal elements of the above tensors vanish. In that event $\hat{A}_3^* = \hat{A}_3$ for the operators. We then neglect diamagnetism and write for the gyromagnetic effects in this experimental arrangement

$$\langle \hat{M}_3 \rangle = \chi_{33} H + \Theta_{33}\omega, \tag{4-114}$$
$$\langle \hat{J}_3 \rangle = \Theta'_{33} H + \eta_{33}\omega. \tag{4-115}$$

One should also note that any angular momentum given up by the field is $O(H^2)$, and can therefore be neglected.

In the Barnett experiments an angular velocity ω_B is impressed upon the system and a resulting polarization observed. This is compared to an equivalent field H_B producing the same polarization, resulting in the definition of the *Barnett coefficient*:

$$\frac{e}{2mc} g'_B \equiv \frac{\omega_B}{H_B} = \frac{\langle \hat{M}_3 \rangle / \Theta_{33}}{\langle \hat{M}_3 \rangle / \chi_{33}} = \frac{\chi_{33}}{\Theta_{33}}. \tag{4-116}$$

Similarly, in the Einstein-de Haas experiment a field H_3 induces a magnetic moment $\langle \hat{M}_3 \rangle$, which is measured, and an electronic angular momentum $\langle \hat{J}_3 \rangle$ is also generated. This latter is then balanced by a macroscopic rotation $Q\omega_E = -\langle \hat{J}_3 \rangle$, where Q is the moment of inertia of the system about the 3-axis. Measurements then yield the *Einstein-de Haas coefficient*:

$$\frac{e}{2mc}g'_{\rm E} \equiv \frac{\langle \hat{M}_3 \rangle}{Q\omega_{\rm E}} = \frac{\langle \hat{M}_3 \rangle}{\langle \hat{J}_3 \rangle} = \frac{\chi_{33}}{\Theta'_{33}}. \tag{4-117}$$

The first symmetry relation in Eq.(4-112) implies that

$$g'_{\rm B} = g'_{\rm E} \equiv g', \tag{4-118}$$

an Onsager-like reciprocity relation.

Several calculations of g' have been carried out for real systems (e.g., Frank, 1932; Gorter and Kahn, 1940). In the rare-earth salts it is found that the Landé g-factor gives an approximation in quite good agreement with most data:

$$g' \simeq g = \frac{3}{2} + \frac{S(S+1) - L(L+1)}{2J(J+1)}. \tag{4-119}$$

For ferromagnetic materials an approximate calculation by Van Vleck (1950) yielded

$$g' \simeq \frac{g}{g-1}, \tag{4-120}$$

which seems to agree satisfactorily with data on Fe, Ni, Co, and their alloys.

E. Galvanomagnetic Effects

In the preceding section we encountered a set of coupled processes arising when two separate external dynamical sources are allowed to interact with the system. There are, in fact, numerous other phenomena that occur when a magnetic field is imposed upon a system in addition to any other processes which may be taking place, and we examine some of these in this section. These effects are most readily discussed in terms of the equilibrium situation resulting after a static and uniform magnetic field has been impressed upon the system. The zero-field ensemble is described by the statistical operator

$$\hat{\rho}_0 = \frac{1}{Z_0}e^{-\beta \hat{H}_0}, \quad Z_0 = {\rm Tr}\, e^{-\beta \hat{H}_0}, \quad \hat{H}_0 \neq \hat{H}_0(t), \tag{4-121}$$

whereas the equilibrium situation after the field has been established is described by

$$\hat{\rho}_1 = \frac{1}{Z_1}e^{-\beta \hat{H}}, \quad Z_1 = {\rm Tr}\, e^{-\beta \hat{H}}, \quad \hat{H} \equiv \hat{H}_0 + \hat{H}_1 \neq \hat{H}(t). \tag{4-122}$$

Here we have written $\hat{H} = \hat{H}_0 + \hat{H}_1 = \hat{H}_0 - \hat{\mathbf{M}} \cdot \mathbf{H}$, where $\hat{\mathbf{M}}$ is the total magnetic-dipole operator for the system and $\mathbf{H}$ is the external magnetic field.

In the context of the equilibrium ensemble described by Eq.(4–122) the linear response of the system to another external perturbation is

$$\langle \hat{B} \rangle_t = \langle \hat{B} \rangle_1 + \int_{t_0}^{t} \phi_{BA}^{(1)}(t - t') F(t') \, dt' , \qquad (4\text{–}123\text{a})$$

where

$$\phi_{BA}^{(1)}(t - t') \equiv \frac{1}{i\hbar} \langle [\hat{A}, \hat{B}(t - t')] \rangle_1$$

$$= \beta \langle \overline{\hat{A}} \hat{B}(t - t') \rangle_1 . \qquad (4\text{–}123\text{b})$$

These functions are generalizations of the isothermal susceptibilities discussed earlier.

Perhaps the most common application of these equations is to the electrical conductivity of a medium in the presence of a uniform magnetic field. The straightforward generalization of Eq.(4–71) to the present case is

$$\sigma_{ij}(\omega, \mathbf{H}) = \beta \int_0^\infty e^{-i\omega t} \langle \overline{\hat{\jmath}_i} \hat{\jmath}_j(t) \rangle_1 \, dt , \qquad (4\text{–}124)$$

where the expectation values and the Kubo transform are calculated in the ensemble of Eq.(4–122). This is known as the *galvanomagnetic tensor*. In linear approximation one can expand the exponentials by means of perturbation techniques discussed earlier and write

$$\sigma_{ij}(\omega, \mathbf{H}) \simeq \sigma_{ij}(\omega) + H_k \chi_{kij} , \qquad (4\text{–}125)$$

where χ_{kij} is a correlation function involving two components of the electric current operator and one of the magnetic dipole-moment operator.

Under most circumstances it is possible to invert the matrix representation of σ_{ij} to obtain the *resistivity tensor*:

$$\rho_{ij}(\omega, \mathbf{H}) \equiv \sigma_{ij}^{-1}(\omega, \mathbf{H}) . \qquad (4\text{–}126)$$

Those terms in the diagonal elements of ρ_{ij} which are quadratic in the magnetic field describe the phenomenon of magnetoresistance (in weak fields), whereas the off-diagonal elements are proportional to $\mathbf{H}$ and represent the Hall effect (Hall, 1878). In the simplest experimental arrangement the induced Hall current is perpendicular to the plane formed by the electric and magnetic field vectors. At high fields and low temperatures one finds field oscillations in the magnetoresistance, as first seen by Shubnikov and de Haas (1930) in Bismuth, the theory of which has been reviewed by Kubo and Miyake (1965). A general investigation of the high-field galvanomagnetic tensor has been carried out by Animalu (1972).

Careful experiments have confirmed the existence of an effect that we overlooked in the previous discussion of linear response in physical systems. These refer to magneto-electric media, which become magnetized when placed in an electric field. The effects were predicted by Curie long ago (e.g., Curie, 1894), but their observation has proved very difficult. The reason for this difficulty is that the effects seem to arise only in time-asymmetric media (Landau and Lifshitz, 1960). Thus, the material under study must either be in a nonequilibrium state or be in motion—or be a magnetic crystal (spin-ordered structure).

In standard treatments of magneto-electricity (e.g., O'Dell, 1970) these effects are usually described by the constitutive relations

$$P_i = \chi^e_{ij} E^j + \chi^{em}_{ij} B^j \,, \tag{4-127a}$$

$$M_i = \chi^{me}_{ij} e^j + \chi^m_{ij} B^j \,, \tag{4-127b}$$

where χ^e_{ij} and χ^m_{ij} are the electric and magnetic susceptibility tensors, respectively. The quantities χ^{em}_{ij} and χ^{me}_{ij} are then the magneto-electric coefficients. We have seen earlier, though, that the fields **E** and **B** are the *total* fields within matter and are usually somewhat difficult to calculate. Rather, one constructs a theory involving coefficients relating the applied external fields to the observed polarizations. These coefficients must then be related carefully to the actual material coefficients for each experimental arrangement, as indicated earlier. Therefore, we shall here only derive formal linear expressions relating the external fields to the observable polarizations.

Consider an isotropic material in which it is supposed that magneto-electric effects will occur. Because of the kind of media which must be considered candidates, we can not presume that there are no spontaneous polarizations. It turns out that most electric polarization experiments to measure χ^e_{ij} are performed using time-varying fields, whereas the measurement of magnetic susceptibilities is usually done in a static magnetic field. Thus, we shall employ a combination of static and alternating fields to develop a straightforward extension of linear response theory. In particular, we refer to previous sections for the appropriate expressions.

With the application of an external electric field $E(t)$ and the use of the 'magnetic' ensemble (4–122), the predicted expectation value of the total electric dipole moment is

$$\langle \hat{M}_i(t) \rangle = \langle \hat{M}_i \rangle_1 + \frac{1}{i\hbar} \int_{t_0}^t \langle [\hat{M}_j(t_0), \hat{M}_i(t,t')] \rangle_1 E_j(t') \, dt'$$

$$\simeq \beta \left[\langle \overline{\hat{M}_j \hat{M}_i} \rangle_0 - \langle \hat{M}_j \rangle_0 \langle \hat{M}_i \rangle_0 \right] H_j$$

$$+ \frac{1}{i\hbar} \int_{t_0}^t \langle [\hat{M}_j(t_0), \hat{M}_i(t-t')] \rangle_0 E_j(t') \, dt' \,. \tag{4-128}$$

The first term on the right-hand side of this result is to be identified as the magneto-electric coefficient χ^{em}_{ij}, and the second is clearly the electric polarizability tensor, $\alpha_{ij}(t)$.

Instead of the alternating field, suppose now that we apply a static electric field **E** and predict the expectation value of the total magnetic moment. Then,

$$\langle \hat{M}_i \rangle = \langle \hat{M}_i \rangle_1 + \beta \left[\langle \overline{\hat{M}_j \hat{M}_i} \rangle_0 - \langle \hat{M}_j \rangle_0 \langle \hat{M}_i \rangle_0 \right] E_j$$

$$\simeq \beta \left[\langle \overline{\hat{M}_j \hat{M}_i} \rangle_0 - \langle \hat{M}_j \rangle_0 \langle \hat{M}_i \rangle_0 \right] H_j$$
$$+ \beta \left[\langle \overline{\hat{M}_j \hat{M}_i} \rangle_0 - \langle \hat{M}_j \rangle_0 \langle \hat{M}_i \rangle_0 \right] E_j . \tag{4-129}$$

Thus, we obtain a set of coupled equations similar to those of Eqs.(4–127):

$$\langle \hat{M}_i(t) \rangle = \alpha_{ij}(t) E_j + \chi_{ij}^{em} H_j , \tag{4-130a}$$
$$\langle \hat{M}_i \rangle = \chi_{ij}^{me} E_j + \chi_{ij}^{m} H_j , \tag{4-130b}$$

where in the first line E_j is the amplitude of the applied alternating field $\mathbf{E}(t)$ and is presumed equal to that of the static field.

From Eqs.(4–128) and (4–129), and the symmetry properties of the correlation functions (in which the dipole-moment operators are odd under PT), we find the following reciprocity relations for the magneto-electric coefficients:

$$\chi_{ij}^{em} \equiv \beta \left[\langle \overline{\hat{M}_j \hat{M}_i} \rangle_0 - \langle \hat{M}_j \rangle_0 \langle \hat{M}_i \rangle_0 \right] = \chi_{ji}^{me} . \tag{4-131}$$

Therefore, because it turns out to be much easier to perform the experiment leading to Eq.(4–128) than to Eq.(4–129), it is sufficient to measure and calculate χ_{ij}^{em}. Thus, Eq.(4–131) provides a complete formal description of the magneto-electric effects.

F. Nonlinear Electromagnetic Response

With the advent of the laser it has become possible to apply very intense fields to dielectric substances and thereby study a number of nonlinear effects, such as second harmonics. Therefore, we outline here the formal derivation of statistical-mechanical expressions for the higher-order polarizabilities, and this in turn provides an introduction to the application of higher-order dynamical response theory. We shall focus attention completely on phenomena observable at optical frequencies, following to large extent the review provided by Franken and Ward (1963).

It is convenient first to summarize briefly the theory of higher-order dynamical response, which is based on the Hamiltonian

$$\hat{\mathsf{H}}_t = \hat{\mathsf{H}}_0 + \hat{\mathsf{H}}_1 = \hat{\mathsf{H}}_0 - \hat{A}_j F_j(t) , \quad t \geq t_0 , \tag{4-132}$$

and we again employ the summation convention over repeated indices. The Heisenberg operators $\hat{A}_j$ couple the external perturbation to the system, and through second order in the perturbation the response is written as

$$\langle \hat{B}_i(t) \rangle - \langle \hat{B}_i \rangle_0 = \int_{t_0}^{t} \phi_{ij}(t - t') F_j(t') \, dt'$$
$$+ \int_{t_0}^{t} dt_1 \int_{t_0}^{t_1} dt_2 \, \phi_{ijk}(t - t_2, t_1 - t_2) F_k(t_1) F_j(t_2) , \tag{4-133}$$

where the first- and second-order linear response functions are given, respectively, by

$$\phi_{ij}(t - t') = \frac{1}{i\hbar}\langle[\hat{A}_j, \hat{B}_i(t - t')]\rangle_0 , \tag{4-134a}$$

$$\phi_{ijk}(t - t_2, t_1 - t_2) = \frac{1}{(i\hbar)^2}\langle[\hat{A}_j, [\hat{A}_k(t_1 - t_2), \hat{B}_i(t - t_2)]]\rangle_0 . \tag{4-134b}$$

Now consider the perturbation turned on at $t = t_0$ to be the electric field $\mathbf{E}\sin(\omega t)$, coupled to the total electric-dipole operator $e\hat{\mathbf{r}}$. One makes the identifications

$$\hat{A}_j \equiv e\hat{r}_j , \qquad F_j(t) = E_j^\omega \sin(\omega t) , \tag{4-135}$$

and the perturbation Hamiltonian takes the explicit form

$$\hat{H}_1 = -e\hat{\mathbf{r}} \cdot \mathbf{E}^\omega \sin(\omega t)$$
$$= -e\hat{r}_j E_j^\omega \sin(\omega t) . \tag{4-136}$$

The superscript ω is used to identify the frequency with which the field amplitude is to be associated. We envision only dipole coupling in this sample calculation, so that we can approximate $\mathcal{P}$ by the dipole density $\mathbf{P}$, and we shall also presume that there is no initial spontaneous polarization. Then, Eq.(4–133) takes the form

$$\langle\hat{P}_i\rangle = \int_{t_0}^{t} \phi_{ij}(t - t')E_j^\omega \sin(\omega t') \, dt'$$
$$+ \int_{t_0}^{t} dt_1 \int_{t_0}^{t_1} dt_2 \, \phi_{ijk}(t - t_2, t_1 - t_2)E_k^\omega \sin(\omega t_1)E_j^\omega \sin(\omega t_2) . \tag{4-137}$$

This expression can be written in a more revealing form if we now set $t_0 = -\infty$ and change integration variables, including an adiabatic-switching factor if necessary. By employing a number of trigonometric identities to expand the resulting harmonic functions of sums and differences of variables, we find that Eq.(4–137) can be recast as follows:

$$\langle\hat{P}_i\rangle = E_j^\omega \sin(\omega t) \int_0^\infty \phi_{ij}(t_1) \cos(\omega t_1) \, dt_1$$
$$- E_j^\omega \cos(\omega t) \int_0^\infty \phi_{ij}(t_1) \sin(\omega t_1) \, dt_1$$
$$+ \tfrac{1}{2}E_k^\omega E_j^\omega \int_0^\infty dt_1 \int_0^\infty dt_2 \, \phi_{ijk}(t_1 + t_2, t_2) \cos(\omega t_2)$$
$$- \tfrac{1}{2}E_k^\omega E_j^\omega \sin(2\omega t) \int_0^\infty dt_1 \int_0^\infty dt_2 \, \phi_{ijk}(t_1 + t_2, t_2) \sin(2\omega t_1 + \omega t_2)$$
$$- \tfrac{1}{2}E_k^\omega E_j^\omega \cos(2\omega t) \int_0^\infty dt_1 \int_0^\infty dt_2 \, \phi_{ijk}(t_1 + t_2, t_2) \cos(2\omega t_1 + \omega t_2) .$$

$$\tag{4-138}$$

The first two terms correspond to ordinary optical polarization, the third is a dc polarization effect, and the fourth and fifth represent an oscillation at twice the driving frequency.

An even more compact notation can be achieved by rewriting Eq.(4–138) in the form

$$\langle \hat{P}_i(t) \rangle = p_i^0 + p_i^\omega + p_i^{2\omega} , \tag{4–139}$$

where

$$p_i^0 = \chi_{ijk}^0 E_k^\omega E_j^\omega , \tag{4–140}$$

$$
\begin{aligned}
p_i^\omega &= p_i'^\omega \sin(\omega t) + p_i''^\omega \cos(\omega t) \\
&= \chi_{ij}'^\omega E_j^\omega \sin(\omega t) + \chi_{ij}''^\omega E_j^\omega \cos(\omega t) ,
\end{aligned}
\tag{4–141}
$$

$$
\begin{aligned}
p_i^{2\omega} &= p_i'^{2\omega} \sin(2\omega t) + p_i''^{2\omega} \cos(2\omega t) \\
&= \chi_{ijk}'^{2\omega} E_k^\omega E_j^\omega \sin(2\omega t) + \chi_{ijk}''^{2\omega} E_k^\omega E_j^\omega \cos(2\omega t) ,
\end{aligned}
\tag{4–142}
$$

and we have defined tensors

$$\chi_{ij}'^\omega \equiv \int_0^\infty \phi_{ij}(t_1) \cos(\omega t_1) \, dt_1 , \tag{4–143a}$$

$$\chi_{ij}''^\omega \equiv - \int_0^\infty \phi_{ij}(t_1) \sin(\omega t_1) \, dt_1 , \tag{4–143b}$$

$$\chi_{ijk}^0 \equiv \tfrac{1}{2} \int_0^\infty dt_1 \int_0^\infty dt_2 \, \phi_{ijk}(t_1 + t_2, t_2) \cos(\omega t_2) , \tag{4–143c}$$

$$\chi_{ijk}'^{2\omega} \equiv -\tfrac{1}{2} \int_0^\infty dt_1 \int_0^\infty dt_2 \, \phi_{ijk}(t_1 + t_2, t_2) \sin(2\omega t_1 + \omega t_2) , \tag{4–143d}$$

$$\chi_{ijk}''^{2\omega} \equiv -\tfrac{1}{2} \int_0^\infty dt_1 \int_0^\infty dt_2 \, \phi_{ijk}(t_1 + t_2, t_2) \cos(2\omega t_1 + \omega t_2) . \tag{4–143e}$$

In Eq.(4–139) we recognize p_i^0 as a dc effect representing optical rectification, and p_i^ω is the contribution from ordinary optical polarization. The quantity $p_i^{2\omega}$ corresponds to second-harmonic generation, as first seen in the optical region by Franken, *et al* (1961). These results can be considerably generalized by adding more terms to the Hamiltonian and then considering linear response functions of order higher than the second. Prior to discussing these possibilities, though, it is useful to comment on some of the general features of Eqs.(4–139)–(4–143).

The tensors defined in Eqs.(4–143) do not have as many independent components as might be thought at first glance, and some are often zero. This reduction occurs owing to the symmetry properties of the medium. For example, in an isotropic substance, or one possessing inversion symmetry (e.g., calcite), significant even-harmonic production can not occur, because **P** must reverse sign with **E** regardless of the complexity of their tensor relation. In this case one expects $\chi_{ijk}^{2\omega}$ to vanish, say. Lack of inversion symmetry is exactly what gives rise to piezoelectricity, in which even powers of **E** contribute. But the magnitudes of piezoelectric and second-harmonic tensors are not at all related.

A further symmetry is evident in Eqs.(4–140) and (4–142), where it is seen that the factors E^ω_k and E^ω_j are physically indistinguishable. Thus, the tensors actually occur in pairs:

$$\chi^{2\omega}_{ijk} + \chi^{2\omega}_{ikj}, \qquad (4\text{–}144)$$

and the factors of $1/2$ in Eqs.(4–143) correctly account for the case $j = k$. This symmetry will not occur if the frequencies of the two fields are different, as in the phenomenon of 'sum-frequency' production, for example. Of course, we recognize the pairing in Eq.(4–144) as arising basically from the definition of generalized susceptibilities as derivatives, or response coefficients. That is, in the general nonlinear expression for the response,

$$\frac{\partial^2 \langle \hat{B}_i(t) \rangle}{\partial F_k \partial F_j}\bigg|_{F=0} = \chi_{ijk} + \chi_{ikj}. \qquad (4\text{–}145)$$

The explicit forms of the response functions introduced above are as follows:

$$\phi_{ij}(t_1) = \frac{1}{i\hbar}\frac{e}{V}\langle [\hat{r}_j, \hat{M}_i(t_1)] \rangle_0, \qquad (4\text{–}146)$$

$$\phi_{ijk}(t_1 + t_2, t_2) = \frac{1}{(i\hbar)^2}\frac{e^2}{V}\langle [\hat{r}_j, [\hat{r}_k(t_2), \hat{M}_i(t_1 + t_2, t_2)]] \rangle_0, \qquad (4\text{–}147)$$

where $\hat{M} \equiv V\hat{P}$ is the total dipole-moment operator for the system, and the time-development operators contain only the unperturbed Hamiltonian $\hat{H}_0$. These response functions relate the polarization to the *external* field, and not to the total local field within the medium. Thus, as Ward (1965) has pointed out, the actual calculation of $\langle \hat{P}(t) \rangle$ and subsequent relation to observation involves three steps: (1) the external fields must be related to the local fields in a region of the medium; (2) the polarization produced by these local fields in that region must be calculated; and, (3) the radiation properties of that polarization must be determined. Clearly, it is only step (2) that is under consideration here, and $\mathbf{E}$ must be interpreted as the local field in a small (but macroscopic) region of the system.

We now outline briefly the generalization of the preceding discussion so as to include many of the nonlinear optical effects accessible to observation with current experimental techniques. In order to generate a fairly comprehensive example, suppose the perturbation to be a plane wave with fields

$$\mathbf{E} = \mathbf{E}^\omega \sin(\omega t - \mathbf{k} \cdot \mathbf{r}), \qquad (4\text{–}148\text{a})$$

$$\mathbf{B} = \mathbf{B}^\omega \sin(\omega t - \mathbf{k} \cdot \mathbf{r})$$

$$= \frac{\mathbf{k} \times \mathbf{E}^\omega}{k} \sin(\omega t - \mathbf{k} \cdot \mathbf{r}). \qquad (4\text{–}148\text{b})$$

In fact, one can also consider a second plane wave with a different frequency ω'. Then a representative perturbation Hamiltonian takes the form

$$\hat{H}_1 = -e\hat{r} \cdot \mathbf{E}^\omega \sin(\omega t) + \tfrac{1}{2}e(\hat{r} \cdot \mathbf{E}^\omega)(\mathbf{k} \cdot \mathbf{r})\cos(\omega t)$$

$$+ \frac{e}{2mc}(\hat{L} + 2\hat{S}) \cdot \mathbf{B}^\omega \sin(\omega t) - e(\mathbf{E}^\omega \cdot \hat{r})(\mathbf{E}^\omega \cdot \mathbf{r})\sin^2(\omega t)$$

$$- e\hat{r} \cdot \mathbf{E}^{\omega'} \sin(\omega' t) - e\mathbf{E}^0 \cdot \hat{r}. \qquad (4\text{–}149)$$

The first term on the right-hand side of this last equation has already been studied, and the second arises from the spatial expansion of the electric field:

$$\mathbf{E}(\mathbf{r}) = \mathbf{E}(0) + \mathbf{r} \cdot \nabla \mathbf{E} + \cdots, \qquad (4\text{--}150)$$

which represents an electric-quadrupole interaction. To leading order, this accounts for field inhomogeneities. The third term is the magnetic-dipole interaction and we shall use it below only in the zero-frequency, or dc form. The leading-order contribution from quadrupole polarization is described by the fourth term, and represents a coupling of the external field to the electric-quadrupole operator $e\hat{\mathbf{r}}\hat{\mathbf{r}}$. Application of a second electric field with frequency ω' is described by the fifth term, which allows for sum-difference frequency generation at $(\omega \pm \omega')$, and in the last term we have allowed for additional application of an external dc electric field. Obviously, arbitrarily more terms of a higher-order multipole character could be added, but these will suffice for the present discussion.

A considerable extension of the calculation carried out above, which includes the third-order linear response function, yields the following expression for the total polarization:

$$\langle \hat{\boldsymbol{P}}(t) \rangle = \mathbf{p}^0 + \mathbf{p}^\omega + \mathbf{p}^{2\omega} + \mathbf{p}^{\omega \pm \omega'} + \mathbf{p}^{3\omega} + \nabla \cdot \mathbf{q}^{2\omega} + \cdots, \qquad (4\text{--}151)$$

where the various terms can be written in an obvious notation as follows:

$$p_i^0 = \chi_{ij}^0 E_j + \chi_{ijk}^0 E_k^\omega E_j^\omega, \qquad (4\text{--}152)$$

$$
\begin{aligned}
p_i^\omega = {} & \chi_{ij}^\omega E_j + {}_1\chi_{ijk}^\omega E_j^0 E_k^\omega + {}_2\chi_{ijk}^\omega \overline{k}_j^\omega E_k^\omega \\
& + {}_1\chi_{ijk\ell}^\omega E_j^0 E_k^0 E_\ell^\omega + {}_2\chi_{ijk\ell}^\omega \varepsilon_{jkm} B_m^0 E_\ell^\omega \\
& + {}_3\chi_{ijk\ell}^\omega E_j^\omega E_k^\omega E_\ell^\omega + {}_4\chi_{ijk\ell}^\omega B_j^0 B_k^0 E_\ell^\omega,
\end{aligned} \qquad (4\text{--}153)
$$

$$
\begin{aligned}
p_i^{2\omega} = {} & \chi_{ijk}^{2\omega} E_j^\omega E_k^\omega + {}_1\chi_{ijk\ell}^{2\omega} \overline{k}_j^\omega E_k^\omega E_\ell^\omega \\
& + {}_2\chi_{ijk\ell}^{2\omega} E_j^0 E_k^\omega E_\ell^\omega + \chi_{ijk\ell m}^{2\omega} \varepsilon_{jkn} B_n^0 E_\ell^\omega E_m^\omega,
\end{aligned} \qquad (4\text{--}154)
$$

$$p^{\omega \pm \omega'} = \chi_{ijk}^{\omega \pm \omega'} E_j^\omega E_k^{\omega'}, \qquad (4\text{--}155)$$

$$p_i^{3\omega} = \chi_{ijk\ell}^{3\omega} E_j^\omega E_k^\omega E_\ell^\omega, \qquad (4\text{--}156)$$

$$q_{ij}^{2\omega} = \eta_{ijk\ell}^{2\omega} E_k^\omega E_\ell^\omega, \qquad (4\text{--}157)$$

where $\overline{k}_i^\omega$ is a unit propagation vector and ε_{jkm} is the Levi-Civitá symbol. One can identify the various tensors as previously from response theory.

The physical phenomena associated with each of the above contributions to the total polarization are readily identified. In Eq.(4–152) the contributions correspond, respectively, to an ordinary linear dc polarization, and optical rectification. Equation (4–153) represents, in order, the ordinary linear polarizability, the linear electro-optic effect, optical activity, the Kerr effect and electric double refraction, the Faraday and Voight effects, intensity-dependent refractive index, and the Cotton-Mouton effect. In Eq.(4–154) we find second-harmonic generation, magnetic-dipole absorption and second-harmonic emission, an electric-field induced second harmonic, and a magnetic-field induced second harmonic, respectively. Equation (4–155) describes sum-difference frequency production, whereas Eq(4–156) corresponds to third-harmonic generation. Finally, Eq.(4–157) describes quadratic polarization.

Ward (1965) has developed a diagrammatic procedure for systematically evaluating the relevant tensors in all these contributions to the polarization. He has provided extensive tables in terms of the dipole operators and fields, which should prove quite useful in actual calculations.

Problems

4.1 Verify the symmetry relations exhibited in Eqs.(4–112).

4.2 Derive an explicit expression for the correlation function X_{kij} appearing in Eq.(4–125).

4.3 Prove the symmetry property of Eq.(4–131).

REFERENCES

Animalu, A.O.E.: 1972, 'High-Field Magnetoresistance of Metals by Kubo-Mott Formula', *Ann. Phys.* **70**, 150.

Argyres, P.N., and P.L.Kelley: 1964, 'Theory of Spin Resonance and Relaxation', *Phys. Rev.* **134**, A98.

Barnett, S.J.: 1915, 'Magnetization by Rotation', *Phys. Rev.* **6**, 239.

Böttcher, C.J.F.: 1973, *Theory of Electric Polarization*, 2nd ed., Elsevier, Amsterdam.

Curie, P.: 1894, 'Sur la Symétrie dans les Phénomènes Physiques, Symétrie d'un Champ Électrique et d'un Champ Magnétique', *J. de Phys.* **3**, 393.

Debye, P.: 1945, *Polar Molecules*, Dover, New York.

DuBois, D.F., V. Galinsky, and M.G. Kivelson: 1963, 'Propagation of Electromagnetic Waves in Plasmas', *Phys. Rev.* **129**, 2376.

Einstein, A., and W.J. de Haas: 1915, 'Experimenteller Nachweis der Ampéreschen Molekularströme', *verhandl. Deut. Physik Ges.* **17**, 152.

Esaki, L.: 1962, 'New Phenomenon in Magnetoresistance of Bismuth at Low Temperature', *Phys. Rev. Letters* **8**, 4.

Frank, A.: 1932, 'Temperature Variation of the Magnetic Susceptibility, Gyromagnetic Ratio, and Heat Capacity in Sm^{+++} and Eu^{+++}', *Phys. Rev.* **39**, 119.

Franken, P.A., and J.F. Ward: 1963, 'Optical Harmonics and Nonlinear Phenomena', *Rev. Mod. Phys.* **35**, 23.

Franken, P.A., A.E. Hill, C.W. Peters, and G. Weinreich: 1961, 'Generation of Optical Harmonics', *Phys. Rev. Letters* **7**, 118.

Friedmann, H.L.: 1964a, 'A Cluster Expansion for the Electrical Conductance of Solutions', *Physica* **30**, 509.

Friedmann, H.L.: 1964b, 'On the Limiting Law for Electrical Conductance in Ionic Solutions', *Physica* **30**, 537.

Fröhlich, H.: 1949, *Theory of Dielectrics*, Oxford Univ. Press, Oxford.

Glarum, S.H.: 1960, 'Dielectric Relaxation of Polar Liquids', *J. Chem. Phys.* **33**, 1371.

Gorter, C.J., and B. Kahn: 1940, 'On the Theory of the Gyromagnetic Effects', *Physica* **7**, 753.

Hall, E.H.: 1878, 'On a New Action of the Magnet on Electric Currents', *Am. J. Math.* **2**, 287.

Heims, S.P., and E.T. Jaynes: 1962, 'Theory of Gyromagnetic Effects and Some Related Magnetic Phenomena', *Rev. Mod. Phys.* **34**, 143.

Izuyama, T.: 1961, 'An Expansion Theorem for the Electric Conductivity of Metals. I', *Prog. Theor. Phys.* **25**, 964.

Kubo, R., and K. Tomita: 1954, 'A General Theory of Magnetic Resonance', *J. Phys. Soc. Japan* **9**, 888.

Kubo, R., and S.J. Miyake: 1965, 'Quantum Theory of Galvanomagnetic Effect', in F. Seitz and D. Turnbull (eds.), *Solid State Physics, Vol.17*, Academic Press, New York, p.270.

Kubo, R., M. Toda, and N. Hashitsume: 1985, *Statistical Physics II*, Springer-Verlag, Berlin.

Landau, L.D., and E.M. Lifshitz: 1960, *Electrodynamics of Continuous Media*, Pergamon, Oxford.

Langer, J.S.: 1960, 'Theory of Impurity Resistance in Metals', *Phys. Rev.* **120**, 714.

Langer, J.S.: 1961, 'Theory of Impurity Resistance in Metals. II', *Phys. Rev.* **124**, 1003.

Langer, J.S.: 1962a, 'Evaluation of Kubo's Formula for the Impurity Resistance of an Interacting Electron Gas', *Phys. Rev.* **127**, 5.

Langer, J.S.: 1962b, 'Thermal Conductivity of a System of Interacting Electrons', *Phys. Rev.* **128**, 110.

Miyake, S.J., and Kubo, R.: 1962, 'Nonlinear Magnetoresistance in Bismuth', *Phys. Rev. Letters* **9**, 62.

O'Dell, T.H.: 1970, *The Electrodynamics of Magneto-Electric Media*, North-Holland, Amsterdam.

Oppenheim, I., M. Bloom, and H.C. Torrey: 1964, 'Nuclear Spin Relaxation in Gases and Liquids', *Can. J. Phys.* **42**, 70.

Shubnikov, L., and W.J. de Haas: 1930, 'Magnetische Widerstandvergrosserung in Einkristallen von Wismut bei tiefen Temperaturen', *Proc. Acad. Amsterdam* **33**, 130.

Van Vleck, J.H.: 1950, 'Concerning the Theory of Ferromagnetic Resonance Absorption', *Phys. Rev.* **78**, 266.

Ward, J.F.: 1965, 'Calculation of Nonlinear Optical Susceptibilities Using Diagrammatic Perturbation Theory', *Rev. Mod. Phys.* **37**, 1.

Chapter 5

General Nonequilibrium Processes

In Chapter 2 we developed the general algorithm for obtaining an initial-state statistical operator from given data for systems far from equilibrium. Subsequently we studied only a very special, albeit very important case of the general formalism, consisting of a weak dynamical perturbation in the form of known modifications to the system Hamiltonian. But the more interesting applications of the theory are to irreversible processes which can not be described unambiguously by adding terms to the equilibrium Hamiltonian, and in this chapter we initiate the further development of the theory needed to investigate in detail arbitrary nonequilibrium phenomena.

It is first useful to summarize for convenient reference the general equations of the theory. The physical observables of interest are presumed to be described by a number of Heisenberg operators $\hat{F}_k(\mathbf{x}, t)$, whose time development is governed by

$$\hat{F}_k(\mathbf{x}, t) = \hat{U}^\dagger(t, t_0)\hat{F}_k(\mathbf{x}, t)\hat{U}(t, t_0), \qquad (5\text{--}1)$$

and $\hat{U}(t, t_0)$ is found from the well-known operator equations of motion, as described in Chapter 2 and Appendix A.

If information in the form $\langle \hat{F}_k(\mathbf{x}, t) \rangle$ is available throughout a space-time region R_k, then the PME ensemble encompassing this—and only this—information is characterized by the statistical operator

$$\hat{\rho} = \frac{1}{Z} \exp\left[\sum_k \int_{R_k} \lambda_k(\mathbf{x}, t)\hat{F}_k(\mathbf{x}, t)\, d^3x\, dt \right], \qquad (5\text{--}2)$$

where

$$Z[\{\lambda_k(\mathbf{x}, t)\}] \equiv \operatorname{Tr} \exp\left[\sum_k \int_{R_k} \lambda_k(\mathbf{x}, t)\hat{F}_k(\mathbf{x}, t)\, d^3x\, dt \right], \qquad (5\text{--}3)$$

is the partition functional. The Lagrange-multiplier functions are determined from the set of functional differential equations relating the measured or given data to the ensemble:

$$\begin{aligned}
\langle \hat{F}_k(\mathbf{x}, t) \rangle &= \operatorname{Tr}[\hat{\rho}\hat{F}_k(\mathbf{x}, t)] \\
&= \frac{\delta}{\delta\lambda_k(\mathbf{x}, t)} \log Z[\{\lambda_k(\mathbf{x}, t)\}], \quad (\mathbf{x}, t) \in R_k.
\end{aligned} \qquad (5\text{--}4)$$

It must be emphasized that in R_k the quantity $\langle \hat{F}_k(\mathbf{x},t) \rangle$ is a known number. Owing to the cyclic invariance of the trace the expectation value of any other operator in either the Heisenberg or Schrödinger picture is

$$\langle \hat{J}(\mathbf{x},t) \rangle = \mathrm{Tr}[\hat{\rho}\hat{J}(\mathbf{x},t)] = \mathrm{Tr}[\hat{\rho}(t)\hat{J}(\mathbf{x})]\,, \qquad (5\text{--}5)$$

and often $\hat{J}$ will be independent of $\mathbf{x}$ as well. Again we emphasize the complete generality and strong nonlinearity of these expressions—they describe arbitrarily large departures from equilibrium. Let us next discuss a further special case of these equations.

A. Theory of the Steady State

We distinguish among systems which are: completely *isolated*; *closed* systems which can exchange energy with their surroundings, particularly under controlled conditions; and, *open* systems which can exchange energy, particles, etc., with other systems. Just as the equilibrium situation represents the time-invariant state of a closed system, a steady-state, or stationary process characterizes the time-invariant state of an open system. In a sense this latter situation is perhaps the simplest type of irreversible process—although currents can persist in the medium, they are constant in time. In this section we construct that modification of the above formalism arising from additional information that the system is undergoing a steady-state process. What is commonly meant experimentally by such a process is that the expectation values of all relevant operators are constant in time. Gradients will be present, of course, but they are constant, as are the various possible rate processes. The important point, though, is that by either measurement or design one must *determine* that the system is in a steady state. Such determination then constitutes additional data which must be incorporated into the initial ensemble.

Statistical mechanical treatments of steady-state processes have been formulated previously, by Bergmann and Lebowitz (1955) [also, Lebowitz and Bergmann (1957)], and by McLennan (1959), but have been somewhat restricted in ways discussed earlier. A very general description of stationary processes has been worked out in the present context by Scalapino (1961), although the mathematical criterion had been discussed earlier by Fano (1957), among others.

In order to realize the steady state one must require as a minimum that expectation values of relevant operators not change in time:

$$\frac{d}{dt}\langle \hat{A} \rangle = \frac{d}{dt}\,\mathrm{Tr}(\hat{\rho}\hat{A}) = \frac{1}{i\hbar}\,\mathrm{Tr}\big([\hat{\rho},\hat{\mathrm{H}}]\hat{A}\big) = 0\,, \qquad (5\text{--}6)$$

where it is presumed that neither $\hat{\mathrm{H}}$ nor $\hat{A}$ depend explicitly on the time. Although this weak condition is necessary, it is by no means sufficient. Owing to cyclic invariance of the trace, it is a simple matter to show that Eq.(5–6) is also valid if either $\hat{A}$ is a constant of the motion or if $\hat{\rho}$ and $\hat{A}$ commute. Thus, one would have to make such a specification about every operator $\hat{A}$.

A more stringent requirement is to demand that $\hat{\rho}$ be stationary, so that the expectation value of any dynamical variable will also be stationary. This leads us to define a system undergoing a steady-state process as one described by a statistical operator satisfying the *strong stationarity condition*:

$$[\hat{\rho}, \hat{\mathrm{H}}] = 0, \tag{5--7}$$

of which the equilibrium situation is just a special case. This criterion has also been employed by Nakajima (1958).

In order to incorporate these ideas into construction of the initial statistical operator, we rederive Eqs.(5--2)-(5--4) under the condition (5--7). For this purpose it is convenient to omit temporarily explicit reference to space-time variables in the measurement process. Hence, one again presumes a collection of initial data in the form

$$\langle \hat{F}_i \rangle = \mathrm{Tr}(\hat{\rho}\hat{F}_i), \tag{5--8}$$

and the character of these expectation values implies the additional condition (5--7). Now maximize the entropy subject to these constraints.

In solving this variational problem we find it convenient to write a small variation in $\hat{\rho}$ as

$$\delta\hat{\rho} = \hat{\rho}_s + \epsilon\hat{\eta}, \tag{5--9}$$

where $\hat{\rho}_s$ is the desired stationary statistical operator, $[\hat{\rho}_s, \hat{\mathrm{H}}] = 0$, and $\hat{\eta}$ represents an arbitrary variation. The result of this calculation is the condition

$$\mathrm{Tr}\left\{ \hat{\eta}\left[\ln\hat{\rho}_s + \sum_i \lambda_i \hat{F}_i \right] \right\} = 0. \tag{5--10}$$

We have here omitted a constant term, for $\hat{\rho}_s$ will presently be normalized. Were $\hat{\eta}$ completely arbitrary we would obtain the usual expression (5--2) for $\hat{\rho}$, but this arbitrariness is now severely restricted owing to the additional constraint

$$[\hat{\eta}, \hat{\mathrm{H}}] = 0. \tag{5--11}$$

In turn, this will lead to additional restrictions on the forms of the $\hat{F}_i$ appearing in $\hat{\rho}_s$.

We choose a representation in which $\hat{\mathrm{H}}$ is diagonal,

$$\hat{\mathrm{H}}\,|E\nu\rangle = E|E\nu\rangle, \tag{5--12}$$

where ν represents all other quantum numbers except the system energy. The matrix elements of the constraint equation (5--11) in this representation are

$$\langle E\nu|[\hat{\mathrm{H}}, \hat{\eta}]|E'\nu'\rangle = 0,$$

or

$$(E - E')\langle E\nu|\hat{\eta}|E'\nu'\rangle = 0. \tag{5--13}$$

When $E \neq E'$ the matrix elements must vanish, but within any degenerate manifold corresponding to the spectrum of $\hat{H}$ they are arbitrary. From Eq.(5–10) we then obtain

$$\langle E\nu| \ln \hat{\rho}_s + \sum_i \lambda_i \hat{F}_i |E\nu'\rangle = 0. \qquad (5\text{–}14)$$

Now define the *diagonal part* of an operator as

$$\langle E\nu| \hat{F}_d |E'\nu'\rangle \equiv \delta_{EE'} \langle E\nu| \hat{F} |E\nu'\rangle. \qquad (5\text{–}15)$$

Because $\hat{\rho}_s$ is already diagonal in this representation (it commutes with $\hat{H}$), one can rewrite Eq.(5–14) as

$$\langle E\nu| \ln \hat{\rho}_s + \sum_i \lambda_i \hat{F}_{id} |E'\nu'\rangle = 0. \qquad (5\text{–}16)$$

Finally, with the presumption that the state vectors form a complete set, we see that Eq.(5–16) generates the following description of the *stationary ensemble*:

$$\hat{\rho}_s = \frac{1}{Z_s} \exp\left[-\sum_i \lambda_i \hat{F}_{id}\right], \qquad Z_s = \text{Tr} \exp\left[-\sum_i \lambda_i \hat{F}_{id}\right], \qquad (5\text{–}17)$$

where we have induced the normalization $\text{Tr}\,\hat{\rho}_s = 1$. Thus, the stationary statistical operator contains only the diagonal parts of the defining operators, and $[\hat{H}, \hat{F}_d] = 0$.

The usefulness of this initial ensemble depends, of course, on having an explicit representation for the diagonal part of an operator. In order to obtain such an expression we write

$$\hat{F}_d \equiv \hat{F} - \hat{B}, \qquad (5\text{–}18a)$$

where $\hat{B}$ is the off-diagonal part of $\hat{F}$, and then express $\hat{B}$ in terms of another operator $\hat{A}$:

$$\hat{B} = \frac{1}{\hbar}[\hat{H}, \hat{A}]. \qquad (5\text{–}18b)$$

Now require that

$$\langle E\nu| \hat{F} |E'\nu'\rangle = \frac{i}{\hbar} \langle E\nu| [\hat{H}, \hat{A}] |E'\nu'\rangle, \qquad (5\text{–}19)$$

for $E \neq E'$. Hence,

$$\langle E\nu| \hat{A} |E'\nu'\rangle = \frac{\langle E\nu| \hat{F} |E'\nu'\rangle}{i(E - E')/\hbar}, \qquad (5\text{–}20)$$

for $E \neq E'$. One verifies by direct substitution that $\hat{A}$ has the formal representation

$$\hat{A} = \lim_{\epsilon \to 0} \int_{-\infty}^{0} e^{\epsilon t}\, e^{it\hat{H}/\hbar}\, \hat{F} e^{-it\hat{H}/\hbar}\, dt, \qquad (5\text{–}21)$$

and $\epsilon > 0$ is a convergence factor. This is a necessary device because of convergence difficulties at the lower limit of integration. Owing to the presumption of time-independent $\hat{H}$, we can make the identification

$$\hat{F}(t) \equiv e^{it\hat{H}/\hbar}\, \hat{F} e^{-it\hat{H}/\hbar}. \qquad (5\text{–}22)$$

If one combines these equations with the equation of motion for the Heisenberg operator $\hat{F}(t)$, the desired representation for the diagonal part of an operator is obtained:

$$\hat{F}_\mathrm{d} = \hat{F} - \int_{-\infty}^0 e^{\epsilon t} \dot{\hat{F}}(t) \, dt \,, \tag{5-23}$$

where the limit $\epsilon \to 0^+$ is implied. An alternative form emerges by integrating once by parts,

$$\hat{F}_\mathrm{d} = \lim_{\epsilon \to 0^+} \epsilon \int_{-\infty}^0 \hat{F}(t) e^{\epsilon t} \, dt \,, \tag{5-24}$$

which has also been found useful in similar contexts by Kubo (1959).

Physical insight into the diagonal part of an operator can be gained from the form (5–24). In Problem 5.2 we find that $\hat{F}_\mathrm{d}$ can be interpreted as a time average of $\hat{F}(t)$. In addition, suppose that $\hat{F}(t)$ can be written as the sum of a dc component and a time-varying part, such that

$$\hat{F}(t) = \hat{F}_0 + \hat{F}_v(t) \,. \tag{5-25}$$

Substitution of Eq.(5–25) into Eq.(5–24) yields the identification $\hat{F}_\mathrm{d} = \hat{F}_0$. Thus, $\hat{F}_\mathrm{d}$ is just that part of $\hat{F}$ remaining constant under a unitary transformation generated by $\hat{H}$, which is precisely one's expectation for a steady-state process.

It has been stated occasionally that it may not be possible to formulate a steady-state ensemble for systems exposed to external conservative forces (See, e.g., Peterson, 1967). For example, an isolated piece of conducting material in the presence of a static electric field can not support a steady current for long, and strong fields will also induce significant Joule heating. One is free to *presume* the existence of a steady state in such cases, of course, but at one's own peril. The entire point of the present approach is that one should not make any unwarranted presumptions concerning the initial state of the system, but construct the statistical operator based only on what is actually known. Should observation imply that the system is actually in a stationary state, this leads directly to the formalism outlined above. If, in addition, the full Hamiltonian is also known, so much the better. We shall return to this point presently.

B. The Linear Approximation

Although the algorithm described by Eqs.(5–2)-(5–4) can describe any system arbitrarily far from equilibrium, its very generality presents formidable mathematical difficulties. The statistical operator is completely nonlinear, as well as nonlocal and memory retaining. As a consequence it is not only useful, but almost necessary at this point to examine in some detail the simplest approximation to the general theory. In this section, therefore, we develop the linear approximation, where by 'linear' we mean only terms linear in the departure from equilibrium are retained. Though only a special case, it is that which has been studied most extensively by other methods, and therefore provides useful comparisons. Moreover, the mathematical problems posed by nonlinear phenomena present special difficulties for this as well as for any other theory and require further developments.

It is convenient, as well as sufficient for illustration to consider only a single Heisenberg operator $\hat{F}(t)$. A homogeneous closed system in thermal equilibrium is described by the statistical operator

$$\hat{\rho}_0 = \frac{1}{Z_0(\beta)} e^{-\beta\hat{H}} , \qquad Z_0(\beta) = \text{Tr}\, e^{-\beta\hat{H}} , \qquad (5\text{–}26)$$

which maximizes the entropy for the given datum $E = \langle\hat{H}\rangle_0$. For any other operator $\hat{F}$,

$$\langle\hat{F}\rangle_0 \equiv \text{Tr}(\hat{\rho}\hat{F}) . \qquad (5\text{–}27)$$

Now suppose that information is obtained in the form of the expectation value $\langle\hat{F}(t)\rangle$ throughout an information-gathering interval $-\tau \le t \le 0$, say. The statistical operator incorporating this new information into a description of the system is then given by

$$\hat{\rho} = \frac{1}{Z} \exp\left\{ -\beta\hat{H} + \int_{-\tau}^{0} \lambda(t)\hat{F}(t)\, dt \right\}, \qquad (5\text{–}28a)$$

$$Z[\beta, \lambda(t)] = \text{Tr} \exp\left\{ -\beta\hat{H} + \int_{-\tau}^{0} \lambda(t)\hat{F}(t)\, dt \right\}. \qquad (5\text{–}28b)$$

It must be emphasized strongly that the Lagrange multiplier $\beta = (\kappa T)^{-1}$ has already been identified above, and refers to the temperature of the *equilibrium* system—it is not defined as such otherwise. Thus, $\hat{\rho}_0$ actually serves as a prior probability in obtaining Eqs.(5–28), so that the latter really arise from a principle of minimum cross-entropy (see, e.g., Volume I, Chapter 2). The reader is asked to carry out this variational calculation in Problem 5.1, but at the moment the important point is to realize that β is now *fixed* by the energy of the initial equilibrium system. If one believes the perturbation to have altered the system temperature—as it well might—then this can only be verified by allowing the system to return to equilibrium and then making a new measurement. We shall consider this point in further detail in Chapter 7.

It is presumed throughout that $\hat{H}$ is time independent; this is not actually necessary, but it corresponds to the usual physical situation. Because there is no information about any possible change in the Hamiltonian owing to external influences, the time-evolution operator takes the simple form

$$\hat{U}(t) = \hat{U}_0(t) \equiv e^{-it\hat{H}/\hbar} , \qquad (5\text{–}29)$$

and

$$\hat{F}(t) = \hat{U}_0^\dagger(t)\hat{F}\hat{U}_0(t) . \qquad (5\text{–}30)$$

Given the initial ensemble (5–28), we wish to predict the expectation value of another operator $\hat{C}(t)$ at some later time $t > 0$. Hence, we presume the departure from equilibrium to be small in some sense and apply perturbation theory to the exponential operators in Eqs.(5–28). Note that this procedure does *not* imply that $\lambda(t)\hat{F}(t)$ is necessarily small, but only that the integral is small.

A complete perturbation theory for exponential operators has been presented by Heims and Jaynes (1962), and we merely summarize the results here. Let ϵ be a small positive number representing a perturbation parameter, and let $\hat{A}$ and $\hat{B}$ be arbitrary linear operators on an appropriate space. Then the identity

$$e^{\hat{A}+\hat{B}} = e^{\hat{A}}\left[\hat{1} + \int_0^1 e^{-x\hat{A}}\,\hat{B}e^{x(\hat{A}+\hat{B})}\,dx\right] \qquad (5\text{–}31)$$

from Appendix A can be applied repeatedly to generate the series

$$e^{\hat{A}+\epsilon\hat{B}} = e^{\hat{A}}\left[\hat{1} + \sum_{n=1}^{\infty}\epsilon^n\hat{S}_n\right], \qquad (5\text{–}32)$$

where

$$\hat{S}_n \equiv \int_0^1 dx_1 \int_0^{x_1} dx_2 \cdots \int_0^{x_{n-1}} dx_n\,\hat{B}(x_1)\hat{B}(x_2)\cdots\hat{B}(x_n), \qquad (5\text{–}33)$$

and

$$\hat{B}(x) \equiv e^{-x\hat{A}}\,\hat{B}e^{x\hat{A}}. \qquad (5\text{–}34)$$

With the notation

$$\hat{\rho}_0 = \frac{1}{Z_0}e^{\hat{A}}, \quad Z_0 = \operatorname{Tr} e^{\hat{A}}, \quad \langle\hat{C}\rangle_0 = \operatorname{Tr}(\hat{\rho}_0\hat{C}), \qquad (5\text{–}35a)$$

$$\hat{\rho} = \frac{1}{Z}e^{\hat{A}+\epsilon\hat{B}}, \quad Z = \operatorname{Tr} e^{\hat{A}+\epsilon\hat{B}}, \quad \langle\hat{C}\rangle = \operatorname{Tr}(\hat{\rho}\hat{C}), \qquad (5\text{–}35b)$$

the expectation value of an operator $\hat{C}$ in the ensemble (5–35b) can be written

$$\langle\hat{C}\rangle - \langle\hat{C}\rangle_0 = \sum_{n=1}^{\infty}\frac{\epsilon^n}{n!}\left(\frac{d^n\langle\hat{C}\rangle}{d\epsilon^n}\right)_{\epsilon=0}$$

$$= \sum_{n=1}^{\infty}\epsilon^n[\langle\hat{Q}_n\hat{C}\rangle_0 - \langle\hat{Q}_n\rangle_0\langle\hat{C}\rangle_0], \qquad (5\text{–}36)$$

where $\hat{Q}_1 \equiv \hat{S}_1$ and

$$\hat{Q}_n \equiv \hat{S}_n - \sum_{k=1}^{n-1}\langle\hat{Q}_k\rangle_0\hat{S}_{n-k}, \qquad n > 1. \qquad (5\text{–}37)$$

Note that all expectation values on the right-hand side of Eq.(5–36) are in terms of $\hat{\rho}_0$.

In the linear, or first-order approximation Eq.(5–36) yields, when we set $\epsilon = 1$,

$$\langle \hat{C} \rangle - \langle \hat{C} \rangle_0 \simeq \int_0^1 dx \, \langle e^{-x\hat{A}} \, \hat{B} e^{x\hat{A}} \, \hat{C} \rangle_0 - \langle \hat{B} \rangle_0 \langle \hat{C} \rangle_0 . \tag{5–38}$$

It is useful to define again the *Kubo transform* of the operator $\hat{B}$ with respect to $\hat{A}$ as

$$\overline{B} \equiv \int_0^1 e^{-x\hat{A}} \, \hat{B} e^{x\hat{A}} \, dx . \tag{5–39}$$

If we now return to the scenario described by Eq.(5–26), then the time-evolution operator has the form of Eq.(5–29) for time-independent $\hat{H}$, we can set $\hat{A} = -\beta \, \hat{H}$, and then the Kubo transform can be written alternatively as

$$\overline{\hat{B}} = \beta^{-1} \int_0^\beta \hat{B}(t - i\hbar u) \, du . \tag{5–40}$$

The notation is defined by Eqs.(5–29) and (5–30). If the transform is defined with respect to the operator $\hat{A}$ appearing in $\hat{\rho}$, as is almost always the case, then $\langle \overline{\hat{B}} \rangle_0 = \langle \hat{B} \rangle_0$.

The right-hand side of Eq.(5–38) is just the quantum-mechanical covariance of the two operators $\hat{B}$ and $\hat{C}$, and so we find it also useful to define the *covariance function* as

$$K_{CB} \equiv \langle \overline{\hat{B}} \hat{C} \rangle_0 - \langle \hat{B} \rangle_0 \langle \hat{C} \rangle_0 . \tag{5–41}$$

The general properties of covariance functions will be explored in some detail in the following chapter, but it is useful to exhibit two of the simpler results here and now. Along with cyclic invariance of the trace, a change of integration variables allows one to prove the reciprocity relation

$$K_{CB} = K_{BC} , \tag{5–42}$$

where it must be remembered that it is always the second subscript that refers to the Kubo transform. In a similar manner, if $\hat{B}$ has the form

$$\hat{B}(u) \equiv e^{-u\hat{A}} \, \hat{B}(0) e^{u\hat{A}} , \tag{5–43}$$

where u is an arbitrary parameter, then straightforward differentiation yields the very useful relationship

$$\frac{dK_{CB}}{du} = \langle [\hat{C}, \hat{B}(u)] \rangle_0 . \tag{5–44}$$

With this brief outline of the perturbation theory for exponential operators, let us return to the problem specified by Eqs.(5–26)-(5–28). If the departure from equilibrium is presumed small, the deviation of the expectation value of $\hat{C}$ at time t from its equilibrium value is just

$$\langle \hat{C}(t) \rangle - \langle \hat{C} \rangle_0 = \int_{-\tau}^{0} K_{CF}(t,t') \lambda(t')\, dt' , \qquad (5\text{--}45)$$

where

$$K_{CF}(t,t') \equiv \overline{\langle \hat{F}(t')\hat{C}(t) \rangle_0} - \langle \hat{F} \rangle_0 \langle \hat{C} \rangle_0 . \qquad (5\text{--}46)$$

We have observed that the time independence of $\hat{\text{H}}$ confers the same property on the single-operator expectation values, which is readily proved by appeal to the cyclic invariance of the trace. Similarly, this time independence guarantees time-translation invariance:

$$K_{CF}(t,t') = K_{CF}(t - t') . \qquad (5\text{--}47)$$

In this case the reciprocity relation (5–42) becomes

$$K_{CF}(t) = K_{FC}(-t) , \qquad (5\text{--}48)$$

and the Kubo transform can be written in the alternative form of Eq.(5–40).

It is important to note that the covariance function of Eq.(5–46) depends only on equilibrium properties of the system. Hence, the message conveyed by Eq.(5–45) is that, quite generally, small departures from equilibrium caused by *anything* are governed principally by equilibrium fluctuations. When Eq.(5–45) provides a valid description, it makes manifest the stability of the equilibrium state.

In order for Eq.(5–45) to be useful one must be able to determine the Lagrange-multiplier function $\lambda(t)$ from the prescription (5–4). On the one hand, we shall see in subsequent applications that it is usually possible to identify the functions $\lambda(t)$ on physical or intuitive grounds as the independent variables of the linear theory. On the other hand, it is necessary as a matter of principle to be able to determine the Lagrange-multiplier functions exactly, although direct application of the prescription (5–4) is not usually practicable. Rather, we now illustrate that $\lambda(t)$ can be determined completely within the desired approximation of the theory, linear or otherwise.

We again employ the perturbation expansion (5–32) to expand the partition functional through second order:

$$Z[\beta, \lambda(t)] = Z_0(\beta) \left[1 + \langle \hat{F} \rangle_0 \int_{-\tau}^{0} \lambda(t)\, dt + \int_{-\tau}^{0} dt_1 \int_{-\tau}^{0} dt_2\, \lambda(t_1)\lambda(t_2) L(t_1,t_2) + \cdots \right] , \qquad (5\text{--}49)$$

where we have defined a function

$$L(t_1,t_2) \equiv \beta^{-2} \int_{0}^{\beta} du_1 \int_{0}^{u_1} du_2\, \langle \hat{F}(t_1 - i\hbar u_1)\hat{F}(t_2 - i\hbar u_2) \rangle_0$$
$$= L(t_1 - t_2) . \qquad (5\text{--}50)$$

The parameter dependence of all operators in these expressions is that of Eq.(5–43), with $\hat{A} = i\hat{H}/\hbar$. Hence, a change of integration variables reveals that Eq.(5–50) can equally well be written

$$L(t) = \beta^{-2} \int_0^\beta (\beta - u)\langle \hat{F}(t - i\hbar u)\hat{F}(0)\rangle_0 \, du \, . \qquad (5\text{–}51)$$

Now take the logarithm of Z and employ the expansion of $\log(1 + x)$ for small x. Through second order,

$$\log Z = \log Z_0 + \langle \hat{F}\rangle_0 \int_{-\tau}^0 \lambda(t) \, dt + \int_{-\tau}^0 dt_1 \int_{-\tau}^0 dt_2 \, L(t - t_2)\lambda(t_1)\lambda(t_2)$$

$$- \tfrac{1}{2}\langle \hat{F}\rangle_0^2 \int_{-\tau}^0 \lambda(t_1) \, dt_1 \int_{-\tau}^0 \lambda(t_2) \, dt_2 + \cdots . \qquad (5\text{–}52)$$

If we define a function

$$K(t_1 - t_2) \equiv L(t_1 - t_2) + L(t_2 - t_1) - \langle \hat{F}\rangle_0^2 = K(t_2 - t_1) \, , \qquad (5\text{–}53)$$

then through second order we have the approximation

$$\log Z \simeq \log Z_0 + \langle \hat{F}\rangle_0 \int_{-\tau}^0 \lambda(t) \, dt + \tfrac{1}{2} \int_{-\tau}^0 dt_1 \int_{-\tau}^0 dt_2 \, K(t_1 - t_2)\lambda(t_1)\lambda(t_2) \, . \qquad (5\text{–}54)$$

A functional differentiation as prescribed by Eq.(5–4) yields

$$\langle \hat{F}(t)\rangle = \langle \hat{F}\rangle_0 + \int_{-\tau}^0 K(t - t')\lambda(t') \, dt' \, , \qquad -\tau \leq t \leq 0 \, . \qquad (5\text{–}55)$$

Thus, in the linear approximation $\lambda(t)$ is determined by a Fredholm integral equation with driving equal to the departure from equilibrium, $\langle \hat{F}(t)\rangle - \langle \hat{F}\rangle_0$. This last quantity is completely known in the information-gathering interval, which is the *only* place where $\lambda(t)$ is defined.

The kernel of the integral equation, as defined in Eq.(5–53), appears to be rather complicated. But let us return to Eq.(5–45) and set $\hat{C}(t) = \hat{F}(t)$. Then

$$\langle \hat{F}(t)\rangle - \langle \hat{F}\rangle_0 = \int_{-\tau}^0 K_{FF}(t - t')\lambda(t') \, dt' \, , \qquad (5\text{–}56)$$

giving the predicted value of the expectation of $\hat{F}$ at *any* time t. If, however, one restricts t to the interval $-\tau \leq t \leq 0$, then this last equation bears a strong resemblance to the integral equation (5–55). That the two expressions are in fact equivalent follows from an identity involving the function $f(t, u) \equiv \langle \hat{F}(t - i\hbar u)\hat{F}(0)\rangle_0$. That is,

$$\beta \int_0^\beta f(t, u) \, du = \int_0^\beta (\beta - u)[f(t, u) + f(-t, u)] \, du \, , \qquad (5\text{–}57)$$

which is verified by writing out the expectation values explicitly in a representation in which $\hat{H}$ is diagonal. This is the first in a hierarchy of identities which can be generated from successive approximations to the theory. It is now left to the reader to conclude that

$$K(t) \equiv K_{FF}(t) = K_{FF}(-t). \qquad (5\text{--}58)$$

In retrospect an identity such as this should have been expected from inspection of Eq.(5–45), but it was first necessary to establish an equivalence with the prescription of Eq.(5–4). Moreover, the discussion has exposed an additional subtlety in the theory and we are now in a position to uncover a truly striking aspect of the formalism.

From the manner of its derivation it appears that Eq.(5–55), is valid only in the information-gathering interval $-\tau \le t \le 0$, whereas Eq.(5–45) is presumably valid for all t. But the equivalence demonstrated in Eq.(5–58) shows that Eq.(5–55), or Eq.(5–56), is in fact valid for all t. Thus, when $t > 0$ Eq.(5–56) gives the *predicted future* of $\hat{F}(t)$; when $-\tau \le t \le 0$ it provides the linear Fredholm integral equation determining $\lambda(t)$; and when $t < -\tau$ the equation yields the *retrodicted past* of $\hat{F}(t)$. This last observation underscores the fact that $K_{FF}(t)$ is not a causal function, unless so used in that specific context. Although physical influences must propagate forward in time, logical inferences about the present can affect our knowledge of the past as well as of the future. These points have been discussed in more detail elsewhere (Jaynes, 1979).

Among other things, Eq.(5–56) is a linear Fredholm integral equation of the first kind from which to determine the Lagrange-multiplier function. Its theory is understood well enough to realize that such equations are in general quite difficult to solve, unless special conditions are met. For example, if the initial information is provided throughout the infinite past, such that $\tau = \infty$, the equation can be studied through Fourier analysis. An example of this technique will be given presently. In the general case of finite τ one has recourse to the Hilbert-Schmidt and Riesz-Fischer theorems, from which are found both solutions and conditions for their existence in terms of the eigenvalues and eigenfunctions of the kernel (e.g., Tricomi, 1957).

At this point it is useful to mention several necessary generalizations of the linear theory. First of all, the preceding results are essentially unchanged if the equilibrium situation is described by the grand canonical ensemble. One need only introduce the Gibbs function $G\hat{1} \equiv \hat{G} = \mu\hat{N}$, where μ is the chemical potential and N the total number operator (which commutes with $\hat{H}$). In terms of the operator

$$\hat{K} \equiv \hat{H} - \hat{G}, \qquad (5\text{--}59)$$

Eq.(5–26) is replaced by

$$\hat{\rho}_0 = \frac{1}{Z_0} e^{-\beta\hat{K}}, \qquad Z_0 = \operatorname{Tr} e^{-\beta\hat{K}}, \qquad (5\text{--}60)$$

and one need only replace $\hat{H}$ by $\hat{K}$ in all the preceding equations. Although there are no formal changes, it should be noted that $\hat{G}$ must commute with *both* $\hat{H}$ and $\hat{F}(t)$ in order to write the Kubo transform in the alternative form (5–40).

A second generalization is necessary when the perturbed system is spatially nonuniform, a possibility already anticipated in Eqs.(5–1) and (5–4). This description is realized theoretically by utilizing the second-quantized formalism and field operators, as will be discussed in some detail presently. We merely note here that Eqs.(5–45) and (5–46) become, respectively,

$$\langle \hat{C}(\mathbf{x},t)\rangle - \langle \hat{C}(\mathbf{x})\rangle_0 = \int_R K_{CF}(\mathbf{x},t;\mathbf{x}',t')\lambda(\mathbf{x}',t')\, d^3x'\, dt' , \tag{5–61}$$

$$K_{CF}(\mathbf{x},t;\mathbf{x}',t') = \langle \hat{F}(\mathbf{x}',t')\hat{C}(\mathbf{x},t)\rangle_0 - \langle \hat{F}(\mathbf{x}')\rangle_0\langle \hat{C}(\mathbf{x})\rangle_0 . \tag{5–62}$$

Often the single-operator expectation values are also independent of $\mathbf{x}$. The generalization to include a number of operators $\hat{F}_k(\mathbf{x},t)$ is straightforward.

It is interesting to note that the series expansion (5–36) can be put into a more revealing form through the observation from Eq.(5–61) that the covariance function can be obtained by taking a functional derivative of $[\langle \hat{C}(\mathbf{x},t)\rangle - \langle \hat{C}(\mathbf{x})\rangle_0]$ with respect to $\lambda(\mathbf{x},t)$. Equivalence with the derivation of Eq.(5–38) is established by means of the following identity found in Appendix A:

$$\frac{\partial}{\partial \lambda}e^{\hat{A}+\lambda \hat{B}}\bigg|_{\lambda=0} = e^{\hat{A}}\,\overline{\hat{B}} . \tag{5–63}$$

One can generate a countably infinite hierarchy of such identities corresponding to derivatives of all orders, and then generalize the results to functional differentiation. As a result we find that the series of Eq.(5–36) can be rewritten in the generalized form

$$\langle \hat{C}(\mathbf{x},t)\rangle - \langle \hat{C}(\mathbf{x})\rangle_0 = \sum_{n=1}^{\infty} \int_R d^3x'_1\, dt'_1 \cdots \int_R d^3x'_n\, dt'_n$$

$$\lambda(\mathbf{x}'_1,t'_1)\cdots \lambda(\mathbf{x}'_n,t'_n)\left[\frac{\delta^n\langle \hat{C}(\mathbf{x},t)\rangle}{\delta\lambda(\mathbf{x}',t')^n}\right]_{\lambda=0} \tag{5–64}$$

and the $n = 1$ term yields Eq.(5–61).

Of course, from the preceding discussion it is also evident that these quantities are related to the functional derivatives of the logarithm of the partition functional. For example, in the linear theory the covariance function of Eq.(5–62) is given by the second functional derivative:

$$K_{F_i F_j}(\mathbf{x},t;\mathbf{x}',t') = \frac{\delta^2}{\delta\lambda_i(\mathbf{x},t)\,\delta\lambda_j(\mathbf{x}',t')}\log Z , \tag{5–65}$$

where $\hat{F}_i$ and $\hat{F}_j$ are two of the Heisenberg operators on which initial data are available, as in Eqs.(5–2) and (5–3).

This last expression illustrates that K is analogous to the covariance matrix A of the equilibrium theory [e.g., Eq.(2–79) in Volume I]. Indeed, in the case that the $\{\hat{F}_k\}$ are constants of the motion K reduces to A, which describes ordinary fluctuations in the equilibrium system.

Continuing the analogy, we see that the near-equilibrium theory is equally well described in terms of the entropy functional $S[F_k(\mathbf{x}, t)]$, dependent only on the measured values $F_k(\mathbf{x}, t) \equiv \langle \hat{F}_k(\mathbf{x}, t) \rangle \in R_k(\mathbf{x}, t)$. The Lagrange-multiplier functions—the basic variables or 'potentials' of the theory—are given by the functional derivatives

$$\lambda_k(\mathbf{x}, t) = \frac{\delta S}{\delta F_k(\mathbf{x}, t)}. \tag{5-66}$$

In a region where S is locally convex the second functional derivatives then generate a set of space-time functions

$$G_{ij}(\mathbf{x}, t; \mathbf{x}', t') \equiv \frac{\delta S}{\delta F_i(\mathbf{x}, t)\, \delta F_j(\mathbf{x}', t')}, \tag{5-67}$$

and the convexity criterion is

$$\sum_k \int_{R_k} \delta \lambda_k(\mathbf{x}, t)\, \delta F_k(\mathbf{x}, t)\, d^3 x\, dt \quad < 0. \tag{5-68}$$

Again in analogy with the equilibrium theory [e.g., Eq.(2–80) in Volume I], the G_{ij} are functional inverses of the K_{ij}. If the $\{\hat{F}_k\}$ refer to the usual thermodynamic variables and we are in fact at an equilibrium point, then the matrix G simply defines the quadratic form in the expansion of the entropy about that point:

$$S = S_0 - \tfrac{1}{2} G_{ij} \delta F_i \delta F_j + \cdots, \tag{5-69}$$

thereby making contact with the Onsager theory of Section 1C. In the following chapter we shall establish a number of general reciprocity relations for the covariance functions, which in the appropriate limits are precisely the so-called 'Onsager relations'.

COMPARISON WITH LINEAR DYNAMICAL RESPONSE

Although it arises as only a special case of the general theory of irreversible processes, the theory of linear dynamical response is clearly much better known and, in fact, is taken for granted as a basic tool of modern physical theory. Over the years it has been applied to a broad spectrum of dynamical problems and, in particular, there exist calculations of the linear response function for numerous models. It seems appropriate, therefore, to exhibit the explicit relation between the linear response function and the covariance function. A more practical motivation, of course, is that known results for the former may facilitate in some cases calculation of the latter.

The mathematical relationship between the two functions follows almost immediately from the identity (5–44). By setting $u = i(t - t')/\beta\hbar$ in that equation, one obtains from the definitions

$$\frac{d}{dt'}K_{JF}(t - t') = \beta^{-1} \phi_{JF}(t - t') , \qquad (5\text{–}70)$$

with ϕ_{JF} defined in Eq.(3–10). Thus, K_{JF} appears here in the role of Kubo's relaxation function discussed in Eqs.(3–18) and (3–20). Note, however, that Eq.(5–70) is an *exact* identity and that K_{JF} and R_{JF} are by no means equivalent functions—the latter is defined only for static forces at any rate. It is clear that the covariance function contains much more information than the linear response function, and the context in which the former is derived gives it a much deeper significance than the latter. The suggestion is strong that relaxation processes are to be described by covariance functions.

In order to examine the specific relationship between the two functions more closely, let us construct a hypothetical experiment which is viewed differently by two separate observers. (We shall omit spatial variables in the rest of this chapter without loss of generality.) Both observers possess the same information about a system while it is in complete thermal equilibrium, and both know that in this state it is described by Hamiltonian $\hat{H}$. At time $t = -\tau$ observer A applies an external dynamical field that couples to the system operator $\hat{F}$, and then removes the external force at time $t = 0$. During this period the system Hamiltonian is

$$\hat{H}_t = \hat{H} - \hat{F}v(t) , \qquad -\tau \leq t \leq 0 , \qquad (5\text{–}71)$$

where temporarily we take $v(t)$ to vanish smoothly at $t = -\tau$ and $t = 0$. At some future time $t > 0$ observer A predicts the value of another operator $\hat{J}$ to be

$$J_{\mathrm{A}}(t) \equiv \langle \hat{J}(t) \rangle - \langle \hat{J} \rangle_0 = \int_{-\tau}^{0} \phi_{JF}(t - t')v(t') \, dt' . \qquad (5\text{–}72)$$

Now observer B is completely unaware of what has been done to the system by A, but he monitors the observable characterized by $\hat{F}$ through measurement of the expectation value throughout the interval $-\tau \leq t \leq 0$. From this information B constructs an initial-state statistical operator from which he also predicts the expectation value of $\hat{J}$ at the same time t as did A. Observer B finds that

$$J_{\mathrm{B}}(t) = \int_{-\tau}^{0} K_{JF}(t - t')\lambda(t') \, dt' , \qquad (5\text{–}73)$$

which he can evaluate after first solving the following integral equation for $\lambda(t)$:

$$\langle \hat{F}(t) \rangle - \langle \hat{F} \rangle_0 = \int_{-\tau}^{0} K_{FF}(t - t')\lambda(t') \, dt' , \qquad -\tau \leq t \leq 0 . \qquad (5\text{–}74)$$

The two observers are now asked to compare their predictions, although they clearly possess different information. Although this is only meant as a thought experiment, the procedure obviously can be carried out.

In order for observers A and B to compare their predictions, observer A utilizes Eq.(5–70) and integrates once by parts in Eq.(5–72). Owing to his complete control over the experiment observer A knows that $v(-\tau) = v(0) = 0$, and therefore concludes that his expression (5–72) can be written in the alternative form

$$J_A(t) = -\beta \int_{-\tau}^{0} K_{JF}(t - t') \frac{dv(t')}{dt'} \, dt' . \tag{5–75}$$

Physically we must have $J_A(t) = J_B(t)$ and, if the two theories are equally valid as applied to this experiment, it follows that

$$\lambda(t) = -\beta \frac{dv(t)}{dt}, \qquad -\tau \le t \le 0 . \tag{5–76}$$

Although we shall discuss relaxation processes of this kind in more detail below, in Chapter 7, it is worth pursuing here this connection between the two types of observations, for it illustrates several important points.

If the perturbation Hamiltonian is actually known, $\lambda(t)$ can be obtained directly from Eq.(5–76). That is, in the situation where the two aspects of the theory lead to equivalent results the Lagrange-multiplier function is completely determined, as it should be. Note, however, that observer B could have begun his measurement at some time $t > -\tau$, and thus not had the full range of information available to A. If it then happens that experiment confirms the prediction $J_A(t) \neq J_B(t)$, observer B would learn that there must have been additional constraints upon the system of which he was not aware, although he did his best with the information available to him. But, according to Eq.(5–74), he could now retrodict the previous behavior. These are rather general and beneficial features of the PME formalism.

The usual scenario envisioned by the theory of linear dynamic response does not, however, require the boundary condition that $v(t)$ vanish *smoothly* at $t = 0$. Rather, the force is removed abruptly in step-function fashion, so that $v(t')$ should be replaced by $\theta(-t')v(t')$ in order to correctly account for the discontinuity in the driving force. That is, it is essential to note that the upper limit of integration in Eq.(5–75) must be considered as 0^+. Consequently, in this case the predicted response and relaxation is

$$J_A(t) = 2\beta v(0) K_{JF}(t) - \beta \int_{-\tau}^{0} K_{JF}(t - t') \frac{dv(t')}{dt'} \, dt' , \tag{5–77}$$

owing to the step-function arising from differentiation of the δ-function. In turn the determination (5–76) becomes

$$\lambda(t) = 2\beta \delta(t) v(t) - \beta \theta(-t) \frac{dv(t)}{dt} . \tag{5–78}$$

There is thus a *double* contribution from the upper limit of integration. (The reader may wish to re-examine Problem 3.1 in light of this result!) One suspects that the factor of 2 is anomalous, for it would not be expected to appear in the solution to the integral equation for $\lambda(t)$ obtained by observer B, Eq.(5–74). In order to demonstrate that this is indeed the case we shall carry out an explicit calculation in which the role of observer B has been filled by Jaynes (1979), except for details.

For convenience we presume $\langle \hat{F}(t) \rangle$ to be specified over the interval $[-\infty, 0]$, but make no presumption at all about the driving mechanism. We thus set $\tau = \infty$ in Eq.(5–74), and then reformulate the problem as follows. Solve the equation

$$\int_{-\infty}^{\infty} k(t - t')\lambda(t')\,dt' = f(t)\,, \tag{5–79a}$$

subject to the subsidiary condition

$$\lambda(t) = 0\,, \qquad t > 0\,, \tag{5–79b}$$

using only past values of $f(t)$, and employ this solution to obtain the future prediction for $f(t)$ which also satisfies Eq.(5–74) with $\tau = \infty$. Here we have written $k(t) \equiv K_{FF}(t)$, $f(t) \equiv \langle \hat{F}(t) \rangle - \langle \hat{F} \rangle_0$. This is now a Wiener-Hopf integral equation, and thus solvable by known techniques of analytic continuation.

First make the decomposition

$$f(t) = f_+(t) + f_-(t)\,, \tag{5–80a}$$

where

$$f_+(t) \equiv f(t)\theta(t) = \begin{cases} f(t), & t > 0 \\ 0, & t \le 0 \end{cases},$$

$$f_-(t) \equiv f(t)\theta(-t) = \begin{cases} 0, & t > 0 \\ f(t), & t \le 0 \end{cases}. \tag{5–80b}$$

Fourier transformation and continuation into the complex ω-plane yields

$$F_+(\omega) + F_-(\omega) = K(\omega)\Lambda(\omega)\,, \tag{5–81}$$

where

$$F_\pm(\omega) \equiv \int_{-\infty}^{\infty} f_\pm(t)e^{i\omega t}\,dt\,, \tag{5–82}$$

and similarly for the other functions. Now $F_-(\omega)$ and $K(\omega)$ are known, but $F_+(\omega)$ and $\Lambda(\omega)$ are unknown. We shall assert, however, that $F_+(\omega)$ is to be regular in the upper half-plane (uhp) $\text{Im}(\omega) > 0$, $\Lambda(\omega)$ is regular in the lower half-plane (lhp) $\text{Im}(\omega) < 0$, and that is just enough information to solve the problem. One next carries out the Wiener-Hopf factorization (Wiener and Hopf, 1931),

$$K(\omega) \equiv K_+(\omega)K_-(\omega)\,, \tag{5–83}$$

where K_+ is chosen to be regular and have no zeros in the uhp, K_- has the same behavior in the lhp, and on the real axis $K_-(\omega) = K_+^*(\omega)$. In general (e.g., Roos, 1969), $K_+(\omega)$ is uniquely determined in the uhp by

$$K_+(\omega) = \exp\left[\frac{1}{2\pi i}\int_{-\infty}^{\infty}\frac{\log K(z)}{z - \omega}\,dz\right], \qquad (5\text{-}84)$$

in the lhp by analytic continuation, and the integral converges if

$$\left|\int_{-\infty}^{\infty}\frac{\log K(\omega)}{1 + \omega^2}\,d\omega\right| < \infty. \qquad (5\text{-}85)$$

This latter condition is also the Paley-Wiener criterion for solvability of the integral equation (5–79)—(Paley and Wiener, 1934).

Now write, from Eqs.(5–81) and (5–83),

$$\frac{F_+(\omega)}{K_+(\omega)} + \frac{F_-(\omega)}{K_+(\omega)} = K_-(\omega)\Lambda(\omega), \qquad (5\text{-}86)$$

and introduce a second decomposition of the one term which is unknown:

$$H(\omega) \equiv \frac{F_-(\omega}{K_+(\omega)} = H_+(\omega)H_-(\omega). \qquad (5\text{-}87)$$

Here H_+ is to be regular in the uhp and H_- in the lhp. Equation (5–86) is then

$$\frac{F_+(\omega)}{K_+(\omega)} + H_+(\omega) = K_-(\omega)\Lambda(\omega) - H_-(\omega). \qquad (5\text{-}88)$$

Each side of this equation is $(\pm)$ the representation of an entire function $V(\omega)$ in their respective half-planes of regularity. Determination of this entire function, and hence the uniqueness of the solutions, can only be accomplished by imposing additional physical conditions.

One evident physical requirement not yet included in the statement of the problem concerns continuity conditions on $f(t)$ and its derivatives at $t = 0$. For example, one would expect the predicted value of $f(t)$ an arbitrarily short time after $t = 0$ to be nearly equal to its *specified* value at $t = 0$. Such requirements can be imposed mathematically by noting that the entire function $V(\omega)$ can be expanded as

$$V(\omega) = \sum_{n=0}^{\infty} V_n\,(-i\omega)^n. \qquad (5\text{-}89)$$

The Fourier transform is

$$v(t) = \sum_{n=0}^{\infty} V_n\,\delta^{(n)}(t). \qquad (5\text{-}90)$$

With these observations it is a short step to find that the essentially unique solutions to Eq.(5–88) are

$$\Lambda(\omega) = \frac{H_-(\omega) + V(\omega)}{K_-(\omega)}, \qquad (5\text{–}91a)$$

$$F_+(\omega) = -K_+(\omega)H_+(\omega) - K_+(\omega)V(\omega). \qquad (5\text{–}91b)$$

Inverse Fourier transformation yields

$$\lambda(t) = \int_{-\infty}^{\infty} \frac{d\omega}{2\pi} e^{i\omega t} \frac{H_-(\omega) + V(\omega)}{K_-(\omega)}, \qquad (5\text{–}92a)$$

$$f_+(t) = -\int_{-\infty}^{\infty} \frac{d\omega}{2\pi} e^{i\omega t} K_+(\omega)[H_+(\omega) + V(\omega)]. \qquad (5\text{–}92b)$$

Thus, depending on how $K_+(\omega)$ ultimately decays for large ω, we can choose the coefficients V_n so as to ensure continuity of $f(t)$ and any of its derivatives at $t = 0$ by subtracting away the discontinuous parts.

As a definite example, suppose that the covariance function is the simple exponential

$$k(t - t') = K_{FF}(t - t') \equiv \tfrac{a}{2} e^{-a|t-t'|}, \qquad (5\text{–}93a)$$

with Fourier transform

$$K(\omega) = \frac{a^2}{a^2 + \omega^2}. \qquad (5\text{–}93b)$$

In this case the factorizations can be carried out by inspection:

$$\begin{aligned} K(\omega) &= K_+(\omega)K_-(\omega) \\ &= \frac{a}{a - i\omega} \frac{a}{a + i\omega}, \end{aligned} \qquad (5\text{–}94)$$

$$H_+(\omega) = 0, \qquad H_-(\omega) = \frac{a - i\omega}{a} F_-(\omega), \qquad (5\text{–}95)$$

because $H(\omega) = a^{-1}(a - i\omega)F_-(\omega)$ is itself regular in the lhp. Hence,

$$\begin{aligned} f_+(t) &= -\int_{-\infty}^{\infty} \frac{d\omega}{2\pi} e^{-i\omega t} K_+(\omega) \sum_{n=0}^{\infty} V_n (-i\omega)^n \\ &= \sum_{n=0}^{\infty} V_n \frac{d^n}{dt^n} k_+(t), \end{aligned} \qquad (5\text{–}96)$$

with

$$k_+(t) = \begin{cases} \int_{-\infty}^{\infty} \frac{d\omega}{2\pi} e^{-i\omega t} K_+(\omega), & t \geq 0 \\ 0, & t < 0 \end{cases}. \qquad (5\text{–}97)$$

As discussed above, we can now choose $V(\omega)$ so as to enforce continuity of $f(t)$ at $t = 0$. According to the physical description of the problem, though, if the driving is going 'up' as $t \to 0^-$, it is doubtless going 'down' at $t = 0^+$, and thus we would not expect $f'(t)$ to be continuous at $t = 0$. Consequently, we shall retain only V_0 and choose $V_n \equiv 0$ for all $n > 0$. Equations (5–92) then become

$$\lambda(t) = \int_{-\infty}^{\infty} \frac{d\omega}{2\pi} e^{i\omega t} \left[\frac{a^2 + \omega^2}{a^2} F_-(\omega) + \frac{a + i\omega}{a} V(\omega) \right]$$

$$= f_-(t) - \frac{1}{a^2} f''(t) + v(t) - \frac{1}{a} v'(t), \tag{5–98}$$

$$f_+(t) = -\int_{-\infty}^{\infty} \frac{d\omega}{2\pi} e^{i\omega t} \frac{a}{a - i\omega} V(\omega). \tag{5–99}$$

One readily verifies that

$$k_+(t) = a e^{-at} \tag{5–100}$$

and

$$V_0 = -\frac{1}{a} f(0), \qquad v(t) = -\frac{1}{a} f(0)\delta(t). \tag{5–101}$$

Hence, the final solutions are

$$\lambda(t) = \left[1 - \frac{1}{a^2} \frac{d^2}{dt^2} \right] f_-(t) - \frac{f(0)}{a^2} \left[a\delta(t) - \delta'(t) \right], \tag{5–102}$$

$$f_+(t) = f(0)e^{-at}. \tag{5–103}$$

We see, therefore, that a concrete example behaves very much as we would have expected intuitively. Removal of the driving force allows the dynamical variable to relax back to its equilibrium value—although our choice of model covariance function had a great deal to do with just how this occurs. Indeed, the covariance function of Eq.(5–93a) appears to be 'Markovian', in that the predicted future of $f(t)$ depends only on the specified departure from equilibrium at $t = 0$, and not on information about its past history. Subsequently we shall see that this is always true in this type of problem, independent of exactly how the driving is done, and the linear response depends *only* on the parameters of the initial nonequilibrium state. Also, note that the relaxation illustrated in Eq.(5–103) must be attributed as much to lack of new information in this scenario as to dynamics.

Finally, observe the appearance of δ-functions in the solution (5–102) for $\lambda(t)$, confirming that the limits of integration in the original equation *must* be taken as $[-\infty, 0^+]$. Yet, there are no factors of 2 here! In fact, it is difficult to imagine any reasonable driving term $v(t)$ in Eq.(5–78) which would produce the result (5–102). We shall return to these points when we discuss the theory of *thermal* driving in Chapter 7.

C. Higher-Order Terms

Owing to the mathematical complexity of the statistical operator of Eqs.(5–28), it appears extremely difficult to avoid making approximations in any but the most trivial nonequilibrium calculations. Nevertheless, one can extend the linear theory to higher orders, and this in fact becomes necessary in applications to hydrodynamics, say. It is of some value, therefore, to illustrate in detail how the Lagrange-multiplier function is determined through second order—that is, when leading-order nonlinear corrections are retained.

If the partition functional is expanded as in Eq.(5–49), but through third order in $\lambda(t)$, we find that

$$
Z[\beta, \lambda(t)] = Z_0(\beta) \left[1 + \langle \hat{F} \rangle_0 \int_{-\tau}^{0} \lambda(t)\, dt \right.
$$

$$
+ \int_{-\tau}^{0} dt_1 \int_{-\tau}^{0} dt_2\, L(t_1, t_2)\lambda(t_1)\lambda(t_2)
$$

$$
+ \int_{-\tau}^{0} dt_1 \int_{-\tau}^{0} dt_2 \int_{-\tau}^{0} dt_3\, L_1(t_1, t_2, t_3)\lambda(t_1)\lambda(t_2)\lambda(t_3)
$$

$$
\left. + \cdots \right],
\tag{5–104}
$$

where $L(t_1, t_2)$ is defined in Eq.(5–50), and

$$
L_1(t_1, t_2, t_3) \equiv \frac{1}{\beta^3} \int_{0}^{\beta} du_1 \int_{0}^{u_1} du_2 \int_{0}^{u_2} du_3
$$

$$
\times \langle \hat{F}(t_1 - i\hbar u_1)\hat{F}(t_2 - i\hbar u_2)\hat{F}(t_3 - i\hbar u_3) \rangle_0 . \tag{5–105}
$$

By means of a judicious sequence of interchanges of integration orders, such as in Eq.(5–51), this function can also be written in the form

$$
L_1(t_1, t_2, t_3) = \frac{1}{\beta^3} \int_{0}^{\beta} du_1 \int_{0}^{u_1} du_2\, (\beta - u_1)\langle \hat{F}(t_1 - i\hbar u_1)\hat{F}(t_2 - i\hbar u_2)\hat{F}(t_3) \rangle_0 .
$$

$$
\tag{5–106}
$$

We now take the logarithm of Z and expand through third order in $\lambda(t)$:

$$
\log Z = \log Z_0 + \langle \hat{F} \rangle_0 \int_{-\tau}^{0} \lambda(t)\, dt + \int_{-\tau}^{0} dt_1 \int_{-\tau}^{0} dt_2\, L(t_1, t_2)\lambda(t_1)\lambda(t_2)
$$

$$
+ \int_{-\tau}^{0} dt_1 \int_{-\tau}^{0} dt_2 \int_{-\tau}^{0} dt_3\, L_1(t_1, t_2, t_3)\lambda(t_1)\lambda(t_2)\lambda(t_3)
$$

$$
- \tfrac{1}{2}\langle \hat{F} \rangle_0^2 \int_{-\tau}^{0} \lambda(t_1)\, dt_1 \int_{-\tau}^{0} \lambda(t_2)\, dt_2
$$

$$
- \langle \hat{F} \rangle_0 \int_{-\tau}^{0} \lambda(t_1)\, dt_1 \int_{-\tau}^{0} dt_2 \int_{-\tau}^{0} dt_3\, L(t_2, t_3)\lambda(t_2)\lambda(t_3)
$$

$$
+ \tfrac{1}{3}\langle \hat{F} \rangle_0^3 \int_{-\tau}^{0} \lambda(t_1)\, dt_1 \int_{-\tau}^{0} \lambda(t_2)\, dt_2 \int_{-\tau}^{0} \lambda(t_3)\, dt_3 + \cdots .
$$

$$
\tag{5–107}
$$

In analogy with Eq.(5–53) we define another function

$$K_1(t_1, t_2, t_3) \equiv L_1(t_3, t_1, t_2) + L_1(t_1, t_2, t_3)$$
$$+ L_1(t_2, t_3, t_1) - \langle \hat{F} \rangle_0 L(t_3, t_2) . \qquad (5\text{–}108)$$

Although K_1 is defined in terms of the cyclic permutations of variables in L_1, it is *not* invariant under cyclic permutations. Rather, it appears to have no obvious symmetry properties.

If we now perform the functional differentiation prescribed by Eq.(5–4), we find that

$$\langle \hat{F}(t) \rangle - \langle \hat{F} \rangle_0 = \int_{-\tau}^{0} K(t, t') \lambda(t')\, dt'$$

$$- \langle \hat{F} \rangle_0 \int_{-\tau}^{0} dt_1\, \lambda(t_1) \int_{-\tau}^{0} dt_2\, K(t, t_2) \lambda(t_2)$$

$$+ \int_{-\tau}^{0} dt_1 \int_{-\tau}^{0} dt_2\, K_1(t, t_1, t_2) \lambda(t_1) \lambda(t_2) , \qquad (5\text{–}109)$$

with the definition (5–53) of $K(t, t')$. Note that this expression is valid in the interval $-\tau \le t \le 0$.

The function $K(t, t')$ has already been identified with the covariance function $K_{FF}(t, t')$ in Eq.(5–58). If the expectation value in Eq.(5–5) were expanded to second order in $\lambda(t)$, with $\hat{J} = \hat{F}$, one would obtain an expression containing three terms, the first two of which correspond to the first two terms on the right-hand side of Eq.(5–109). The third term contains a function

$$K_{FFF}(t, t_1, t_2) \equiv \overline{\langle \hat{F}(t_1)\hat{F}(t_2)\hat{F}(t) \rangle}_0 - \overline{\langle \hat{F}(t_1)\hat{F}(t_2) \rangle}_0 \langle \hat{F} \rangle_0 , \qquad (5\text{–}110)$$

where the *double Kubo transform* is defined as

$$\overline{\hat{F}(t_1)\hat{G}(t_2)} \equiv \int_0^1 x_1\, dx_1 \int_0^1 dx_2\, e^{\beta \hat{H} x_1}\, \hat{F}(t_1) e^{-\beta \hat{H} x_1(1-x_2)}\, \hat{G}(t_2) e^{-\beta \hat{H} x_1 x_2}$$

$$= \frac{1}{\beta^2} \int_0^{\beta} du_1 \int_0^{u_1} du_2\, \hat{F}(t_1 - i\hbar u_1)\hat{G}(t_2 - i\hbar u_2) , \qquad (5\text{–}111)$$

the last line holding for time-independent $\hat{H}$. Thus, we must show that $K_1 \equiv K_{FFF}$. Because of the identity (5–58), this means we must prove that

$$\frac{1}{\beta^2} \int_0^{\beta} du_1 \int_0^{u_1} du_2\, \langle \hat{F}(t_1 - i\hbar u_1)\hat{F}(t_2 - i\hbar u_2)\hat{F}(t) \rangle_0 =$$

$$\frac{1}{\beta^3} \int_0^{\beta} du_1 (\beta - u_1) \int_0^{u_1} du_2\, [\langle \hat{F}(t_1 - i\hbar u_1)\hat{F}(t_2 - i\hbar u_2)\hat{F}(t) \rangle_0$$

$$+ \langle \hat{F}(t_2 - i\hbar u_1)\hat{F}(t - i\hbar u_2)\hat{F}(t_1) \rangle_0$$

$$+ \langle \hat{F}(t - i\hbar u_1)\hat{F}(t_1 - i\hbar u_2)\hat{F}(t_2) \rangle_0] . \qquad (5\text{–}112)$$

In turn, this is equivalent to proving that

$$\int_0^\beta u_1 \, du_1 \int_0^{u_1} du_2 \, \langle \hat{F}(t_1 - i\hbar u_1)\hat{F}(t_2 - i\hbar u_2)\hat{F}(t)\rangle_0 =$$

$$\int_0^\beta du_1 \, (\beta - u_1) \int_0^{u_1} du_2 \, [\langle \hat{F}(t_2 - i\hbar u_1)\hat{F}(t - i\hbar u_2)\hat{F}(t_1)\rangle_0$$

$$+ \langle \hat{F}(t - i\hbar u_1)\hat{F}(t_1 - i\hbar u_2)\hat{F}(t_2)\rangle_0]. \tag{5-113}$$

The identity (5–112) is one immediately following that of Eq.(5–57) in a hierarchy of similar identities arising in all higher-order terms. One proves Eq.(5–113) by writing out both sides explicitly in a representation in which $\hat{H}$ is diagonal and then performing the u-integrations. As verified in Problem 5.4, the algebra is tedious but straightforward, and the result is that, indeed, $K_1 = K_{FFF}$.

This analysis of the leading-order nonlinear term is summarized by writing for the expectation value of $\hat{C}(t)$ the following expression:

$$\langle \hat{C}(t)\rangle - \langle \hat{C}\rangle_0 = \int_{-\tau}^0 K_{CF}(t, t')\lambda(t') \, dt'$$

$$- \langle \hat{F}\rangle_0 \int_{-\tau}^0 dt_1 \int_{-\tau}^0 dt_2 \, K_{CF}(t, t_2)\lambda(t_1)\lambda(t_2)$$

$$+ \int_{-\tau}^0 dt_1 \int_{-\tau}^0 dt_2 \, K_{CFF}(t, t_1, t_2)\lambda(t_1)\lambda(t_2). \tag{5-114}$$

The predicted future and retrodicted past of $\langle \hat{C}(t)\rangle$ are given by this equation in the intervals $t > 0$ and $t < -\tau$, respectively. When $\hat{C} = \hat{F}$ this provides a nonlinear integral equation for determining $\lambda(t)$ in the interval $-\tau \leq t \leq 0$, as specified by Eq.(5–109). This nonlinear integral equation is rather difficult to solve and the theory is not very well developed. Nevertheless, equations of this type have been studied (e.g., Schmidt, 1908), and the problem has been reduced to one of working out the requisite calculational procedures.

We re-emphasize that the intrinsically nonlinear situation occurs when the statistical operator of Eq.(5–2) describes a truly nonequilibrium initial state. One is not even sure in this case what sense to give to the meaning of a linear approximation, and the available mathematical tools are few indeed. The formulation of the problem, however, is at least unambiguous in the context of the present theory. We shall return to this fascinating aspect of the full theory eventually, but for the next several chapters we shall focus on an analysis of the linear approximation, because it is here where the most immediate results can be obtained.

Problems

5.1 Let $\langle \hat{F}(t)\rangle$ be specified throughout the interval $-\tau \leq t \leq 0$ for a system in thermal equilibrium with Hamiltonian $\hat{H}$ for $t < -\tau$. Construct the ensemble of Eqs.(5–28) by minimizing the cross-entropy

$$H \equiv \kappa \, \mathrm{Tr} \, \hat{\rho}[\ln \hat{\rho} - \ln \hat{\rho}_0]$$

subject to this constraint, as well as $\text{Tr}\,\hat{\rho} = \text{Tr}\,\hat{\rho}_0 = 1$.

5.2 Show that the diagonal part of an operator can also be interpreted as a time average, in the sense that we can write

$$\hat{F}_d = \lim_{\tau \to \infty} \frac{1}{\tau} \int_0^\tau \hat{F}(-t)\,dt\,.$$

5.3 Carry out the detailed derivation of the series expansion in Eq.(5–64).

5.4 Instead of the simple exponential model of the covariance function given in Eq.(5–93a), find the solutions to the problem being studied there for a model covariance function given by

$$k(t - t') = \frac{1/2}{(a\tau)^{1/2}(1 + a^2\tau^2)^{1/2}}\,,$$

where a is a constant and $\tau \equiv |t - t'|$.

REFERENCES

Bergmann, P.G., and J.L. Lebowitz: 1955, 'New Approach to Nonequilibrium Processes', *Phys. Rev.* **99**, 578.

Fano, U.: 1957, 'Description of States in Quantum Mechanics by Density Matrix and Operator Techniques', *Rev. Mod. Phys.* **29**, 74.

Heims, S.P., and E.T.Jaynes: 1962, 'Theory of Gyromagnetic Effects and Some Related Magnetic Phenomena', *Rev. Mod. Phys.* **34**, 143.

Jaynes, E.T.: 1979, 'Where Do We Stand on Maximum Entropy?', in R.D. Levine and M. Tribus (eds.), *The Maximum Entropy Formalism*, M.I.T. Press, Cambridge, MA.

Kubo, R.: 1959, 'Some Aspects of the Statistical-Mechanical Theory of Irreversible Processes', in W.E. Brittin and Lita G. Dunham (eds.), *Lectures in Theoretical Physics*, *Vol.I*, Interscience, New York.

Lebowitz, J.L., and P.G. Bergmann: 1957, 'Irreversible Gibbsian Ensembles', *Ann. Phys. (N.Y.)* **1**, 1.

McLennan, J.A., Jr.: 1959, 'Statistical Mechanics of the Steady State', *Phys. Rev.* **115**, 1405.

Nakajima, S.: 1958, 'On Quantum Theory of Transport Phenomena', *Prog. Theor. Phys.* **20**, 948.

Paley, R.E.A.C., and N. Wiener: 1934, *Fourier Transforms in the Complex Domain*, Am. Math. Soc., Providence, RI, p.16.

Peterson, R.L.: 1967, 'Formal Theory of Nonlinear Response', *Rev. Mod. Phys.* **39**, 69.

Roos, B.W.: 1969, *Analytic Functions and Distributions in Physics and Engineering*, Wiley, New York.

Scalapino, D.J.: 1961, 'Irreversible Statistical Mechanics and the Principle of Maximum Entropy', *Ph.D. thesis*, Stanford Univ. (unpublished).

Schmidt, E.: 1908, 'Über die Auflösung der nichtlinearen Integralgleichung und die verzweigung ihrer Lösungen', *Math. Ann.* **65**, 370.

Tricomi, F.G.: 1957, *Integral Equations*, Interscience, New York.

Wiener, N., and E. Hopf: 1931, 'Über eine Klasse singulärer Integralgleichungen', *Sitz. Ber. Preuß. Akad. Wiss.* **30**, 696.

Chapter 6

Analysis of Covariance Functions
for
Simple Fluids

Because the covariance functions play such a central role in the linear theory it is important to analyze their structure and properties in considerable detail. In order to carry out this analysis it is necessary to choose a somewhat specific example, which is still broad enough in scope so as to apply to a wide range of common problems. Therefore, in view of later applications, we focus here on simple fluids, a more exact definition of which will emerge presently. Application to more complex systems, such as lattices, then follows easily, for most general properties remain unchanged. Although covariance functions arise here from a definite origin, and thus in a completely different context than elsewhere, they clearly have the structure of correlation functions and their role in physical theory has been in the air for some time. Consequently, one can refer to the vast literature on that subject in order to deduce many similar properties.

Let us consider for simplicity a single-component fluid of N particles, either fermions or bosons, for which the commutator of any two linear operators will be written

$$[\hat{A}, \hat{B}] \equiv \hat{A}\hat{B} - \varepsilon\hat{B}\hat{A}, \qquad (6\text{-}1)$$

with $\varepsilon = +1\,(-1)$ for bosons (fermions). As described in Volume I, say, and elsewhere (e.g., Fetter and Walecka, 1971), a formulation of N-body quantum mechanics in the Fock representation involves creation operators $\hat{a}_i^\dagger$ for a particle in the single-particle state i, and annihilation operators $\hat{a}_i$ for the same state. For time independent $\hat{H}$, time evolution is described by

$$\hat{a}_i(t) = e^{it\hat{H}/\hbar}\,\hat{a}_i e^{-it\hat{H}/\hbar}\,, \qquad (6\text{-}2)$$

and similarly for creation operators.

The Hermitian product

$$\hat{N}_i \equiv \hat{a}_i^\dagger \hat{a}_i \qquad (6\text{-}3)$$

defines the number operator for the ith state, and has as eigenvalues occupation numbers n_i, which specify the total number of particles in the various single-particle states $|\lambda_i\rangle$. The total-number operator is then

$$\hat{N} \equiv \sum_i \hat{N}_i = \sum_i \hat{a}_i^\dagger \hat{a}_i\,, \qquad (6\text{-}4)$$

and the sum goes over the complete set of single-particle states. (Because we shall never include spin-dependent interactions in the sequel, it will be convenient to suppress all spin indices when possible.)

The general N-body Hamiltonian has the form $\hat{H}^{(N)} = \hat{H}_0^{(N)} + \hat{V}^{(N)}$, where $\hat{H}_0^{(N)}$ is the N-body free-particle Hamiltonian, and $\hat{V}^{(N)}$ a general N-body interaction. We shall usually presume that $\hat{V}^{(N)}$ is well approximated by a sum over all two-body interactions, and that any external field is described by a one-body interaction that can be incorporated into $\hat{H}_0^{(N)}$. Thus,

$$\hat{H}^{(N)} = \hat{H}_0^{(N)} + \hat{V}_1 + \hat{V}_2$$
$$= \sum_{k=1}^{N} \hat{H}_k + \tfrac{1}{2} \sum_{i \neq j}^{N} \hat{V}_{ij} , \qquad (6\text{--}5)$$

where $\hat{V}_{ij}$ is an arbitrary two-body potential and $\hat{H}_k = \hat{H}_0(k) + \hat{V}_1(k)$ is the single-particle 'free' Hamiltonian for the kth particle.

The Fock-space representations for these operators are

$$\hat{H}_0 + \hat{V}_1 = \sum_{\lambda_i \lambda_k} \hat{a}_i^\dagger \hat{a}_k \langle \lambda_i | \hat{H}_0 + \hat{V}_1 | \lambda_k \rangle , \qquad (6\text{--}6)$$

$$\hat{V}_2 = \tfrac{1}{2} \sum_{\substack{\lambda_i \lambda_j \\ \lambda_k \lambda_\ell}} \hat{a}_k^\dagger \hat{a}_\ell^\dagger \langle \lambda_k \lambda_\ell | \hat{V}_2 | \lambda_i \lambda_j \rangle \hat{a}_j \hat{a}_i . \qquad (6\text{--}7)$$

If $|\lambda_i\rangle$ is an eigenstate of $\hat{H}_0$, as is generally presumed, then

$$\langle \lambda_i | \hat{H}_0 | \lambda_j \rangle = \delta_{ij} \omega_i , \qquad (6\text{--}8)$$

where

$$\omega_i \equiv \frac{p_i^2}{2m} = \frac{\hbar^2 k_i^2}{2m} \qquad (6\text{--}9)$$

is a free-particle energy for particles of mass m. For local interactions, which we shall always presume here, the matrix elements above are written in terms of single-particle wavefunctions:

$$\langle \lambda_i | \hat{V}_1 | \lambda_j \rangle = \int \phi_{\lambda_i}^*(x) V_1(x) \phi_{\lambda_k}(x) \, d^3 x , \qquad (6\text{--}10)$$

$$\langle \lambda_k \lambda_\ell | \hat{V}_2 | \lambda_i \lambda_j \rangle = \int \int \phi_{\lambda_k}^*(x_1) \phi_{\lambda_\ell}^*(x_2) V_2(x_1, x_2) \phi_{\lambda_i}(x_1) \phi_{\lambda_j}(x_2) \, d^3 x_1 \, d^3 x_2 .$$
$$(6\text{--}11)$$

Various processes in fluids are most readily described in the coordinate representation, so we introduce field operators into Fock space as follows:

$$\hat{\psi}(x) \equiv \sum_i \phi_i(x) \hat{a}_i , \qquad \hat{\psi}^\dagger(x) \equiv \sum_i \phi_i^*(x) \hat{a}_i^\dagger . \qquad (6\text{--}12)$$

The coefficients are just single-particle wavefunctions normalized to the system volume and belonging to a complete set. The sums range over that complete set and it should be remembered that implicitly the ϕ_i are actually $(2s+1)$-component wavefunctions for particles of spin s. The field operators satisfy the following commutation relations:

$$[\hat{\psi}(\mathbf{x}), \hat{\psi}(\mathbf{x}')] = [\hat{\psi}^\dagger(\mathbf{x}), \hat{\psi}^\dagger(\mathbf{x}')] = 0, \quad [\hat{\psi}(\mathbf{x}), \hat{\psi}^\dagger(\mathbf{x}')] = \delta(\mathbf{x} - \mathbf{x}'). \qquad (6\text{--}13)$$

One can employ the classical expressions for various quantities in order to deduce their Fock-space forms. For example, from the classical expression for the number density,

$$n(\mathbf{x}) = \sum_{i=1}^{N} \delta(\mathbf{x} - \mathbf{x}_i), \qquad (6\text{--}14)$$

we deduce the following form for the number-density operator:

$$\hat{n}(\mathbf{x}) \equiv \hat{\psi}^\dagger(\mathbf{x})\hat{\psi}(\mathbf{x}), \qquad (6\text{--}15)$$

and the total-number operator becomes

$$\hat{N} \equiv \int \hat{\psi}^\dagger(\mathbf{x})\hat{\psi}(\mathbf{x}) \, d^3x. \qquad (6\text{--}16)$$

Time dependence of all field operators can be deduced from Eq.(6–2), as in the following example:

$$\hat{n}(\mathbf{x}, t) = \hat{\psi}^\dagger(\mathbf{x}, t)\hat{\psi}(\mathbf{x}, t) = \sum_{ij} \phi_i^*(\mathbf{x})\phi_j(\mathbf{x})\hat{a}_i^\dagger(t)\hat{a}_j(t). \qquad (6\text{--}17)$$

Similarly, the energy-density operator is defined as

$$\hat{h}(\mathbf{x}, t) \equiv \frac{\hbar^2}{2m}(\nabla\hat{\psi}^\dagger) \cdot (\nabla\hat{\psi}) + \frac{1}{2}\int V_2(\mathbf{x}, \mathbf{x}')\hat{\psi}^\dagger(\mathbf{x}, t)\hat{n}(\mathbf{x}', t)\hat{\psi}(\mathbf{x}, t) \, d^3x', \qquad (6\text{--}18)$$

so that the Hamiltonian is

$$\hat{H} = \int \hat{h}(\mathbf{x}, t) \, d^3x, \qquad (6\text{--}19)$$

which is actually independent of the time. Any one-body operator $V_1(\mathbf{x})$ describing an external field, say, can be included by adding a term $\hat{n}(\mathbf{x}, t)V_1(\mathbf{x})$ to the right-hand side of Eq.(6–18). In fact, this procedure is what allows one to introduce nonlocal couplings into the theory of dynamical response in the form

$$\int \hat{n}(\mathbf{x}, t)V_1(\mathbf{x}) \, d^3x. \qquad (6\text{--}20)$$

Finally, the particle current-density operator will be written

$$\hat{\jmath}(\mathbf{x}, t) \equiv \frac{i\hbar}{2m}[(\nabla\hat{\psi}^\dagger)\hat{\psi} - \hat{\psi}^\dagger(\nabla\hat{\psi})]. \qquad (6\text{--}21)$$

Fock-space operators in the Heisenberg picture satisfy the usual equation of motion, $i\hbar(d\hat{A}/dt) = [\hat{A}, \hat{H}]$, under the presumption of no explicit time dependence. In turn, the density operators defined above have the following explicit equations of motion:

$$\partial_t \hat{n}(\mathbf{x}, t) = \tfrac{i}{\hbar}[\hat{H}, \hat{n}(\mathbf{x}, t)],\tag{6-22}$$

$$m\partial_t \hat{\jmath}(\mathbf{x}, t) = \tfrac{i}{\hbar}[\hat{H}, m\hat{\jmath}(\mathbf{x}, t)],\tag{6-23}$$

$$\partial_t \hat{h}(\mathbf{x}, t) = \tfrac{i}{\hbar}[\hat{H}, \hat{h}(\mathbf{x}, t)],\tag{6-24}$$

where $m\hat{\jmath}$ is the momentum-density operator. The left-hand sides of these equations are also involved in statements of the local microscopic conservation laws for the system, which usually relate time derivatives of the densities to divergences of the corresponding currents. We expect, therefore, that these differential conservation laws can be obtained directly from Eqs.(6–22)-(6–24) by evaluating the commutators is such a way that they take the form of the divergence of another operator. Thus, we rewrite the above equations as

$$\partial_t \hat{n}(\mathbf{x}, t) = -\nabla \cdot \hat{\jmath}(\mathbf{x}, t),\tag{6-25}$$

$$m\partial_t \hat{\jmath}(\mathbf{x}, t) = -\nabla \cdot \hat{\mathbf{T}}(\mathbf{x}, t),\tag{6-26}$$

$$\partial_t \hat{h}(\mathbf{x}, t) = -\nabla \cdot \hat{q}(\mathbf{x}, t).\tag{6-27}$$

Equation (6–25) is readily verified directly, but great care must be exercised in identifying the energy current-density operator $\hat{q}$, and the components of the stress-tensor operator, $\hat{T}_{ij}$. The latter is actually the momentum current density, whereas $\hat{\jmath}$ serves as both the number current and the momentum density. The total-momentum operator is just

$$\hat{\mathbf{P}}(t) \equiv \int m\hat{\jmath}(\mathbf{x}, t)\, d^3x,\tag{6-28}$$

integrated over the entire system volume.

Identification of the operators $\hat{q}$ and $\hat{T}_{ij}$ depends crucially on the specific character of the two-body interaction $\hat{V}_2(\mathbf{x}, \mathbf{x}')$, because it is not at all clear that they can always be identified unambiguously for arbitrary interactions. In the present case, though, we have excluded explicit mention of spin variables in the formalism, and this is only consistent if spin-dependent interactions are also excluded. As a consequence, we are led to consider only spherically-symmetric two-body potentials in what follows: $\hat{V}_2(\mathbf{x}, \mathbf{x}') = \hat{V}(|\mathbf{x} - \mathbf{x}'|)$. More generally, the Hamiltonian is taken to be rotationally invariant, including $\hat{V}_1(\mathbf{x})$.

In addition to rotational invariance, time independence of the Hamiltonian implies that the system is also time-translation invariant. Moreover, in most applications of the linear theory it suffices to consider only initially uniform systems, so that the system is also space-translation invariant. These invariance properties are characterized mathematically by writing

$$\hat{\psi}(\mathbf{x}, t) = e^{it\hat{H}/\hbar}\, \hat{\psi}(\mathbf{x}) e^{-it\hat{H}/\hbar}\,, \tag{6-29}$$

$$\hat{\psi}(\mathbf{x} + \mathbf{x}') = e^{i\mathbf{x}'\cdot\hat{\mathbf{P}}/\hbar}\, \hat{\psi}(\mathbf{x}) e^{-i\mathbf{x}'\cdot\hat{\mathbf{P}}/\hbar}\,, \tag{6-30}$$

and both the total-momentum and total-number operators are conserved. That is,

$$[\hat{H}, \hat{\mathbf{P}}] = [\hat{H}, \hat{N}] = [\hat{\mathbf{P}}, \hat{N}] = \hat{0}\,, \tag{6-31}$$

which are readily proved directly in Problem 6.1. (It is true, of course, that the Hamiltonian can be translation invariant even if the system state is nonuniform, but we shall not explicitly consider such states here.)

Specification of these three invariance properties constitutes a precise definition of what we here call a *simple fluid*. For this system the stress-tensor and energy current-density operators are determined by direct evaluation of the right-hand sides of Eqs.(6–23) and (6–24) by means of the commutation relations (6–13), the properties (6–29)-(6–31), and the following identity from Appendix A for arbitrary $\hat{A}$ and $\hat{S}$:

$$e^{-\hat{S}}\,\hat{A}e^{\hat{S}} = \hat{A} + \int_0^1 e^{-\lambda\hat{S}}\,[\hat{A}, \hat{S}]e^{\lambda\hat{S}}\, d\lambda\,. \tag{6-32}$$

In Problem 6.2 the reader is asked to carry out the tedious algebra required to verify that the two operators are given by the following expressions:

$$\hat{T}_{ij}(\mathbf{x}, t) = -\frac{\hbar^2}{4m}[(\nabla_i - \nabla_i')(\nabla_j - \nabla_j')\hat{\psi}^\dagger(\mathbf{x}', t)\hat{\psi}(\mathbf{x}, t)]_{\mathbf{x}'=\mathbf{x}}$$
$$- \frac{1}{4}\int d^3r'\left[\frac{r_i' r_j'}{r'}\frac{dV_2(r')}{dr'}\right]\int_{-1}^{1} d\lambda\, \hat{\psi}^\dagger\left(\mathbf{x} + \tfrac{1+\lambda}{2}\mathbf{r}', t\right)$$
$$\hat{n}\left(\mathbf{x} - \tfrac{1-\lambda}{2}\mathbf{r}', t\right)\hat{\psi}\left(\mathbf{x} + \tfrac{1+\lambda}{2}\mathbf{r}', t\right), \tag{6-33}$$

$$\hat{q}_i(\mathbf{x}, t) = \frac{1}{i\hbar}\left(\frac{\hbar^2}{2m}\right)^2[(\nabla_i - \nabla_i')(\nabla_j - \nabla_j')\nabla' \cdot \nabla\hat{\psi}^\dagger(\mathbf{x}', t)\hat{\psi}(\mathbf{x}, t)]_{\mathbf{x}'=\mathbf{x}}$$
$$+ \frac{1}{2}\int d^3x'\, V_2(|\mathbf{x} - \mathbf{x}'|)\hat{\psi}^\dagger(\mathbf{x}', t)\hat{j}_i(\mathbf{x}, t)\hat{\psi}(\mathbf{x}', t)$$
$$- \frac{1}{4}\int d^3r'\left[\frac{r_i' r_j'}{r'}\frac{dV_2(r')}{dr'}\right]\int_{-1}^{1} d\lambda\, \hat{\psi}^\dagger\left(\mathbf{x} + \tfrac{1+\lambda}{2}\mathbf{r}', t\right)$$
$$\hat{j}_i\left(\mathbf{x} - \tfrac{1-\lambda}{2}\mathbf{r}', t\right)\hat{\psi}\left(\mathbf{x} + \tfrac{1+\lambda}{2}\mathbf{r}', t\right). \tag{6-34}$$

We employ the summation convention over repeated indices and note that it is the possible noncommutativity of the operators that gives rise to the λ-integrals.

The space-integrated forms of these operators are somewhat simpler, because in that case one can make the replacement

$$\int_{-1}^{1} f(\lambda)\, d\lambda \longrightarrow 2f(1)\,, \tag{6-35}$$

and it is then useful to define

$$\hat{T}_{ij}(t) \equiv \int \hat{T}_{ij}(\mathbf{x}, t) \, d^3x \,, \tag{6-36}$$

$$\hat{q}(t) \equiv \int \hat{q}(\mathbf{x}, t) \, d^3x \,. \tag{6-37}$$

Unlike the situation with $\hat{\mathbf{P}}(t)$ and $\hat{N}(t)$, these operators are *not* conserved, in that they do not commute with the Hamiltonian. The space integration does have beneficial effects, though, as illustrated in the following identities:

$$\hat{T}_{11}(t) = \hat{T}_{22}(t) = \hat{T}_{33}(t) = \tfrac{1}{3}\hat{T}^i{}_i(t) \,, \tag{6-38}$$

the last quantity being one-third the tensor trace of $\hat{T}_{ij}(t)$.

In addition to the symmetry properties discussed above, one notes that $\hat{n}$, $\hat{h}$, and the symmetric tensor $\hat{T}_{ij}$ are even under time reversal, whereas $\hat{\jmath}$ and $\hat{q}$ are odd. These properties, along with the appropriate behavior under Galilean transformations, are sufficient to determine the operators $\hat{q}$ and $\hat{T}_{ij}$ *uniquely* for spin-independent interactions (Puff and Gillis, 1968). Were the two-body interaction not spherically symmetric, $\hat{T}_{ij}$ would not be symmetric. In that case the antisymmetric part would be related to an angular-momentum operator and it would be necessary to develop an explicit microscopic conservation law for angular momentum. We shall not consider this possibility further at this time.

In view of subsequent applications it is useful to exhibit the equilibrium expectation values of the above operators in the grand canonical ensemble. With the use of Eqs.(5–59) and (5–60), along with the above symmetry properties, we see that

$$\langle \hat{n}(\mathbf{x}, t) \rangle_0 = \frac{1}{Z_0} \operatorname{Tr}\!\left[e^{-\beta \hat{K}} \hat{\psi}^\dagger(\mathbf{x}, t) \hat{\psi}(\mathbf{x}, t) \right] = \frac{1}{Z_0} \operatorname{Tr}\!\left[e^{-\beta \hat{K}} \hat{\psi}^\dagger(0) \hat{\psi}(0) \right]$$

$$= n_0 \equiv \frac{N_0}{V} = \frac{\langle \hat{N} \rangle}{V} \,, \tag{6-39}$$

where V is the system volume. In like manner, the equilibrium energy density is

$$\langle \hat{h}(\mathbf{x}, t) \rangle_0 = \frac{\hbar^2}{2m} \langle (\nabla \hat{\psi}^\dagger) \cdot (\nabla \hat{\psi}) \rangle_0 + \tfrac{1}{2} \int d^3r \, V_2(r) \langle \hat{\psi}^\dagger(\mathbf{r}) \hat{\psi}^\dagger(0) \hat{\psi}(0) \hat{\psi}(\mathbf{r}) \rangle_0 \,, \tag{6-40}$$

and the first term on the right-hand side is seen to be the average kinetic-energy density. If there is no external field, then Eq.(6–40) will yield the equilibrium value of the total energy per unit volume:

$$\langle \hat{h}(\mathbf{x}, t) \rangle_0 = h_0 \equiv \frac{E_0}{V} = \frac{\langle \hat{H} \rangle_0}{V} \,. \tag{6-41}$$

The current-density operators $\hat{\jmath}$ and $\hat{q}$ must have zero expectation values in the equilibrium system, of course, so that the only remaining equilibrium expectation value is that for the stress tensor. A short calculation yields:

$$\langle \hat{T}_{ij}(\mathbf{x}, t) \rangle_0 = \tfrac{2}{3} \frac{\hbar^2}{2m} \langle (\nabla \hat{\psi}^\dagger) \cdot (\nabla \hat{\psi}) \rangle_0 \delta_{ij}$$

$$- \tfrac{1}{6} \int d^3r \, [\mathbf{r} \cdot \nabla V_2(r)] \langle \hat{\psi}^\dagger(\mathbf{r}) \hat{\psi}^\dagger(0) \hat{\psi}(0) \hat{\psi}(\mathbf{r}) \rangle_0 \, \delta_{ij} \,. \tag{6-42}$$

The first term on the right-hand side is just 2/3 the average kinetic-energy density, which suggests that Eq.(6–42) is an expression of the virial theorem and that the left-hand side is the equilibrium pressure. This conclusion can be verified in several ways—from Eq.(9–37) in Volume I, or from Puff and Gillis (1968)—so that one can write

$$\langle \hat{T}_{ij}(\mathbf{x}, t)\rangle_0 = P_0\, \delta_{ij}\,. \tag{6–43}$$

As it should, this interpretation depends crucially on rotational invariance of the Hamiltonian. That is, the stress tensor has no shear components in equilibrium and the hydrostatic pressure is related to the normal stresses only.

A possibly more familiar notation is achieved by noting that the expectation value of a product of creation and annihilation operators appearing in Eqs.(6–40) and (6–42) is essentially the *radial distribution function*:

$$g(r) \equiv n_0^{-2}\langle \hat{\psi}^\dagger(\mathbf{r})\hat{\psi}^\dagger(0)\hat{\psi}(0)\hat{\psi}(\mathbf{r})\rangle_0\,. \tag{6–44}$$

The dependence of $g(r)$ only on the magnitude of $\mathbf{r}$ is a further consequence of rotational invariance. Define also

$$\frac{\langle KE\rangle_0}{V} \equiv \frac{\hbar^2}{2m}\int \langle (\nabla\hat{\psi}^\dagger)\cdot(\nabla\hat{\psi})\rangle_0\, d^3r\,. \tag{6–45}$$

The average energy density and pressure in a simple equilibrium fluid can then be written, respectively, as

$$\langle \hat{h}(\mathbf{x}, t)\rangle_0 = \frac{\langle KE\rangle_0}{V} + \tfrac{1}{2}n_0^2\int V_2(r)g(r)\, d^3r\,, \tag{6–46}$$

$$P_0 = \tfrac{2}{3}\frac{\langle KE\rangle_0}{V} - \tfrac{1}{6}n_0^2\int [\mathbf{r}\cdot\nabla V_2(r)]g(r)\, d^3r\,. \tag{6–47}$$

Simple fluids can be completely characterized by the five locally conserved quantities of Eqs.(6–25)-(6–27), in that these microscopic conservation laws lead to five long-lived hydrodynamic modes. Local disturbances of these quantities can not be dissipated locally, but spread out over the entire system. Thus, as we shall see, it is just the existence of these conservation laws that leads to a macroscopic hydrodynamic description. Similar quantities will stand out in other model systems, and their characteristic operators will yield to physical interpretations in much the same way as here. In studying the covariance functions for the present model of a simple fluid, however, we shall generally have in mind the five conserved operators above.

A. General Properties of Covariance Functions

Let us rewrite the covariance function defined in Eq.(5–62) explicitly as

$$
\begin{aligned}
K_{AB}(\mathbf{x}', t'; \mathbf{x}, t) &\equiv \overline{\langle \hat{B}(\mathbf{x}', t')\hat{A}(\mathbf{x}, t)\rangle_0} - \langle \hat{B}(\mathbf{x}', t')\rangle_0\langle \hat{A}(\mathbf{x}, t)\rangle_0 \\
&= \mathrm{Tr}\Big\{\hat{\rho}_0\int_0^1 ds\, e^{s\beta\hat{K}}\, \hat{B}(\mathbf{x}', t')e^{-s\beta\hat{K}}\, \hat{A}(\mathbf{x}, t)\Big\} \\
&\quad - \mathrm{Tr}\{\hat{\rho}_0\hat{B}(\mathbf{x}', t')\}\,\mathrm{Tr}\{\hat{\rho}_0\hat{A}(\mathbf{x}, t)\}\,, \tag{6–48}
\end{aligned}
$$

Operator	P	T
$\hat{H}$	+	+
$\hat{N}(t)$	+	+
$\hat{\mathbf{P}}(t)$	−	−
$\hat{n}(\mathbf{x},t)$	+	+
$\hat{\jmath}(\mathbf{x},t)$	−	−
$\hat{h}(\mathbf{x},t)$	+	+
$\hat{q}(\mathbf{x},t)$	−	−
$\hat{T}_{ij}(\mathbf{x},t)$	+	+

Table 6–1. Inversion properties of the field operators for a simple fluid.

where $\hat{\rho}_0$ is given by Eq.(5–60). Although $\hat{A}$ and $\hat{B}$ may be general linear Heisenberg operators, we understand $\hat{H}$ to exhibit rotational and space-time translational invariance, as discussed above and manifested in Eqs.(6–29) and (6–30). Significant simplification of K_{AB} is immediate in this case because the space and time variables then occur only in the combinations

$$\mathbf{r} \equiv \mathbf{x} - \mathbf{x}', \qquad \tau \equiv t - t'. \qquad (6\text{–}49)$$

That is,

$$K_{AB}(\mathbf{x}',t';\mathbf{x},t) = K_{AB}(0,0;\mathbf{r},\tau) \equiv K_{AB}(\mathbf{r},\tau), \qquad (6\text{–}50)$$

and the single-operator expectation values are actually independent of space and time variables. As with most identities of this type, the proof relies on cyclic invariance of the trace. In addition, we have noted that Eqs.(6–29) and (6–30) remain valid with $\hat{H}$ replaced by $\hat{K}$, as in Eq.(5–59). The physical scenario envisioned here is that for almost all physical applications: it is presumed that the nonequilibrium situation arises from perturbation of an equilibrium system that is spatially and temporally uniform.

When the system exhibits the foregoing invariance properties, the five basic field operators for a simple fluid are well defined. These operators are all Hermitian and possess definite space-reflection (parity) and time-reversal properties as well. In fact, they all exhibit the following behavior: under the parity operation,

$$\hat{A}(\mathbf{r},\tau) \longrightarrow \hat{A}'(\mathbf{r},\tau) = P_A\,\hat{A}(-\mathbf{r},\tau), \quad P_A = \pm 1; \qquad (6\text{–}51)$$

under time reversal,

$$\hat{A}(\mathbf{r},\tau) \longrightarrow \hat{A}'(\mathbf{r},\tau) = T_A\,\hat{A}(\mathbf{r},-\tau), \quad T_A = \pm 1. \qquad (6\text{–}52)$$

These symmetry properties for the operators describing a simple fluid are tabulated in Table 6–1, and we note that $PT = +1$ for all of them. Similar tables are readily constructed for other systems.

Operator properties of the kind indicated in Eqs.(6–51) and (6–52) lead to corresponding properties for the covariance functions. The parity operation is described by a linear, unitary operator, whereas time inversion is governed by an antilinear, antiunitary operator. One finds that

$$K_{AB}(\mathbf{r},\tau) = P_A P_B K_{AB}(-\mathbf{r},\tau) = T_A T_B K_{A\dot{B}}(\mathbf{r},-\tau)$$
$$= P_A P_B T_A T_B K_{AB}(-\mathbf{r},-\tau) = K_{BA}(-\mathbf{r},-\tau), \qquad (6\text{--}53)$$

the last being a generalization of the identity (5–48). If $\hat{A}$ and $\hat{B}$ have symmetry properties of the type indicated in Table 6–1, however, $PT = 1$ and we obtain the complete reciprocity relation

$$K_{AB}(\mathbf{r},\tau) = K_{BA}(\mathbf{r},\tau). \qquad (6\text{--}54)$$

Note that on the left-hand side $\hat{B}$ is independent of space-time variables, whereas on the right-hand side it is $\hat{A}$ that exhibits this independence. As we shall see in Chapter 8, this very general form of reciprocity provides the basis for the so-called Onsager relations discussed in Chapter 1. An identity which remains valid regardless of any particular space-time dependence is that of Eq.(5–44), relating the covariance function to that of the expectation value of a commutator. Further identities of this kind are explored in the problems.

If $\hat{A}$ and $\hat{B}$ are Hermitian, as is always presumed, then K_{AB} is real and

$$K_{AA} \geq 0. \qquad (6\text{--}55)$$

We emphasize, however, that this implies both operators $\hat{A}$ are evaluated at the same space-time point, for $\hat{A}(\mathbf{r},\tau)$ is not necessarily the same operator as $\hat{A}(0,0)$— they do not commute. One proves Eq.(6–55) by studying the product of deviations, $\langle \overline{\delta \hat{A} \delta \hat{A}} \rangle$, in a representation in which $\hat{\rho}_0$ is diagonal. The equality is seen to hold when and only when $\hat{A} = \langle \hat{A} \rangle \hat{1}$.

For Hermitian operators K_{AB} satisfies all the properties for the scalar product of a linear vector space. In particular, when all the operators under consideration are linearly independent one can always find an orthogonal basis for the manifold. We shall find an opportunity later to exploit these observations, but for now we simply note the implication that the covariance function must satisfy the Schwarz inequality:

$$K_{AA}(0,0)K_{BB}(0,0) \geq K_{AB}^2(\mathbf{r},\tau), \qquad (6\text{--}56)$$

with equality if and only if $\hat{B} = c\hat{A}$, where c is a real number. Thus, if we set $\hat{A} = \hat{A}(\mathbf{x},t)$, $\hat{B} = \hat{A}(\mathbf{x}',t')$, then

$$K_{AA}(0,0) \geq |K_{AA}(\mathbf{r},\tau)|, \qquad (6\text{--}57)$$

so that $K_{AA}(\mathbf{r},\tau)$ is bounded above by its initial values. More generally, by considering the covariance function of $[\hat{A}(\mathbf{r},\tau) - \hat{B}(0,0)]$ with itself we obtain the bound

$$|K_{AB}(\mathbf{r},\tau)| \leq \tfrac{1}{2}[K_{AA}(0,0) + K_{BB}(0,0)]. \qquad (6\text{--}58)$$

As discussed earlier in connection with the linear response function, Fourier transformation leads to a closer relation with experimental quantities. In addition, actual calculations of covariance functions are usually much facilitated in this representation. On physical grounds one expects $K_{AB}(\mathbf{r}, \tau)$ to vanish for large $|\mathbf{r}|$, so that the spatial transform can be extended to infinite limits. Alternatively, one can either invoke the infinite-volume limit or adopt periodic boundary conditions. In any event, we utilize the conventions (3–27) and, because it will later be convenient to refer to them explicitly, write the Fourier transforms as

$$K_{AB}(\mathbf{k}, \omega) = \int d^3 r\, e^{-i\mathbf{k}\cdot\mathbf{r}} \int_{-\infty}^{\infty} d\tau\, e^{i\omega\tau} K_{AB}(\mathbf{r}, \tau), \qquad (6\text{--}59)$$

$$K_{AB}(\mathbf{r}, \tau) = \int \frac{d^3 k}{(2\pi)^3} e^{i\mathbf{k}\cdot\mathbf{r}} \int_{-\infty}^{\infty} \frac{d\omega}{2\pi} e^{-i\omega\tau} K_{AB}(\mathbf{k}, \omega). \qquad (6\text{--}60)$$

When $PT = 1$ for all the operators under consideration Eqs.(6–53) and (6–59) generate the following identities:

$$K_{AB}(\mathbf{k}, \omega) = K^*_{AB}(\mathbf{k}, \omega) = K_{AB}(-\mathbf{k}, -\omega) = K_{BA}(\mathbf{k}, \omega)$$
$$= P_A P_B K_{AB}(-\mathbf{k}, \omega) = T_A T_B K_{AB}(\mathbf{k}, -\omega). \qquad (6\text{--}61)$$

One can not take these identities for granted, of course, in other model systems which may exhibit different symmetry properties.

There exist numerous reasonably general properties of covariance functions which can only be proved rigorously from a knowledge of the asymptotic behavior of $K_{AB}(\mathbf{r}, \tau)$. We insert this *caveat* because it is not always made sufficiently explicit. Although it does not appear possible, despite Eqs.(6–57) and (6–58), to prove that $K_{AB}(\mathbf{r}, \tau)$ vanishes in general for large $|\mathbf{r}|$ and $|\tau|$, one certainly expects such behavior on physical grounds (with the exception of special circumstances, such as near critical points). With these words of caution, we adopt the reasonable physical expectation that $K_{AB}(\mathbf{r}, \tau)$ vanishes as either $|\mathbf{r}|$ or $|\tau|$ tends to infinity, although we do not specify the rates of decay. We shall return to this latter question later.

Now recall the series of conservation laws (6–25)-(6–27). In the absence of external fields they all have the form $\partial \hat{D}/\partial t = -\nabla \cdot \hat{\mathbf{J}}$, where $\hat{D}$ is a density and $\hat{\mathbf{J}}$ a current. In terms of covariance functions,

$$\frac{\partial}{\partial \tau} K_{DB}(\mathbf{r}, \tau) = -\nabla \cdot K_{\mathbf{J}B}(\mathbf{r}, \tau), \qquad (6\text{--}62)$$

for *any* operator $\hat{B}$. Upon Fourier transformation, and with presumption of the above asymptotic properties, we obtain the identity

$$K_{DB}(\mathbf{k}, \omega) = \sum_i \frac{k_i}{\omega} K_{J_i B}(\mathbf{k}, \omega). \qquad (6\text{--}63)$$

Rotational invariance of the Hamiltonian describing the equilibrium system leads to further insight into the structure of covariance functions. An essential feature of this invariance property is that there is only one preferred direction in the isotropic system—that defined by the vector **k** introduced by Fourier transformation. As a consequence, all the Fourier-transformed covariance functions can be written in terms of vectors and tensors constructed from the components of **k**, with respect to which all tensors are decomposed into transverse and longitudinal parts. One next employs the identity (6–63) to generate a number of other identities among the various covariance functions. By comparing all these identities one finds that the fifteen possible covariance functions which can be constructed from the five basic operators describing a simple fluid can be written in terms of only ten scalar functions $X_i(k,\omega)$. These are functions of only the magnitude $k = |\mathbf{k}|$ and can be chosen such that they are even functions of ω. All possible relations are tabulated in Table 6–2, where appropriate factors of particle mass have been inserted in order to maintain dimensional uniformity. One should note, though, that these definitions must be re-examined should contributions from the Bose ground state be important.

The functions $X_i(k,\omega)$ are readily identified by exploiting the orthogonality properties of the unit longitudinal and transverse tensors, as outlined in Problem 6.4. Table 6–2 is of limited direct utility, however, because the functions X_i are almost as difficult to calculate as the K_{AB} themselves. Rather, the important implications of this analysis have to do with how these symmetry properties eventually simplify explicit calculations, an assertion which will become clear in the following section. As an example, the covariance functions $K_{AB}(\mathbf{r},\tau)$ in space-time variables must have a structure similar to that exhibited in Table 6–2, because then $\mathbf{r}$ defines the only preferred direction in the isotropic equilibrium system. An analysis similar to that outlined above leads to identification of ten scalar functions $X_i(r,\tau)$, which are *not* the Fourier transforms of the $X_i(k,\omega)$. That is, one can not merely Fourier transform Table 6–2, but must actually re-do the analysis. When this is carried out, however, one finds that the space-integrated covariance function $K_{T_{ij}T_{mn}}(\mathbf{r},\tau)$ has only a small number of nonzero components. In fact, the indices must be equal in pairs, such as $K_{T_{11}T_{22}}$, $K_{T_{13}T_{13}}$, etc. One begins to appreciate the enormous utility of the symmetry analysis.

For the operators of Table 6–2 it is clear that both $K_{AB}(\mathbf{r},\tau)$ and its *full* space-time Fourier transform are real. As was found to be the case for the linear response function in Chapter 3, it is often useful to introduce a complex Fourier-Laplace integral which is not necessarily real:

$$\tilde{K}_{AB}(\mathbf{k},z) \equiv \int_0^\infty e^{izt}\, K_{AB}(\mathbf{k},t)\, dt\,, \quad \mathrm{Im}\, z > 0$$

$$= \frac{1}{2\pi} \int_{-\infty}^\infty d\omega\, \frac{K_{AB}(\mathbf{k},\omega)}{\omega - z}\,, \quad \mathrm{Im}\, z \neq 0\,. \tag{6–64}$$

The function $\tilde{K}_{AB}(\mathbf{k},z)$ is regular in the upper half-plane and can be continued to the lower half-plane; it has a branch cut along the entire real axis. The real function $K_{AB}(\mathbf{k},\omega)$ is essentially the discontinuity in $\tilde{K}_{AB}(\mathbf{k},z)$ across the cut. In a manner

$\hat{A}$	$\hat{B}$	$K_{AB}(\mathbf{k},\omega)$
$\hat{n}$	$\hat{n}$	χ_0
$\hat{h}$	$\hat{h}$	χ_1
$\hat{n}$	$\hat{h}$	χ_2
$\hat{n}$	$\hat{j}_i$	$\frac{\omega}{k} k_i \chi_0$
$\hat{h}$	$\hat{j}_i$	$\frac{\omega}{k} k_i \chi_2$
$\hat{n}$	$\hat{q}_i$	$\frac{\omega}{k} k_i \chi_2$
$\hat{h}$	$\hat{q}_i$	$\frac{\omega}{k} k_i \chi_1$
$\hat{j}_i$	$\hat{j}_j$	$\frac{\omega^2}{k^2} k_{ij} \chi_0 + \kappa_{ij} \chi_3$
$\hat{j}_i$	$\hat{q}_j$	$\frac{\omega^2}{k^2} k_{ij} \chi_2 + \kappa_{ij} \chi_4$
$\hat{q}_i$	$\hat{q}_j$	$\frac{\omega^2}{k^2} k_{ij} \chi_1 + \kappa_{ij} \chi_5$
$\hat{n}$	$\hat{T}_{ij}$	$m\frac{\omega^2}{k^2} k_{ij} \chi_0 + \kappa_{ij} \chi_6$
$\hat{h}$	$\hat{T}_{ij}$	$m\frac{\omega^2}{k^2} k_{ij} \chi_2 + \kappa_{ij} \chi_7$
$\hat{j}_i$	$\hat{T}_{mn}$	$m\frac{\omega^3}{k^3} k_{imn} \chi_0 + \frac{\omega}{k} k_i \kappa_{mn} \chi_6 + [k_m \kappa_{in} + k_n \kappa_{im}] m \frac{\omega}{k} \chi_3$
$\hat{q}_i$	$\hat{T}_{mn}$	$m\frac{\omega^3}{k^3} k_{imn} \chi_2 + \frac{\omega}{k} k_i \kappa_{mn} \chi_7 + [k_m \kappa_{in} + k_n \kappa_{im}] m \frac{\omega}{k} \chi_4$
$\hat{T}_{ij}$	$\hat{T}_{mn}$	$m^2 \frac{\omega^4}{k^4} k_{ijmn} \chi_0 + m\frac{\omega^2}{k^2} [k_{ij}\kappa_{mn} + k_{mn}\kappa_{ij}]\chi_6$
		$\quad + m^2 \frac{\omega^2}{k^2} [k_{in}\kappa_{jm} + k_{jm}\kappa_{in} + k_{im}\kappa_{jn} + k_{jn}\kappa_{im}]\chi_3$
		$\quad + \kappa_{ij}\kappa_{mn} \chi_8 + [\kappa_{im}\kappa_{jn} + \kappa_{in}\kappa_{jm}]\chi_9$

Table 6-2. Covariance functions $K_{AB}(\mathbf{k},\omega)$ in terms of the scalar functions $\chi_i(k,\omega)$, unit vectors $k_i \equiv (\mathbf{k})_i / |\mathbf{k}|$, products of unit vectors $k_{imn\cdots} \equiv k_i k_m k_n \cdots$, and the unit transverse tensor $\kappa_{ij} \equiv \delta_{ij} - k_{ij}$.

completely similar to the analysis carried out for the response function, we find that

$$K_{AB}(\mathbf{k},\omega) = 2\mathrm{Re}\,K_{AB}(\mathbf{k},\omega + i0)\,. \tag{6-65}$$

Now recall Eq.(5-70) relating K_{AB} to the linear response function ϕ_{AB}. If one attributes the appropriate asymptotic properties to K_{AB}, as discussed above, then Fourier transformation yields the equivalent relation

$$i\omega K_{AB}(\mathbf{k},\omega) = \beta^{-1}\phi_{AB}(\mathbf{k},\omega)\,. \tag{6-66}$$

But when $PT = 1$ for all the operators, as in the present discussion, the analysis leading to Eq.(3-37) shows that $\phi_{AB}(\mathbf{k},\omega)$ is related to the dissipation in the system.

That is,

$$K_{AB}(\mathbf{k}, \omega) = \frac{2}{\beta \omega} \chi''_{AB}(\mathbf{k}, \omega), \tag{6-67}$$

where χ''_{AB} is the imaginary part of the physical response, Eq.(3–37). This representation also provides an additional integral representation by substitution into Eq.(6–64):

$$\tilde{K}_{AB}(\mathbf{k}, z) = \frac{1}{\beta \pi i} \int_{-\infty}^{\infty} d\omega \, \frac{\chi''_{AB}(\mathbf{k}, \omega)}{\omega(\omega - z)}$$
$$= \frac{1}{\beta i z} [\tilde{\phi}_{AB}(\mathbf{k}, z) - \tilde{\phi}_{AB}(\mathbf{k}, i0)], \tag{6-68}$$

in terms of the function defined in Eq.(3–35).

Thus, it is the real part of $\tilde{K}_{AB}$, or $K_{AB}(\mathbf{k}, \omega)$ itself which describes dissipation in the general system. Equations (3–41) and (6–67) immediately yield the fluctuation-dissipation theorem

$$K_{AB}(\mathbf{k}, \omega) = \frac{1 - e^{-\beta \hbar \omega}}{\beta \hbar \omega} S_{AB}(\mathbf{k}, \omega), \tag{6-69}$$

where $S_{AB}(\mathbf{k}, \omega)$ is the Fourier transform of the space-time correlation function defined in Eq.(3–39). In the limit $\beta \hbar \omega \ll 1$ the covariance function is simply the ordinary correlation function, which also corresponds to omitting Kubo transforms. Hence, this last expression provides a direct means for deducing properties of covariance functions from those of correlation functions.

In Eq.(6–60) we can perform the inverse spatial transform only and write

$$K_{AB}(\mathbf{k}, \tau) \equiv \int_{-\infty}^{\infty} \frac{d\omega}{2\pi} e^{-i\omega \tau} K_{AB}(\mathbf{k}, \omega). \tag{6-70}$$

Expand the exponential and appeal to uniform convergence to rewrite the series as

$$K_{AB}(\mathbf{k}, \tau) = \omega^0(\mathbf{k}) + \tau \omega^1(\mathbf{k}) + \frac{\tau^2}{2} \omega^2(\mathbf{k}) + \cdots, \tag{6-71}$$

where the nth coefficient is defined in terms of the nth moment:

$$\omega^n(\mathbf{k}) \equiv \int_{-\infty}^{\infty} \frac{d\omega}{2\pi} (-i\omega)^n K_{AB}(\mathbf{k}, \omega). \tag{6-72}$$

For $\beta \hbar \omega \ll 1$ all the odd moments vanish, because in that case $S_{AB}(\mathbf{k}, \omega)$ is an even function of ω. We thus obtain a number of moment sum rules which are often useful in constructing models for correlation functions (e.g., Boon and Yip, 1980), and for examining limiting forms of covariance functions.

It will be useful in later applications to understand the behavior of covariance functions for various limiting values of **k** and ω. Of particular interest is the long wavelength limit $k \equiv |\mathbf{k}| \to 0$, commonly known as the *hydrodynamic limit* (HL). By this is meant wavelengths long compared to all microscopic lengths such as mean free paths, as well as frequencies small compared to microscopic quantities such as collision or excitation frequencies of molecules. Most often one includes in this definition the quantum inequality $\beta\hbar\omega \ll 1$. Note that this limit concerns only the magnitude of **k**, which may differ from the value **k** = 0 in a superfluid, say, and thus require a more careful examination in that event.

By taking the HL we obtain that part of the covariance function which varies slowly in space and time, thereby characterizing processes exhibiting long-time and long-distance behavior. It must be emphasized, though, that this limiting procedure by no means implies a simplification of calculational problems, a point made very clearly by Kadanoff and Martin (1963). Hydrodynamic equations refer to a system in some kind of local 'thermodynamic equilibrium', which is produced and enforced by frequent collisions among the particles. Whereas the hydrodynamic regime is collision dominated, on the one hand, conventional techniques for calculating correlation functions are usually based on expansions in powers of some kind of collision parameter, on the other. As we shall see, this parameter is often the strength of the two-body interaction. Moreover, this type of calculation usually results in an expansion in powers of the particle density in its naive form, so that perturbation methods tend to produce descriptions of dilute systems. These opposing 'forces' eventually cause enormous difficulties for microscopic theories of hydrodynamics.

We begin a study of the HL by noting that Eq.(6–59) yields the formal expression

$$\lim_{k \to 0} K_{AB}(\mathbf{k},\omega) = \int d^3r \int_{-\infty}^{\infty} d\tau\, e^{i\omega\tau}\, K_{AB}(\mathbf{r},\tau)\,, \qquad (6\text{–}73)$$

and one can now consider directly the volume-integrated operator on the right-hand side. Special attention is directed to the case in which the volume-integrated operator $\hat{A}$ commutes with the effective Hamiltonian $\hat{K}$ of Eq.(5–59). This occurs when $\hat{A}$ is one of the densities $\hat{n}$, $\hat{h}$, or $\hat{\jmath}$. In that event, whatever the operator $\hat{B}$, one verifies that $K_{AB}(\mathbf{r},\tau)$ is actually independent of time and the Kubo transform can be omitted. Then, when $\hat{A}$ is a density $\hat{D}$,

$$\lim_{k \to 0} K_{DB}(\mathbf{k},\omega) = 2\pi\delta(\omega) \int d^3r\, K_{DB}(\mathbf{r},0)\,, \qquad (6\text{–}74)$$

and we have a direct method for obtaining explicit information from K_{DB} about the system in the hydrodynamic limit. That is, in these special cases the limit provides precisely the same thermodynamic information in terms of equilibrium fluctuations as does the ordinary space-time correlation function $S_{DB}(\mathbf{r},\tau)$. Integration over frequencies in Eq.(6–74) yields the familiar thermodynamic sum rule

$$\lim_{k \to 0} \int_{-\infty}^{\infty} \frac{d\omega}{2\pi}\, K_{DB}(\mathbf{k},\omega) = \int d^3r\, K_{DB}(\mathbf{r},0)\,. \qquad (6\text{–}75)$$

In this special case the covariance functions are directly related to the fluctuations in the equilibrium system, which in turn can be evaluated in terms of thermodynamic functions. For example, the density-density covariance function is

$$K_{nn}(\mathbf{r}, \tau) = \langle \hat{n}(0, o) \hat{n}(\mathbf{r}, \tau) \rangle_0 - \langle \hat{n} \rangle_0 \langle \hat{n} \rangle_0, \qquad (6\text{--}76a)$$

so that

$$\int d^3r \, K_{nn}(\mathbf{r}, 0) = \langle \hat{n}(0, 0) \hat{N}(0) \rangle_0 - V \langle \hat{n} \rangle_0 \langle \hat{n} \rangle_0$$

$$= \frac{n_0^2 \kappa_T}{\beta}, \qquad (6\text{--}76b)$$

where κ_T is the isothermal compressibility, V is the system volume, and subscripts 0 continue to denote equilibrium quantities. The right-hand side of the first line in Eq.(6–76b) is essentially the variance of $\hat{N}$ in the equilibrium grand canonical ensemble—$\Delta^2 \hat{N}$—so that

$$\lim_{k \to 0} K_{nn}(\mathbf{k}, \omega) = 2\pi \delta(\omega) \frac{\Delta^2 \hat{N}}{V}. \qquad (6\text{--}77)$$

Similarly, we can recover all other equilibrium fluctuations in terms of thermodynamic functions. A number of these vanish:

$$K_{nj}(HL), \ K_{nq}(HL), \ K_{hj}(HL), \ K_{hq}(HL), \ K_{jT}(HL), \ K_{qT}(HL).$$

The remaining nonzero quantities are found to have the following thermodynamic forms:

$$\lim_{k \to 0} K_{nh}(\mathbf{k}, \omega) = 2\pi \delta(\omega) \frac{\Delta^2 (\hat{H} \hat{N})}{V}$$

$$= 2\pi \frac{n_0 \kappa_T}{\beta} \Big[h_0 + P_0 - \frac{\alpha T}{\kappa_T} \Big] \delta(\omega), \qquad (6\text{--}78)$$

$$\lim_{k \to 0} K_{nT_{ij}}(\mathbf{k}, \omega) = 2\pi \delta(\omega) \Delta^2 (\hat{T}_{ij} \hat{N}) = 2\pi \frac{n_0}{\beta} \delta_{ij} \delta(\omega), \qquad (6\text{--}79)$$

$$\lim_{k \to 0} K_{hh}(\mathbf{k}, \omega) = 2\pi \delta(\omega) \frac{\Delta^2 \hat{H}}{V}$$

$$= 2\pi \Big\{ \frac{\kappa T^2 C_V}{V} + \frac{\kappa_T}{\beta} \Big[h_0 + P_0 - \frac{\alpha T}{\kappa_T} \Big]^2 \Big\} \delta(\omega), \qquad (6\text{--}80)$$

$$\lim_{k \to 0} K_{hT_{ij}}(\mathbf{k}, \omega) = 2\pi \delta(\omega) \Delta^2 (\hat{T}_{ij} \hat{H}) = 2\pi \frac{h_0 + P_0}{\beta} \delta_{ij} \delta(\omega), \qquad (6\text{--}81)$$

$$\lim_{k \to 0} K_{j_i j_j}(\mathbf{k}, \omega) = 2\pi\delta(\omega)\frac{\Delta^2(\hat{\jmath}_j \hat{P}_i)}{m} = 2\pi\frac{n_0}{\beta}\delta_{ij}\delta(\omega),\qquad(6\text{--}82)$$

$$\lim_{k \to 0} K_{j_i q_j}(\mathbf{k}, \omega) = 2\pi\delta(\omega)\frac{\Delta^2(\hat{q}_j \hat{P}_i)}{m} = 2\pi\frac{h_0 + P_0}{\beta}\delta_{ij}\delta(\omega),\qquad(6\text{--}83)$$

where α is the coefficient of thermal expansion, and C_V the constant-volume heat capacity. Equations (6–77), (6–78), and (6–80) were already obtained in Chapter 4 of Volume I, say, whereas Eqs.(6–79) and (6–81) are most readily verified by differentiating with respect to the appropriate equilibrium Lagrange multipliers in Eq.(6–43). In Problem 6.5 the reader is asked to verify this last assertion, as well as the expressions (6–82) and (6–83).

We see, therefore, that when $\hat{A}$ is a density the hydrodynamic limit of K_{AB} provides considerable thermodynamic information. Observe, however, that these static limits provide *no* information on the nonequilibrium properties of the system, and that when $\hat{A}(\mathbf{r}, \tau)$ is a density one obtains *only* static results in the hydrodynamic limit. The time dependence of $K_{AB}(\mathbf{r}, \tau)$ is crucial to the nonequilibrium theory, and it is not sensible to substitute the above limiting values into Eq.(5–61), say, in order to study transport properties.

Similar information can be obtained from Eq.(6–74) when either $\hat{A}$ or $\hat{B}$ are either $\hat{q}$ or $\hat{T}_{ij}$, of course, but then their space-integrated forms do *not* commute with $\hat{K} = \hat{H} - \mu\hat{N}$. Formally, we can rewrite Eq.(6–74) as

$$\lim_{k \to 0} K_{AB}(\mathbf{k}, \omega) = \int_{-\infty}^{\infty} d\tau\, e^{i\omega\tau} \int d^3r\, \overline{\langle \delta\hat{B}(0,0)\delta\hat{A}(\mathbf{r}, \tau)\rangle_0},\qquad(6\text{--}84)$$

using the notation of the deviation: $\delta\hat{A} \equiv \hat{A} - \langle\hat{A}\rangle_0\hat{1}$. However, the expectation value in this case is *not* independent of time, and the Kubo transform can *not* be omitted. Let us expand the right-hand side of this equation in a formal manner using the identity (6–32) applied to $\hat{A}(\mathbf{r})$. This yields

$$\lim_{k \to 0} K_{AB}(\mathbf{k}, \omega) = 2\pi\delta(\omega)\int d^3r\, K_{AB}(\mathbf{r}, 0)$$

$$-\frac{i}{\hbar}\int_{-\infty}^{\infty} \tau e^{i\omega\tau}\, d\tau$$

$$\times \int d^3r \overline{\langle \delta\hat{B}(0,0)} \int_0^1 d\lambda e^{i\lambda\tau\,\hat{K}/\hbar}\, [\hat{A}(\mathbf{r}), \hat{K}]e^{-i\lambda\tau\,\hat{K}/\hbar}\rangle_0.$$

$$(6\text{--}85)$$

Hence, in the limit $k \to 0$ we generally obtain a term in $\delta(\omega)$, plus terms which are presumably regular in ω at the origin. Only the former arises if the space-integrated $\hat{A}$ commutes with $\hat{K}$, whereas the latter will possess hydrodynamic coefficients in the form of dissipative functions. Unfortunately, identification of these coefficients requires some kind of macroscopic model, and so can be ambiguous. One such model-dependent identification is provided by Puff and Gillis (1968). In Chapter

8 we shall see how the hydrodynamic coefficients emerge unambiguously in a more complete theory.

There is one further limit, however, which yields model-independent information if $\hat{A}$ and $\hat{B}$ are both $\hat{q}$ or both components of $\hat{T}_{ij}$, and that is the high-frequency limit ($\omega \to \infty$). In this case the limit is obtained by *deliberately* setting $t = 0$ in the volume-integrated K_{AB} for these variables—doing so in other cases is just equivalent to the HL, and still yields only thermodynamic information. Thus, if the high-frequency limit of K_{AB} exists for these variables it is given by

$$\lim_{\substack{k \to 0 \\ \omega \to \infty}} K_{AB}(\mathbf{k}, \omega) = \int d^3r \, K_{AB}(\mathbf{r}, 0) , \qquad (6\text{--}86)$$

and this is related to an equal-time commutator, as we shall see. (Because k and ω are not necessarily related here, there is no contradiction in taking opposing limits.)

Intuitively, the high-frequency limit would seem to apply primarily to a simple solid, say, for which the regular terms contribute and the static forms do not. The useful information would then be found in the elastic moduli K and G— the bulk and shear modulus, respectively—and the two would be related through Cauchy's identity, Eq.(1–36). But a liquid will also respond elastically to a sudden disturbance, as noted long ago by Poisson (1837) and studied mathematically by Maxwell (1867). The phenomenon has been described rather graphically by Forster (1975): if one dives into the water from a springboard the low-frequency response is soft and smooth; but if one falls into the water from an airplane at a few thousand feet, then one's last impression may be that the water was frozen! Thus, we can describe these elastic properties of our simple fluid by noting that the initial response to an abrupt disturbance will be described by the high-frequency limits of the elastic moduli, which will be denoted by K_∞ and G_∞.

In Problem 6.3 we study some relationships between covariance functions and commutators. Consider there the equal-time commutator in Eq.(6–178), multiply the equation by $-ir_i$, and then integrate over all space. Integration by parts and comparison with Eq.(6–86) yields the high-frequency expression

$$\lim_{\substack{k \to 0 \\ \omega \to \infty}} K_{BJ_i}(\mathbf{k}, \omega) = \frac{1}{i\hbar} \int d^3r \, r_i \langle [\hat{D}(\mathbf{r}, \tau), \hat{B}(0, \tau)] \rangle_0 , \qquad (6\text{--}87)$$

where $\hat{D}$ is the density corresponding to the current $\hat{\mathbf{J}}$. The point is that for the variables we are considering one operator is always a current. Specifically, we have

$$\lim_{\substack{k \to 0 \\ \omega \to \infty}} K_{q_i q_j}(\mathbf{k}, \omega) = \frac{1}{i\hbar} \int d^3r \, r_i \langle [\hat{h}(\mathbf{r}, \tau), \hat{q}_j(0, \tau)] \rangle_0 , \qquad (6\text{--}88)$$

$$\lim_{\substack{k \to 0 \\ \omega \to \infty}} K_{T_{\ell n} T_{ij}}(\mathbf{k}, \omega) = \frac{1}{i\hbar} \int d^3r \, r_i \langle [m\hat{j}_j(\mathbf{r}, \tau), \hat{T}_{\ell n}(0, \tau)] \rangle_0 . \qquad (6\text{--}89)$$

The first of these expressions involves three operators in a triplet correlation and will not be discussed further, whereas direct evaluation of the second will yield the high-frequency elastic moduli.

This latter calculation is most readily carried out by first computing the Fourier transform

$$\int d^3 r\, e^{-i\mathbf{k}\cdot\mathbf{r}} \langle [\hat{T}_{ij}(\mathbf{r},\tau), m\hat{j}_\ell(0,\tau)] \rangle_0$$

$$= \tfrac{2}{3}[k_i\delta_{j\ell} + k_j\delta_{i\ell} + k_\ell\delta_{ij}]\frac{\langle KE \rangle}{V} + \tfrac{1}{4}n_0\frac{k_i k_j k_\ell}{m}$$

$$+ n_0^2 \int d^3 r\, g(r)[1 - \cos(\mathbf{k}\cdot\mathbf{r})]\frac{\partial}{\partial r_\ell}\left[\frac{r_i r_j}{r}\frac{V_2'(r)}{(\mathbf{k}\cdot\mathbf{r})}\right], \qquad (6\text{--}90)$$

where the radial distribution function $g(r)$ and the expected equilibrium kinetic energy $\langle KE \rangle$ were defined in Eqs.(6–44) and (6–45), respectively, $V_2'(r)$ is the derivative of the spherically-symmetric two-body potential with respect to its argument, and the reader is asked to verify this expression in Problem 6.6. Differentiate both sides with respect to k_n and let $k \to 0$. A short calculation then yields the general result

$$\frac{\beta}{V}\int_0^\beta ds\, \langle \hat{T}_{ij}(-is)\hat{T}_{mn}(0) \rangle_0 - P_0^2 V \delta_{ij}\delta_{mn}$$

$$= mn_0 \left[\delta_{ij}\delta_{mn}T_1 + (\delta_{im}\delta_{jn} + \delta_{in}\delta_{jm})T_2\right], \qquad (6\text{--}91)$$

where we have inserted an additional volume integration in order to facilitate identification of the elastic moduli, employed the notation of Eq.(6–36) for the volume-integrated operators, written out the Kubo transform explicitly, and defined auxiliary functions

$$T_1 \equiv \frac{2}{3}\frac{\langle KE \rangle_0}{m\langle \hat{N} \rangle_0} + \frac{2\pi n_0}{15m}\int_0^\infty r^3 g(r)[-6V_2'(r) + rV_2''(r)]\, dr\,, \qquad (6\text{--}92)$$

$$T_2 \equiv \frac{2}{3}\frac{\langle KE \rangle_0}{m\langle \hat{N} \rangle_0} + \frac{2\pi n_0}{15m}\int_0^\infty r^3 g(r)[4V_2'(r) + rV_2''(r)]\, dr\,. \qquad (6\text{--}93)$$

Zwanzig and Mountain (1965) calculated expressions for the high-frequency elastic moduli in classical homogeneous and isotropic fluids in which the two-body interactions are spherically symmetric. In order to compare with the appropriate quantum-mechanical generalizations of their results, let us recall Eq.(6–47) for the equilibrium pressure P_0. By adding and subtracting this quantity in the high-frequency expression for the bulk modulus, we obtain their expression for K_∞. This modulus is just the diagonal part of Eq.(6–91), the product of 1/3 the traces, although it must be emphasized that it is the *entire* covariance function that determines the elastic moduli and the term in P_0^2 is not a separate contribution. The high-frequency shear modulus G_∞ corresponds to the off-diagonal part of the covariance function, and is essentially T_2. Thus, the explicit expressions for the high-frequency elastic moduli in a simple fluid are

$$K_\infty = \frac{\beta}{V}\int_0^\beta ds\, \langle \tfrac{1}{9}\hat{T}_{ii}(-is)\hat{T}_{jj}(0) \rangle_0 - P_0^2 V$$

$$= \frac{4}{9} \frac{\langle KE \rangle_0}{V} + P_0 + \frac{2\pi n_0^2}{9} \int_0^\infty r^3 g(r)[V_2'(r) + rV_2''(r)] \, dr \,, \qquad (6\text{–}94)$$

$$G_\infty = \frac{\beta}{V} \int_0^\beta ds \, \langle \hat{T}_{ij}(-is)\hat{T}_{ij}(0) \rangle_0 \,, \qquad i \neq j$$

$$= \frac{2}{3} \frac{\langle KE \rangle_0}{V} + \frac{2\pi n_0^2}{15} \int_0^\infty r^3 g(r)[4V_2'(r) + rV_2''(r)] \, dr \,, \qquad (6\text{–}95)$$

where the summation convention is implied in Eq.(6–94) but *not* in Eq.(6–95). (In the homogeneous isotropic system it does not matter which set of off-diagonal elements is employed in the last equation.)

In addition, we note that combination of Eqs.(6–47), (6–94), and (6–95) yields a generalized Cauchy identity for this system:

$$K_\infty = \tfrac{5}{3} G_\infty + 2\left(P_0 - \tfrac{2}{3} \frac{\langle KE \rangle_0}{V} \right). \qquad (6\text{–}96)$$

Numerous values of K_∞ and G_∞ have been tabulated for potentials of Lennard-Jones type by Zwanzig and Mountain.

The mathematical efficacy of the hydrodynamic limit will be discussed further in Chapter 8, where we shall derive quite general expressions for transport coefficients and elastic moduli over the entire frequency range and for all wavelengths. Discussion of other types of limits, such as the precise forms of spatial and temporal decay modes, will also be postponed until more experience has been gained with the calculation and application of covariance functions.

B. Calculation of Covariance Functions

Application of the linear theory to nontrivial problems requires calculation of the covariance functions to an extent that goes well beyond elementary limiting procedures. As is evident from the literature on time correlation functions (e.g., Berne and Harp, 1971; Copley and Lovesey, 1975; Boon and Yip, 1980), this is a difficult undertaking that remains largely unresolved on the level of first principles. Some of the techniques employed in the above references will be reviewed briefly in Chapter 8, so that here for the most part we shall study a straightforward expansion in powers of the interaction. Our earlier discussion, though, has already revealed that perturbation theory may be inadequate in various applications of interest. For that reason we shall relegate many of the more tedious details to Appendix B.

PERTURBATION THEORY

Irrespective of practical matters, it is still important as a matter of principle to demonstrate that the covariance functions *can* be calculated by means of a systematic perturbation expansion if desired. Moreover, one then has a means of addressing the above difficulties, as well as others, by summing various selected subsets of terms in the expansion. Perturbation techniques of this kind are well known from the theory of Green functions, of course, and the general procedure is rather well understood (e.g, Fetter and Walecka, 1971). Owing to the presence of the Kubo transform in the covariance function, however, some novel features arise in the calculations and it is important to exhibit enough detail so as to reveal these points as they relate to a simple fluid.

It is convenient to change notation slightly—*for the present discussion only*— and write the Hamiltonian as

$$\hat{H} = \hat{H}_0 + \hat{H}_1 \,, \tag{6-97}$$

where $\hat{H}_0$ is the Hamiltonian for free particles and $\hat{H}_1$ represents the two-body interaction. The characteristic operator of the grand canonical ensemble is then

$$\hat{K} = \hat{H} - \mu \hat{N} = \hat{K}_0 + \hat{H}_1 \,, \tag{6-98}$$

and $\hat{K}_0$ has eigenvalues

$$\epsilon_i \equiv \omega_i - \mu \,, \tag{6-99}$$

where $\omega_i \equiv \hbar^2 k_i^2 / 2m$ is a single free-particle energy. Time evolution is described in a modified Heisenberg picture:

$$\hat{A}(\mathbf{x}, t) = e^{it\hat{K}/\hbar} \, \hat{A}(\mathbf{x}) e^{-it\hat{K}\hbar} \,. \tag{6-100}$$

The general equilibrium ensemble is described by

$$\hat{\rho} = \frac{1}{Z} e^{-\beta \hat{K}} \qquad Z = \mathrm{Tr}\, e^{-\beta \hat{K}} \,, \tag{6-101}$$

whereas the free-particle equilibrium situation depends on the statistical operator

$$\hat{\rho}_0 = \frac{1}{Z_0} e^{-\beta \hat{K}_0} \,, \qquad Z_0 = \mathrm{Tr}\, e^{-\beta \hat{K}_0} \,, \tag{6-102}$$

and in this latter case expectation values are denoted by $\langle \cdots \rangle_0$. Because there can be no confusion here about equilibrium and nonequilibrium expectation values, subscript 0 will be used throughout this section to denote free-particle quantities.

There is no extraordinary difficulty in evaluating the single-operator quantities $\langle \hat{A} \rangle$ or $\langle \hat{B} \rangle$, so it is convenient to focus on $\Delta K_{AB} \equiv K_{AB} + \langle \hat{A} \rangle \langle \hat{B} \rangle = \langle \overline{\hat{B}} \hat{A} \rangle$. That is, we study

$$\Delta K_{AB}(\mathbf{x}', t'; \mathbf{x}, t) = \frac{1}{Z} \int_0^1 ds \; \mathrm{Tr}\big\{ e^{-\beta \hat{K}} \, e^{\tau' \hat{K}} \, \hat{B}(\mathbf{x}') e^{-\tau' \hat{K}} \, e^{\tau \hat{K}} \, \hat{A}(\mathbf{x}) e^{-\tau \hat{K}} \big\} \,, \tag{6-103}$$

where

$$\tau' \equiv s\beta + it'/\hbar \,, \qquad \tau \equiv it/\hbar \,. \tag{6-104}$$

Thus, the Kubo-transformed operator develops in time with a complex 'time' parameter.

Now introduce a modified interaction picture by writing

$$\hat{A}(\mathbf{x}, \tau) \equiv e^{\tau \hat{K}_0} \hat{A}(\mathbf{x}) e^{-\tau \hat{K}_0}, \qquad (6\text{--}105)$$

and similarly for $\hat{B}(\mathbf{x}', \tau')$. The characteristic evolution operator of this picture is defined as

$$\hat{W}(\tau', \tau) \equiv e^{\tau' \hat{K}_0} e^{-(\tau' - \tau) \hat{K}} e^{-\tau \hat{K}_0}, \qquad (6\text{--}106)$$

which is *not* unitary. One verifies, though, that the operator does possess the following group properties:

$$\hat{W}(\tau, \tau) = \hat{1}, \qquad \hat{W}^{-1}(\tau', \tau) = \hat{W}(\tau, \tau'),$$

$$\hat{W}(\tau, \tau') \hat{W}(\tau', \tau_1) = \hat{W}(\tau, \tau_1). \qquad (6\text{--}107)$$

From the definition (6–106) one easily verifies that for free particles $\hat{W}(\tau', \tau)$ becomes the unit operator, and the following identity is true:

$$e^{-\beta \hat{K}} = e^{-\beta \hat{K}_0} \hat{W}(\beta, 0), \qquad (6\text{--}108)$$

the trace of which is just the grand partition function. As a consequence of these properties of $\hat{W}$ one can now rewrite Eq.(6–103) in terms of free-particle expectation values:

$$\Delta K_{AB}(\mathbf{x}', t'; \mathbf{x}, t) = \int_0^1 ds \, \langle \hat{W}(\beta, \tau') \hat{B}(\mathbf{x}', \tau') \hat{W}(\tau', \tau) \hat{A}(\mathbf{x}, \tau) \hat{W}(\tau, 0) \rangle_0 \langle \hat{W}(\beta, 0) \rangle_0^{-1}. \qquad (6\text{--}109)$$

That is, we are now working in the ensemble of Eq.(6–102).

Further simplification entails a detailed study of the operator $\hat{W}(\tau', \tau)$ which, by direct differentiation in the definition, is seen to satisfy the differential equation

$$\frac{d}{d\tau'} \hat{W}(\tau', \tau) = -\hat{H}_1(\tau') \hat{W}(\tau', \tau), \qquad (6\text{--}110a)$$

where

$$\hat{H}_1(\tau') \equiv e^{\tau' \hat{K}_0} \hat{H}_1 e^{-\tau' \hat{K}_0}. \qquad (6\text{--}110b)$$

With the initial condition provided by the first of Eqs.(6–107), the differential equation (6–110a) is readily converted into the following integral equation:

$$\hat{W}(\tau', \tau) = \hat{1} - \int_\tau^{\tau'} d\tau_1 \, \hat{H}_1(\tau_1) \hat{W}(\tau_1, \tau), \qquad (6\text{--}111)$$

where the integral is taken along an arbitrary contour C in the complex plane, directed from τ to τ'.

This integral equation can now be iterated along the contour C in a manner completely analogous to the case of real variables (e.g., Chapter 8 of Volume I). In

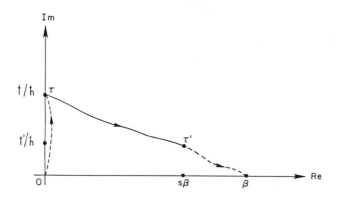

Fig. 6–1. Contour in the complex time-temperature plane along which the integral equation (6–111) is iterated.

the process one must introduce an operator T_C that orders the products of operators along C from τ to τ'. The usual time-ordering operator T is just a special case of T_C, and we note that T_C introduces appropriate factors of $\varepsilon(=\pm 1)$ as it orders the operators. Although not at all necessary, it is convenient to choose the contours such that all integrations are along the same contour C, passing through τ and τ', in that order. Because subsequent applications are solely concerned with the prediction problem, it is always true that $t > t'$ and we choose the contour as in Figure 6–1. That portion of the contour between τ and τ' must be monotonic nonincreasing, but otherwise the contour from 0 to β in the complex plane is rather arbitrary.

Except for modifications introduced by the complex 'time' variable, the subsequent analysis is completely analogous to that made in the usual theory of thermodynamic Green functions. Although complex times have been studied previously by Konstantinov and Perel' (1961), and Mills (1969), further development in the present context differs considerably from that in these references.

Formal iteration of the integral equation (6–111) along the contour C yields the expression

$$\hat{W}(\tau',\tau) = \sum_{n=0}^{\infty} \frac{(-1)^n}{n!} \int_C d\tau_1 \int_C d\tau_2 \cdots \int_C d\tau_n \, T_C[\hat{H}_1(\tau_1) \cdots \hat{H}_1(\tau_n)]. \qquad (6\text{--}112)$$

This is a (perturbation) series expansion for $\hat{W}(\tau',\tau)$, but its importance transcends this fact. If Eq.(6–112) is substituted into Eq.(6–109) for each operator, we find that the $\hat{W}$ operators can be commuted with $\hat{A}$ and $\hat{B}$, exactly as in the case of real variables. This is accomplished by first noting that an ordering operator can be introduced into the expectation value in the numerator in Eq.(6–109) without altering its value, because the operators there are already ordered along the contour C. Then, when the expanded form of $\hat{W}$ is employed, it is possible to interchange $\hat{A}(\mathbf{x}, \tau)$ and $\hat{W}(\tau, 0)$, because T_C will always reposition the operators into their correct locations. Now use the group property of Eq.(6–107) to write $\hat{W}(\tau', \tau)\hat{W}(\tau, 0) = \hat{W}(\tau', 0)$, and then go through the process again, interchanging $\hat{W}(\tau', 0)$ and $\hat{B}(\mathbf{x}', \tau')$. In this manner Eq.(6–109) is considerably simplified, and can be rewritten as

$$\Delta K_{AB} = \int_0^1 ds \, \frac{\langle T_C \, \hat{W}(\beta, 0)\hat{B}(\mathbf{x}', \tau')\hat{A}(\mathbf{x}, \tau)\rangle_0}{\langle \hat{W}(\beta, 0)\rangle_0} . \qquad (6\text{–}113)$$

Thus, *most* of the effects of interactions now reside in the operator $\hat{W}(\beta, 0)$, owing to the definition (6–105). Of course, if $\hat{A}$ or $\hat{B}$ is the energy current, or stress tensor, say, then they will contain the interaction themselves, though linearly. The relevant form of the expansion (6–112) is now

$$\hat{W}(\beta, o) = \sum_{n=0}^{\infty} \frac{(-1)^n}{n!} \int_0^\beta d\tau_1 \cdots d\tau_n T_C[\hat{H}_1(\tau_1) \cdots \hat{H}_1(\tau_n)], \qquad (6\text{–}114)$$

where it is to be emphasized that all the integrals are to be taken from 0 to β along the contour C, passing through τ and τ', in that order. If Eq.(6–114) is now substituted into Eq.(6–113) we obtain

$$\Delta K_{AB} = \int_0^1 ds \, \frac{Num}{Den}, \qquad (6\text{–}115a)$$

where the numerator Num and denominator Den are given by

$$Num = \sum_{n=0}^{\infty} \frac{(-1)^n}{n!} \int_0^\beta d\tau_1 \cdots d\tau_n \, \langle T_C[\hat{H}_1(\tau_1) \cdots \hat{H}_1(\tau_n)]\hat{B}(\mathbf{x}', \tau')\hat{A}(\mathbf{x}, \tau)\rangle_0,$$

$$(6\text{–}115b)$$

$$Den = \sum_{n=0}^{\infty} \frac{(-1)^n}{n!} \int_0^\beta d\tau_1 \cdots d\tau_n \, \langle T_C \, \hat{H}_1(\tau_1) \cdots \hat{H}_1(\tau_n)\rangle_0. \qquad (6\text{–}115c)$$

Note that the ordering operator in the numerator applies to all the variables along C, including τ and τ', and that all but the s-integral are taken along the contour. The s-integration involves only the real part of τ'.

Although Eqs.(6–115) represent the general perturbation expansion of the covariance function that we have been seeking, it is not yet in a really usable form. But, whatever the interaction $\hat{H}_1$, it is always in the form of a Fock-space operator, so that it will be a sum over single-particle states of products of creation and annihilation operators. For example, Eqs.(6–8) and (6–11b) illustrate the form for a local two-body interaction, and Eq.(6–110b) indicates how to construct $\hat{H}_1(\tau_i)$. Similarly, $\hat{A}$ and $\hat{B}$ will be of the same form, though more complicated.

Thus, evaluation of the expressions (6–115) reduces to a calculation of expectation values of ordered products of creation and annihilation operators. These are evaluated with the help of a statistical analog of Wick's theorem, the proof of which is unaffected by the use of complex times. Such an analysis was carried out in Chapter 8 and Appendix G of Volume I, and so will not be repeated here.

The result of Wick's theorem in the present context is that

$$\langle T_C \hat{b}_{\lambda_1}(\tau_1) \cdots \hat{b}_{\lambda_r}(\tau_r)\rangle_0 = \sum \left[\begin{array}{c} \text{All Possible} \\ \text{Fully Contracted Products} \end{array} \right], \qquad (6\text{–}116)$$

where $\hat{b}_{\lambda_i}(\tau_i)$ is either a creation or annihilation operator, λ_i refers to the appropriate single-particle state labeling the operator, τ_i can also be τ or τ', and a contracted product of two operators is defined as

$$\left[\hat{b}_{\lambda_i}(\tau_i) \hat{b}_{\lambda_j}(\tau_j) \right] \equiv \text{Tr}\{\hat{\rho}_0 T_C \hat{b}_{\lambda_i}(\tau_i) \hat{b}_{\lambda_j}(\tau_j)\}. \qquad (6\text{–}117)$$

One finds that the only nonzero contractions are the following two:

$$\left[\hat{a}_{\lambda_i}(\tau_i) \hat{a}^\dagger_{\lambda_j}(\tau_j) \right] = e^{-(\tau_i - \tau_j)\epsilon_i} \left[\theta_C(\tau_i - \tau_j) + \varepsilon\langle \hat{a}^\dagger_{\lambda_i} \hat{a}_{\lambda_i}\rangle_0 \right] \delta_{\lambda_i \lambda_j}, \qquad (6\text{–}118\text{a})$$

$$\left[\hat{a}^\dagger_{\lambda_i}(\tau_i) \hat{a}_{\lambda_j}(\tau_j) \right] = \varepsilon e^{(\tau_i - \tau_j)\epsilon_i} \left[\theta_C(\tau_j - \tau_i) + \varepsilon\langle \hat{a}^\dagger_{\lambda_i} \hat{a}_{\lambda_i}\rangle_0 \right] \delta_{\lambda_i \lambda_j}, \qquad (6\text{–}118\text{b})$$

where ϵ_i is defined in Eq.(6–99), and $\varepsilon = \pm 1$ for bosons or fermions. Also, the step-functions are defined along the contour C, and the single-particle state labels implicitly include spin variables. In the momentum representation, for example,

$$\langle \hat{a}^\dagger_{\mathbf{k}_i} \hat{a}_{\mathbf{k}_i}\rangle_0 = \nu(\mathbf{k}_i) \equiv \frac{e^{-\beta\epsilon(\mathbf{k}_i)}}{1 - \varepsilon e^{-\beta\epsilon(\mathbf{k}_i)}}. \qquad (6\text{–}119)$$

Not only does this form of Wick's theorem allow for straightforward evaluation of arbitrary matrix elements of the form given in Eq.(6–116), but it also induces a further simplification of the expression (6–115a). Let us represent the phrase 'all fully contracted products' with the notation AFCP, and recall that T_C orders the operators such that those with τ_i closer to β lie to the left. Then, some thought shows that the numerator, Eq.(6–115b), can be put into the schematic form

$$\left\{ 1 + \sum \left[\begin{array}{c} \text{AFCP} \\ \hat{b}\text{s only} \end{array} \right]_2 + \sum \left[\begin{array}{c} \text{AFCP} \\ \hat{b}\text{s only} \end{array} \right]_4 + \cdots \right\}$$

$$\times \left\{ \sum \left[\begin{array}{c} \text{AFCP} \\ \hat{B} \text{ and } \hat{A} \end{array} \right] + \sum \left[\begin{array}{c} \text{AFCP connecting} \\ \hat{b}\text{s and } \hat{B} \text{ or } \hat{A} \end{array} \right]_2 + \cdots \right\}, \qquad (6\text{–}120)$$

where the subscripts indicate the number of $\hat{b}$-operators involved in the various products, and these $\hat{b}$s refer here only to those in the factors $\hat{H}_1(\tau_i)$. The $\hat{b}$-operators must always occur in products of pairs of creation and annihilation operators, or the product vanishes.

One now observes that the first factor in Eq.(6–120) is identical with the denominator in Eq.(6–115) when the latter is also expanded by means of Wick's theorem. Thus, the denominator is cancelled and Eq.(6–115) for the covariance function can be rewritten as

$$\Delta K_{AB}(\mathbf{x}',t';\mathbf{x},t) = \beta^{-1} \int_0^\beta ds \, \langle \hat{W}(\beta,0)\hat{B}(\mathbf{x}',\tau')\hat{A}(\mathbf{x},\tau)\rangle_0^c$$

$$= \sum_{n=0}^\infty \frac{(-1)^n}{n!} \int_0^\beta \frac{ds}{\beta} \int_0^\beta d\tau_1 \cdots \int_0^\beta d\tau_n$$

$$\times \langle T_C[\hat{H}_1(\tau_1)\cdots\hat{H}_1(\tau_n)]\hat{B}(\mathbf{x}',\tau')\hat{A}(\mathbf{x},\tau)\rangle_0^c, \quad (6\text{--}121)$$

and the first line can only be interpreted as a short-hand notation for the second. Note again that all the τ_i integrals are along C, and that we have changed integration variables so that in place of the definitions (6–104) we now have

$$\tau' \equiv s + \frac{i}{\hbar}t', \qquad \tau \equiv \frac{i}{\hbar}t. \qquad (6\text{--}122)$$

The superscript 'c' indicates that only connected products are to be included in the evaluation of the matrix element when it is expanded by means of Wick's theorem, where a *connected product* is one in which one or more operators in the expansion of $\hat{W}(\beta,0)$ is contracted with an operator contained in either $\hat{A}$ or $\hat{B}$, and no $\hat{H}_1(\tau_i)$ is fully contracted within itself.

Although Eq.(6–121) provides a formal perturbation expansion for the covariance function in powers of the two-body interaction, further progress depends on a specific choice of what class of system is to be studied. In the present case we have been concerned with a model of a simple fluid which is homogeneous in the equilibrium state. Therefore, it is appropriate to choose for the single-particle wavefunctions in Eqs.(6–12) the plane waves

$$\phi_{\mathbf{k}_i}(\mathbf{x}) \equiv \frac{1}{\sqrt{V}} \exp[i\mathbf{k}_i \cdot \mathbf{x}], \qquad (6\text{--}123)$$

normalized to the system volume V. This choice is consistent with either periodic boundary conditions or the infinite-volume limit, but in either case it should be remembered that these functions are implicitly $(2S+1)$-component wavefunctions for particles of spin $\hbar S$.

It is now useful to display the five basic microscopic field operators explicitly in this plane-wave basis. For the number-density operator we have

$$\hat{n}(\mathbf{x},t) = \hat{\psi}^\dagger(\mathbf{x},t)\hat{\psi}(\mathbf{x},t)$$

$$= \frac{1}{V} \sum_{\mathbf{k}_1\mathbf{k}_2} e^{i(\mathbf{k}_2-\mathbf{k}_1)\cdot\mathbf{x}} \hat{a}^\dagger_{\mathbf{k}_1}(t)\hat{a}_{\mathbf{k}_2}(t), \qquad (6\text{--}124)$$

whereas the current-density operator is

$$\hat{\jmath}(\mathbf{x},t) = \frac{i\hbar}{2m}\left\{[\nabla\hat{\psi}^\dagger(\mathbf{x},t)]\hat{\psi}(\mathbf{x},t) - \hat{\psi}^\dagger(\mathbf{x},t)[\nabla\hat{\psi}(\mathbf{x},t)]\right\}$$

$$= \frac{\hbar}{2mV}\sum_{\mathbf{k}_1\mathbf{k}_2}(\mathbf{k}_1 + \mathbf{k}_2)e^{i(\mathbf{k}_2-\mathbf{k}_1)\cdot\mathbf{x}}\,\hat{a}^\dagger_{\mathbf{k}_1}(t)\hat{a}_{\mathbf{k}_2}(t)\,. \tag{6-125}$$

The other three operators are a bit more complicated because they contain the interaction explicitly. They can be split into separate parts, however, and the contributions from interactions isolated. Thus, the energy-density operator is written

$$\hat{h}(\mathbf{x},t) = \hat{h}^0(\mathbf{x},t) + \hat{h}^1(\mathbf{x},t)\,, \tag{6-126}$$

and the total Hamiltonian is

$$\hat{H} = \int \hat{h}^0(\mathbf{x},t)\,d^3x + \int \hat{h}^1(\mathbf{x},t)\,d^3x\,, \tag{6-127}$$

where

$$\hat{h}^0(\mathbf{x},t) \equiv \frac{\hbar^2}{2mV}\sum_{\mathbf{k}_1\mathbf{k}_2}(\mathbf{k}_1 - \mathbf{k}_2)e^{i(\mathbf{k}_2-\mathbf{k}_1)\cdot\mathbf{x}}\,\hat{a}^\dagger_{\mathbf{k}_1}(t)\hat{a}_{\mathbf{k}_2}(t)\,, \tag{6-128a}$$

$$\hat{h}^1(\mathbf{x},t) \equiv \frac{1}{2V^2}\sum_{\substack{\mathbf{k}_1\mathbf{k}_2\\\mathbf{k}_3\mathbf{k}_4}}e^{i(\mathbf{k}_4+\mathbf{k}_3-\mathbf{k}_1-\mathbf{k}_2)\cdot\mathbf{x}}\,v(\mathbf{k}_3 - \mathbf{k}_2)$$

$$\times \hat{a}^\dagger_{\mathbf{k}_1}(t)\hat{a}^\dagger_{\mathbf{k}_2}(t)\hat{a}_{\mathbf{k}_3}(t)\hat{a}_{\mathbf{k}_4}(t)\,. \tag{6-128b}$$

The function $v(\mathbf{k})$ is defined below.

In like manner, the energy-current operator is written

$$\hat{q}(\mathbf{x},t) = \hat{q}^0(\mathbf{x},t) + \hat{q}^1(\mathbf{x},t) + \hat{q}^2(\mathbf{x},t)\,, \tag{6-129}$$

where

$$\hat{q}^0(\mathbf{x},t) \equiv \frac{\hbar^3}{4m^2V}\sum_{\mathbf{k}_1\mathbf{k}_2}e^{i(\mathbf{k}_2-\mathbf{k}_1)\cdot\mathbf{x}}(\mathbf{k}_1\cdot\mathbf{k}_2)(\mathbf{k}_1 + \mathbf{k}_2)\hat{a}^\dagger_{\mathbf{k}_1}(t)\hat{a}_{\mathbf{k}_2}(t) \tag{6-130a}$$

$$\hat{q}^1(\mathbf{x},t) \equiv \frac{\hbar}{4mV^2}\sum_{\substack{\mathbf{k}_1\mathbf{k}_2\\\mathbf{k}_3\mathbf{k}_4}}e^{i(\mathbf{k}_4+\mathbf{k}_3-\mathbf{k}_1-\mathbf{k}_2)\cdot\mathbf{x}}(\mathbf{k}_2 + \mathbf{k}_3)v(\mathbf{k}_4 - \mathbf{k}_1)$$

$$\times \hat{a}^\dagger_{\mathbf{k}_1}(t)\hat{a}^\dagger_{\mathbf{k}_2}(t)\hat{a}_{\mathbf{k}_3}(t)\hat{a}_{\mathbf{k}_4}(t)\,, \tag{6-130b}$$

$$\hat{q}^2_i(\mathbf{x},t) \equiv \frac{\hbar}{8mV^2}\sum_{\substack{\mathbf{k}_1\mathbf{k}_2\\\mathbf{k}_3\mathbf{k}_4}}e^{i(\mathbf{k}_4+\mathbf{k}_3-\mathbf{k}_1-\mathbf{k}_2)\cdot\mathbf{x}}(\mathbf{k}_2 + \mathbf{k}_3)_j V_{ij}(\mathbf{k}_1,\mathbf{k}_2,\mathbf{k}_3,\mathbf{k}_4)$$

$$\times \hat{a}^\dagger_{\mathbf{k}_1}(t)\hat{a}^\dagger_{\mathbf{k}_2}(t)\hat{a}_{\mathbf{k}_3}(t)\hat{a}_{\mathbf{k}_4}(t)\,, \tag{6-130c}$$

and the summation convention is implied in the last expression. The function $V_{ij}(\mathbf{k}_1, \mathbf{k}_2, \mathbf{k}_3, \mathbf{k}_4)$ is also defined below.

The stress-tensor operator is

$$\hat{T}_{ij}(\mathbf{x}, t) = \hat{T}_{ij}^0(\mathbf{x}, t) + \hat{T}_{ij}^1(\mathbf{x}, t), \tag{6-131}$$

where

$$\hat{T}_{ij}^0(\mathbf{x}, t) \equiv \frac{\hbar^2}{4mV} \sum_{\mathbf{k}_1 \mathbf{k}_2} e^{i(\mathbf{k}_2 - \mathbf{k}_1) \cdot \mathbf{x}} [k_{2i}k_{2j} + k_{2i}k_{1j} + k_{2j}k_{1i} + k_{1i}k_{1j}]$$

$$\times \hat{a}_{\mathbf{k}_1}^\dagger(t) \hat{a}_{\mathbf{k}_2}(t), \tag{6-132a}$$

$$\hat{T}_{ij}^1(\mathbf{x}, t) \equiv -\frac{1}{4V^2} \sum_{\substack{\mathbf{k}_{-1} \mathbf{k}_2 \\ \mathbf{k}_3 \mathbf{k}_4}} e^{i(\mathbf{k}_4 + \mathbf{k}_3 - \mathbf{k}_1 - \mathbf{k}_2) \cdot \mathbf{x}} V_{ij}(\mathbf{k}_1, \mathbf{k}_2, \mathbf{k}_3, \mathbf{k}_4)$$

$$\times \hat{a}_{\mathbf{k}_1}^\dagger(t) \hat{a}_{\mathbf{k}_2}^\dagger(t) \hat{a}_{\mathbf{k}_3}(t) \hat{a}_{\mathbf{k}_4}(t). \tag{6-132b}$$

In addition, complete evaluation of the covariance function from Eq.(6–121) requires explicit consideration of the operator

$$\hat{H}_1(\tau) = \frac{1}{2V^2} \sum_{\substack{\mathbf{k}_1 \mathbf{k}_2 \\ \mathbf{k}_3 \mathbf{k}_4}} V \delta_{\mathbf{k}_2 + \mathbf{k}_4, \mathbf{k}_1 + \mathbf{k}_3} v(\mathbf{k}_2 - \mathbf{k}_1)$$

$$\times \hat{a}_{\mathbf{k}_3}^\dagger(\tau) \hat{a}_{\mathbf{k}_1}^\dagger(\tau) \hat{a}_{\mathbf{k}_2}(t) \hat{a}_{\mathbf{k}_4}(\tau). \tag{6-133}$$

The scalar function $v(\mathbf{k})$ and the tensor V_{ij} contain the particle-particle interaction explicitly, and in this sense play a role similar to the collision integrals of the classical kinetic theory based on the Boltzmann equation. These functions are defined as follows:

$$v(\mathbf{k}) \equiv \int e^{i\mathbf{k} \cdot \mathbf{r}} V_2(r) \, d^3r, \tag{6-134}$$

$$V_{ij}(\mathbf{k}_1, \mathbf{k}_2, \mathbf{k}_3, \mathbf{k}_4) \equiv \int d^3r \left[\frac{r_i r_j}{r} \frac{dV_2(r)}{dr} \right] e^{(i/2)(\mathbf{k}_2 + \mathbf{k}_4 - \mathbf{k}_1 - \mathbf{k}_3) \cdot \mathbf{r}}$$

$$\times \int_{-1}^{1} d\lambda \, e^{(i/2)\lambda(\mathbf{k}_3 + \mathbf{k}_4 - \mathbf{k}_1 - \mathbf{k}_2) \cdot \mathbf{r}}, \tag{6-135}$$

and one must remember the simplification of Eq.(6–35) when a complete volume integration is carried out.

In order to analyze the perturbation expansion (6–121) in a systematic way it is customary—indeed, almost indispensable—to develop diagrammatic calculational procedures. Even for a simple fluid, though, this becomes rather tedious in the present situation. A major complication, for example, arises from the intrinsic dependence of three of the basic operators on the two-body interaction as part of their definition. A detailed construction of the diagrammatic scheme is presented in Appendix B, where further analysis of the perturbation expansion is also carried out, and the efficacy of density expansions in general is discussed in Appendix C. The possibility of logarithmic terms in such expansions will be considered in Chapter 8. Here we shall be content for the moment with a brief study of the zero-order, or free-particle covariance functions for a simple fluid.

FREE-PARTICLE COVARIANCE FUNCTIONS

With reference to Eq.(6–121), we define the zero-order contribution to ΔK_{AB} as

$$\Delta K^0_{AB} \equiv \int_0^\beta \frac{ds}{\beta} \langle \hat{B}\mathbf{x}',\tau')\hat{A}(\mathbf{x},\tau)\rangle_0 , \tag{6–136}$$

where the 'connected' superscript can be omitted because this expectation value must include terms in which contracted products among only operators in $\hat{B}$ and only those in $\hat{A}$ occur. Also, because τ always precedes τ' along the contour C, there is no need to carry the ordering operator T_C explicitly in Eq.(6–136).

It is important to emphasize that this last expression can be thought of as providing a free-particle covariance function only if both $\hat{A}$ and $\hat{B}$ refer to the number-density or current-density operators. If other operators are involved, then a free-particle covariance function emerges only when consideration of the non-interaction terms is implied, as specified in Eqs.(6–128a), (6–130a), and (6–132a), say. We shall presume such a restriction throughout this subsection.

The simplest example of a zero-order covariance function is provided by the density-density covariance, in which $\hat{A} = \hat{B} = \hat{n}$. This is a true free-particle quantity, and from Eqs.(6–124) and (6–136) we have

$$\begin{aligned}
\Delta K^0_{nn} &= \int_0^\beta \frac{ds}{\beta} \langle \hat{n}(\mathbf{x}',\tau')\hat{n}(\mathbf{x},\tau)\rangle_0 \\
&= \int_0^\beta \frac{ds}{\beta} \frac{1}{V^2} \sum_{\substack{\mathbf{k}_1\mathbf{k}_2 \\ \mathbf{k}_3\mathbf{k}_4}} e^{i(\mathbf{k}_2-\mathbf{k}_1)\cdot\mathbf{x}'} e^{i(\mathbf{k}_4-\mathbf{k}_4)\cdot\mathbf{x}} \\
&\qquad \times \langle \hat{a}^\dagger_{\mathbf{k}_1}(\tau')\hat{a}_{\mathbf{k}_2}(\tau')\hat{a}^\dagger_{\mathbf{k}_3}(\tau)\hat{a}_{\mathbf{k}_4}(\tau)\rangle_0 .
\end{aligned} \tag{6–137}$$

Application of Wick's theorem via Eqs.(6–116)-(6–119) then yields

$$\begin{aligned}
\Delta K^0_{nn} &= \frac{1}{V^2} \sum_{\mathbf{k}_1} \nu(\mathbf{k}_1) \sum_{\mathbf{k}_2} \nu(\mathbf{k}_2) \\
&\quad + \int_0^\beta \frac{ds}{\beta} \frac{1}{V^2} \sum_{\mathbf{k}_1\mathbf{k}_2} e^{i(\mathbf{k}_1-\mathbf{k}_2)\cdot(\mathbf{x}-\mathbf{x}')} e^{s(\epsilon_1-\epsilon_2)} \\
&\qquad \times e^{(i/\hbar)(t'-t)(\epsilon_1-\epsilon_2)} \nu(\mathbf{k}_1)[1 + \varepsilon\nu(\mathbf{k}_2)] .
\end{aligned} \tag{6–138}$$

One recognizes the first line as the product of single-operator expectation values, so that we immediately identify the covariance function itself.

It is now convenient to take the thermodynamic limit and introduce the spatial and temporal notation of Eq.(6–49). Thus,

$$\begin{aligned}
K^0_{nn}(\mathbf{r},\tau) &= \frac{(2S+1)}{(2\pi)^6} \int_0^\beta \frac{ds}{\beta} \int d^3k_1 \int d^3k_2 \, e^{i(\mathbf{k}_1-\mathbf{k}_2)\cdot\mathbf{r}} \\
&\qquad \times e^{(s-i\tau/\hbar)(\hbar^2/2m)(\mathbf{k}_1^2-\mathbf{k}_2^2)} \nu(\mathbf{k}_1)[1 + \varepsilon\nu(\mathbf{k}_2)] ,
\end{aligned} \tag{6–139}$$

where we have carried out the sums over spin states, S being the particle spin in units of $\hbar$. Note carefully that $\tau \equiv t - t'$, and is *NOT* a complex variable here!

This is the general result for K_{nn}^0, but the integrals do not appear tractable as they stand. If we take Fourier transforms, though, considerable progress can be made with the resulting integrals. Following the convention of Eqs.(6–59) and (6–60), one notes that the **r**- and τ-integrals produce δ-functions; the former is used to perform the $\mathbf{k}_2$-integration, and the latter to do the radial part of the $\mathbf{k}_1$ integral. The notation is eased somewhat by defining new variables q and p:

$$q^2 \equiv \tfrac{1}{2}\beta m \frac{\omega^2}{k^2}, \quad p^2 \equiv \tfrac{1}{4}\beta \frac{\hbar^2 k^2}{2m}, \quad qp = \tfrac{1}{4}\beta\hbar\omega. \qquad (6\text{--}140)$$

The Fourier-transformed covariance function then reduces to

$$K_{nn}^0(\mathbf{k},\omega) = \frac{\varepsilon m^2(2S+1)}{2\pi\beta^2\hbar^4\omega k} 2(q+p)^2 \int_0^1 \frac{dx}{x^3} \frac{\varepsilon z e^{-(q+p)^2/x^2}}{1 - \varepsilon z e^{-(q+p)^2/x^2}}$$
$$\times \frac{e^{\beta\hbar\omega} - 1}{1 - \varepsilon z e^{\beta\hbar\omega} e^{-(q+p)^2/x^2}}, \qquad (6\text{--}141)$$

where $z \equiv e^{\beta\mu}$ is the fugacity.

A pleasing feature of this calculation is that the integrand in Eq.(6–141) is precisely the derivative of the function

$$\log\left[\frac{1 - \varepsilon z e^{-(q+p)^2/x^2}}{1 - \varepsilon z e^{\beta\hbar\omega} e^{-(q+p)^2/x^2}} \right].$$

We shall use this result after a small digression. The expression (6–141) is actually incomplete if the particles are bosons, because we have not accounted for the Bose-Einstein condensation in the very-low-temperature system. This will introduce a macroscopic occupation of the zero-momentum single-particle quantum state in the system at rest.

The most convenient procedure to follow in this regard is to make the Bogoliubov replacement of creation and annihilation operators for the zero-momentum state by c-numbers. We denote the number density in the Bose ground state by ρ_0 and rewrite the field operators as

$$\hat{\psi}(\mathbf{x}) = \rho_0^{1/2} + \frac{1}{\sqrt{V}} \sum_{\mathbf{k}}' e^{i\mathbf{k}\cdot\mathbf{x}} \hat{a}_{\mathbf{k}}, \qquad (6\text{--}142)$$

where the prime on the summation indicates that the value $\mathbf{k} = 0$ is to be omitted. With a similar expression for $\hat{\psi}^\dagger$, the expectation value in Eq.(6–137) becomes a sum of terms (excluding those that vanish identically):

$$\rho_0^2 + \rho_0 \langle \hat{a}_{\mathbf{k}_1}^\dagger \hat{a}_{\mathbf{k}_2} \rangle_0 + \rho_0 \langle \hat{a}_{\mathbf{k}_3}^\dagger \hat{a}_{\mathbf{k}_4} \rangle_0 + \rho_0 \langle \hat{a}_{\mathbf{k}_1}^\dagger \hat{a}_{\mathbf{k}_4} \rangle_0$$
$$+ \rho_0 \langle \hat{a}_{\mathbf{k}_2} \hat{a}_{\mathbf{k}_3}^\dagger \rangle_0 + \langle \hat{a}_{\mathbf{k}_1}^\dagger \hat{a}_{\mathbf{k}_2} \hat{a}_{\mathbf{k}_3}^\dagger \hat{a}_{\mathbf{k}_4} \rangle_0', \qquad (6\text{--}143)$$

and the last term excludes values $\mathbf{k}_i = \mathbf{0}$, so that it is equivalent to the contribution leading to Eq.(6–141). These additional terms can be identified by introducing a factor $(1 + \varepsilon)/2$—collecting them together, we find the additional contribution to Eq.(6–139) to be

$$\tfrac{1}{2}(1 + \varepsilon)\frac{\rho_0}{(2\pi)^3} \int_0^\beta \frac{ds}{\beta} \int d^3k_1 \left\{ e^{i\mathbf{k}_1 \cdot \mathbf{r}} e^{(s - i\tau/\hbar)(\hbar^2 k_1^2/2m)} \nu(\mathbf{k}_1) \right.$$

$$\left. + e^{-i\mathbf{k}_1 \cdot \mathbf{r}} e^{-(s - i\tau/\hbar)(\hbar^2 k_1^2/2m)} [1 + \varepsilon\nu(\mathbf{k}_1)] \right\}, \quad (6\text{–}144)$$

where for convenience we have set $S = 0$ for bosons.

This last result can be Fourier transformed in a straightforward manner, and we set $z = 1$ in the Bose ground state. The final result for the density-density covariance function is then

$$K_{nn}^0(\mathbf{k},\omega) = (1 + \varepsilon)\frac{\pi\rho_0}{\beta\omega} \left[\delta\!\left(\hbar\omega - \tfrac{\hbar^2 k^2}{2m}\right) - \delta\!\left(\hbar\omega + \tfrac{\hbar^2 k^2}{2m}\right)\right]$$

$$+ 2\pi\frac{\varepsilon(2S + 1)}{\omega k\lambda_T^4} \log\left[\frac{1 - \varepsilon z e^{-(q+p)^2}}{1 - \varepsilon z e^{-(q-p)^2}}\right], \quad (6\text{–}145)$$

and the *thermal wavelength* is defined as $\lambda_T \equiv (2\pi\hbar^2\beta/m)^{1/2}$.

It is useful to note that the second term on the right-hand side of Eq.(6–145) can be expanded in powers of the fugacity as follows:

$$\frac{4\pi\varepsilon(2S + 1)}{\omega k\lambda_T^4} \sum_{n=1}^\infty \frac{(\varepsilon z)^n}{n} e^{-n(\beta\hbar^2 k^2/8m)} e^{-n(\beta m\omega^2/2k^2)} \sinh(n\beta\hbar\omega/2). \quad (6\text{–}146)$$

Thus, at high temperatures and low densities we can extract the Boltzmann limit of $K_{nn}^0(\mathbf{k},\omega)$ by keeping only the first term in the series and noting that the contribution from the Bose ground state is negligible. In addition, this limit corresponds to the requirements

$$\beta\frac{\hbar^2 k^2}{2m} \ll 1, \quad \beta\hbar\omega \ll 1, \quad z \simeq \frac{n_0\lambda_T^3}{2S + 1} \ll 1, \quad (6\text{–}147)$$

n_0 being the equilibrium number density in the free-particle system. The Boltzmann limit of Eq.(6–145) is then

$$K_{nn}^0(\mathbf{k},\omega)_\mathrm{B} = n_0\frac{(2\pi\beta m)^{1/2}}{k} e^{-\frac{1}{2}\beta m\omega^2/k^2}. \quad (6\text{–}148)$$

This is also seen to be a low-frequency limit, corresponding to omission of the Kubo transform.

Next consider the current-current covariance function, for which we have $\hat{A} = \hat{B} = \hat{\mathbf{j}}$. The calculation proceeds in essentially the same way as above, with a few modifications. First of all, from the expression (6–125) for the current-density operator we see that the integrals for the second-rank tensor function K_{ij}^0 will contain a factor $[\mathbf{k_1k_1} + \mathbf{k_2k_2} + \mathbf{k_1k_2} + \mathbf{k_2k_1}]$. Thus, it becomes convenient in the calculation to project all vectors along and transverse to $\mathbf{k}$ in order to keep track of components. This gives rise to quantities

$$\mathbf{e_x e_x} + \mathbf{e_y e_y} = \hat{\mathbf{1}} - \mathbf{e_z e_z}, \qquad (6\text{–}149\text{a})$$

where the $\mathbf{e}_i$ are unit vectors in the coordinate system of integration and $\hat{\mathbf{1}}$ is here the dyadic idemfactor. This latter quantity is invariant under coordinate transformations and in tensor notation is just δ_{ij}. Therefore, at the end of the calculation generality is achieved by making the replacements

$$\mathbf{e_z} \longrightarrow \frac{k_i}{k}, \quad \mathbf{e_x e_x} + \mathbf{e_y e_y} \longrightarrow \delta_{ij} - \frac{k_i k_j}{k^2}. \qquad (6\text{–}149\text{b})$$

Secondly, we shall presume throughout that below the Bose-Einstein transition temperature the boson condensate is uniform. In this case there will be *no* contribution to the current-current covariance function from the Bose ground state—indeed, there will be no contributions of this kind to any of the remaining covariance functions.

Finally, in the calculation of this and other covariance functions there arise higher-order integrals of the kind encountered in Eq.(6–141). It is therefore useful to define a set of such integrals as follows:

$$I_n(q,p) \equiv 2(q+p)^2 \int \frac{dx}{x^n} \frac{\varepsilon z e^{-(q+p)^2/x^2}}{1 - \varepsilon z e^{-(q+p)^2/x^2}} \frac{e^{\beta\hbar\omega} - 1}{1 - \varepsilon z e^{\beta\hbar\omega} e^{-(q+p)^2/x^2}}, \qquad (6\text{–}150)$$

in the notation of Eq.(6–140). These integrals are generally needed only for odd positive integers $n \geq 3$, and we note that I_3 is exactly integrable, as pointed out above. The Boltzmann limits are given by

$$I_n(q,p)_{\mathrm{B}} = \varepsilon z \beta\hbar\omega e^{-q^2} \left[1 + \frac{(n-3)}{2(q+p)^2} + \frac{(n-3)(n-5)}{2^2(q+p)^4} + \cdots \right.$$
$$\left. + \frac{(n-3)(n-5)\cdots 2}{2^{(n-3)/2}(q+p)^{n-3}} \right]. \qquad (6\text{–}151)$$

With these observations, the zero-order current-current covariance function is calculated in a straightforward way, and we obtain

$$K_{j_i j_j}^0(\mathbf{k},\omega) = 2\pi \frac{\varepsilon(2S+1)}{\omega k \lambda_T^4} \left[\frac{k_i k_j}{k^2} \frac{\omega^2}{k^2} I_3 + \left(\delta_{ij} - \frac{k_i k_j}{k^2} \right) \frac{(q+p)^2}{\beta m} (I_5 - I_3) \right], \qquad (6\text{–}152)$$

in agreement with the general form of Table 6–2. Application of Eq.(6–151) yields the following form for the Boltzmann limit:

$$K_{j_i j_j}^0(\mathbf{k},\omega)_{\mathrm{B}} = n_0 \frac{(2\pi\beta m)^{1/2}}{k} \left[\frac{k_i k_j}{k^2} \frac{\omega^2}{k^2} + \frac{1}{\beta m} \left(\delta_{ij} - \frac{k_i k_j}{k^2} \right) \right] e^{-\frac{1}{2}\beta m \omega^2/k^2}. \qquad (6\text{–}153)$$

Because they are relatively simple to calculate, there is some value to taking the inverse Fourier transforms of the Boltzmann limits (6–148) and (6–153). In the latter one must again keep careful track of the components, and we find that

$$K_{nn}^0(\mathbf{r},\tau)_{\mathrm{B}} = n_0 \left(\frac{\beta m}{2\pi}\right)^{3/2} \frac{1}{|\tau|^3} e^{-\frac{1}{2}\beta m r^2/\tau^2}$$

$$\xrightarrow[\tau\to 0^+]{} n_0 \delta(\mathbf{r}), \tag{6–154}$$

$$K_{j_i j_j}^0(\mathbf{r},\tau)_{\mathrm{B}} = n_0 \left(\frac{\beta m}{2\pi}\right)^{3/2} \frac{r_i r_j}{|\tau|^5} e^{-\frac{1}{2}\beta m r^2/\tau^2}$$

$$= \frac{r_i r_j}{\tau^2} K_{nn}^0(\mathbf{r},\tau)_{\mathrm{B}}. \tag{6–155}$$

The remaining zero-order covariance functions require a bit more care, for they are not necessarily free-particle quantities. Thus, we consider only those parts of the operators *not* containing the interaction, which often are referred to as the *kinetic parts* of the correlation functions. For example, if $\hat{A} = \hat{B} = \hat{h}^0$, where $\hat{h}^0$ is given by Eq.(6–128a), then a now-familiar calculation yields

$$K_{h^0 h^0}^0(\mathbf{k},\omega) = \frac{\varepsilon(2S+1)(q+p)^2}{2\beta^2\omega k\lambda_T^2}\left[k^2 I_3 - 4\sqrt{\pi}\frac{k}{\lambda_T}(q+p)I_5\right.$$

$$\left. + 4\pi\frac{(q+p)^2}{\lambda_T^2}I_7\right], \tag{6–156}$$

and in the Boltzmann limit

$$K_{h^0 h^0}^0(\mathbf{k},\omega)_{\mathrm{B}} = \frac{1}{\beta^2}\left[2 + \beta m\frac{\omega^2}{k^2} + \tfrac{1}{4}\beta^2 m^2\frac{\omega^4}{k^4}\right] K_{nn}^0(\mathbf{k},\omega)_{\mathrm{B}}. \tag{6–157}$$

In a manner completely analogous to the calculation of K_{jj}^0, the covariance function involving two energy-current operators is found to be

$$K_{q_i^0 q_j^0}^0(\mathbf{k},\omega) = \frac{\omega^2}{k^2}\frac{k_i k_j}{k^2}\frac{\varepsilon(2S+1)(q+p)^2}{2\beta^2\lambda_T^2\omega k}\left[k^2 I_3 - 4\sqrt{\pi}\frac{k}{\lambda_T}(q+p)I_5\right.$$

$$\left. + 4\pi\frac{(q+p)^2}{\lambda_T^2}I_7\right]$$

$$+ \left(\delta_{ij} - \frac{k_i k_j}{k^2}\right)\frac{\varepsilon(2S+1)k^2(q+p)^2}{2m\beta^3\omega k\lambda_T^2}\left[(I_5 - I_3) - 4\sqrt{\pi}\frac{(q+p)}{k\lambda_T}(I_7 - I_5)\right.$$

$$\left. + 4\pi\frac{(q+p)^2}{k^2\lambda_T^2}(I_9 - I_7)\right]. \tag{6–158}$$

In the Boltzmann limit,

$$K_{q_i^0 q_j^0}^0(\mathbf{k},\omega)_{\mathrm{B}} = \frac{\omega^2}{k^2}\frac{k_i k_j}{k^2}K_{h^0 h^0}^0(\mathbf{k},\omega)_{\mathrm{B}}$$

$$+ \left(\delta_{ij} - \frac{k_i k_j}{k^2}\right)\frac{n_0(2\pi\beta m)^{1/2}}{\beta^3 m k}$$

$$\times \left[6 + 2\beta m\frac{\omega^2}{k^2} + \tfrac{1}{4}\beta^2 m^2\frac{\omega^4}{k^4}\right]e^{-\frac{1}{2}\beta m\omega^2/k^2}. \tag{6–159}$$

Although the covariance function involving two stress tensors would appear to be the most tedious to calculate, even in zero order, its symmetry properties simplify the task considerably. Again the calculation is much like that for K_{nn}^0, except that the integrands now contain factors

$$[k_{2m}k_{2n} + k_{2m}k_{1n} + k_{2n}k_{1m} + k_{1m}k_{1n}][k_{2i}k_{2j} + k_{2i}k_{1j} + k_{2j}k_{1i} + k_{1i}k_{1j}]$$

giving a fourth-rank tensor, $K_{T_{ij}^0 T_{mn}^0}^0$. At first glance the product of two second-rank tensors appears to yield 81 different terms. But the tensors are symmetric and have only six independent components each, so that the total number of terms is immediately reduced to 36.

Further simplification occurs by projecting components parallel and transverse to k in the integrals. This procedure yields poly-adic forms of the type

$$e_x e_x e_x e_x + 2e_x e_x e_y e_y + e_y e_y e_y e_y, \tag{6-160a}$$

which transform to products of the type

$$\left(\delta_{ij} - \frac{k_i k_j}{k^2}\right)\left(\delta_{nm} - \frac{k_n k_m}{k^2}\right) \equiv (\delta_{ij} - k_{ij})(\delta_{nm} - k_{nm})$$

$$\equiv \kappa_{ij}\kappa_{nm} \tag{6-160b}$$

in a general coordinate system. As mentioned earlier, there are a number of possible ways to arrange the indices, and these must all be considered. The algebra is rather tedious, but the result is relatively simple. With the notation of Table 6–2, as recalled in Eq.(6–160b), we find

$$K_{T_{ij}^0 T_{mn}^0}^0 (\mathbf{k}, \omega) = \frac{\omega^4}{k^4} k_{ijmn} \frac{2\pi m^2 \varepsilon(2S+1)}{\omega k \lambda_T^4} I_3$$

$$+ \frac{\omega^2}{k^2} [k_j \kappa_{mn} + k_{mn}\kappa_{ij} + k_{im}\kappa_{jn} + k_{jn}\kappa_{im} + k_{in}\kappa_{jm}$$

$$+ k_{jm}\kappa_{in}](12\pi)\frac{\varepsilon(2S+1)}{\omega k \lambda_T^4}\left(\frac{m}{\beta}\right)(q+p)^2(I_5 - I_3)$$

$$+ [\kappa_{ij}\kappa_{mn} + \kappa_{im}\kappa_{jn} + \kappa_{in}\kappa_{jm}](3\pi)\frac{\varepsilon(2S+1)}{\beta^2 \omega k \lambda_T^4}$$

$$\times (q+p)^4(I_7 - 2I_5 + I_3). \tag{6-161}$$

In the Boltzmann limit,

$$K_{T_{ij}^0 T_{mn}^0}^0 (\mathbf{k}, \omega)_{\mathrm{B}} = n_0 m^2 \frac{(2\pi\beta m)^{1/2}}{k} \frac{\omega^4}{k^4} k_{ijmn} e^{-\frac{1}{2}\beta m \omega^2 / k^2}$$

$$+ 6n_0 \left(\frac{m}{\beta}\right)\frac{(2\pi\beta m)^{1/2}}{k}\frac{\omega^2}{k^2}[k_{ij}\kappa_{mn} + k_{mn}\kappa_{ij}$$

$$+ k_{im}\kappa_{jn} + k_{jn}\kappa_{im} + k_{in}\kappa_{jm} + k_{jm}\kappa_{in}]e^{-\frac{1}{2}\beta m \omega^2 / k^2}$$

$$+ 3n_0 \frac{(2\pi\beta m)^{1/2}}{\beta^2 k}[\kappa_{ij}\kappa_{mn} + \kappa_{im}\kappa_{jn} + \kappa_{in}\kappa_{jm}]$$

$$\times e^{-\frac{1}{2}\beta m \omega^2 / k^2} \tag{6-162}$$

Finally, there remain various covariance functions involving cross-products of free-particle operators. Because no new aspects of the calculations are encountered, we merely record the results:

$$K^0_{nj_i}(\mathbf{k},\omega) = \frac{k_i}{k}\frac{\omega}{k}K^0_{nn}(\mathbf{k},\omega)\,, \tag{6-163}$$

$$K^0_{nh^0}(\mathbf{k},\omega) = \frac{\varepsilon(2S+1)}{2\beta\omega k\lambda_T^2}\left[4\pi\frac{(q+p)^2}{\lambda_T^2}I_5 - 2\sqrt{\pi}\frac{k}{\lambda_T}(q+p)I_3\right]\,, \tag{6-164}$$

$$
\begin{aligned}
K^0_{j_i q_j^0}(\mathbf{k},\omega) = {}& \frac{\omega^2}{k^2}k_{ij}\frac{\varepsilon(2S+1)}{2\beta\omega k\lambda_T^2}\left[4\pi\frac{(q+P)^2}{\lambda_T^2}I_5 - 2\sqrt{\pi}\frac{k}{\lambda_T}(q+p)I_3\right] \\
&+ \kappa_{ij}\frac{\varepsilon(2S+1)(q+p)^2}{2\beta^2 m\omega k\lambda_T^2}\left[4\pi\frac{(q+p)^2}{\lambda_T^2}(I_7 - I_5)\right. \\
&\left.\qquad - 2\sqrt{\pi}\frac{k}{\lambda_T}(q+p)(I_5 - I_3)\right]\,,
\end{aligned}
\tag{6-165}
$$

$$
\begin{aligned}
K^0_{j_m T^0_{ij}}(\mathbf{k},\omega) = {}& \frac{\omega^3}{k^3}\frac{k_{mij}}{\omega k}\frac{2\pi m\varepsilon(2S+1)}{\lambda_T^4}I_3 \\
&+ \frac{3}{2}\frac{\varepsilon(2S+1)(q+p)^2}{\beta^2\hbar\omega\lambda_T^2}\left[1 - 4\sqrt{\pi}\frac{q+p}{k\lambda_T}(I_5 - I_3)\right] \\
&\times [k_m\kappa_{ij} + k_i\kappa_{mj} + k_j\kappa_{im}]\,,
\end{aligned}
\tag{6-166}
$$

$$
\begin{aligned}
K^0_{q_m^0 T^0_{ij}}(\mathbf{k},\omega) = {}& \frac{\omega^3}{k^3}k_{mij}\frac{\varepsilon m(2S+1)(q+p)^2}{2\beta\omega k\lambda_T^2}\left[4\pi\frac{(q+p)^2}{\lambda_T^2}I_5 + 2\sqrt{\pi}\frac{k}{\lambda_T}(q+p)I_3\right] \\
&+ \frac{3}{8}\frac{\varepsilon(2S+1)(q+p)^2}{\pi\beta^2\hbar\omega}\left[1 - 4\sqrt{\pi}\frac{q+p}{k\lambda_T}\right][k_m\kappa_{[ij} + k_i\kappa_{mj} + k_j\kappa_{im}] \\
&\times \left[4\pi\frac{(q+p)^2}{\lambda_T^2}(I_7 - I_5) - 2\sqrt{\pi}\frac{k}{\lambda_T}(q+p)(I_5 - I_3)\right]\,,
\end{aligned}
\tag{6-167}
$$

$$K^0_{h^0 j_i}(\mathbf{k},\omega) = \frac{\omega}{k}\frac{k_i}{k}K^0_{nh^0}(\mathbf{k},\omega)\,, \tag{6-168}$$

$$K^0_{nq_i^0}(\mathbf{k},\omega) = K^0_{h^0 j_i}(\mathbf{k},\omega)\,, \tag{6-169}$$

$$K^0_{nT^0_{ij}}(\mathbf{k},\omega) = mK^0_{j_i j_j}(\mathbf{k},\omega)\,, \tag{6-170}$$

$$K^0_{h^0 T^0_{ij}}(\mathbf{k},\omega) = mK^0_{j_i q_j^0}(\mathbf{k},\omega)\,, \tag{6-171}$$

$$K^0_{h^0 q_i^0}(\mathbf{k},\omega) = \frac{\omega}{k}\frac{k_i}{k}K^0_{h^0 h^0}(\mathbf{k},\omega)\,. \tag{6-172}$$

Most limiting forms can be found from other results, so we need only record the following Boltzmann limits:

$$K^0_{nh^0}(\mathbf{k},\omega)_{\mathrm{B}} = n_0\frac{(2\pi\beta m)^{1/2}}{\beta k}\left[1 + \tfrac{1}{2}\beta m\frac{\omega^2}{k^2}\right]e^{-\frac{1}{2}\beta m\omega^2/k^2}\,, \tag{6-173}$$

$$K^0_{j_i q^0_j}(\mathbf{k},\omega)_{\mathrm{B}} = n_0 \frac{(2\pi\beta m)^{1/2}}{\beta k} \frac{\omega^2}{k^2} k_{ij}\left[1 + \tfrac{1}{2}\beta m \frac{\omega^2}{k^2}\right] e^{-\frac{1}{2}\beta m \omega^2/k^2}$$

$$+ 2n_0 \frac{(2\pi\beta m)^{1/2}}{\beta^2 mk}\left[1 + \tfrac{1}{4}\beta m \frac{\omega^2}{k^2}\right]\kappa_{ij} e^{-\frac{1}{2}\beta m \omega^2/k^2}, \qquad (6\text{-}174)$$

$$K^0_{j_m T^0_{ij}}(\mathbf{k},\omega)_{\mathrm{B}} = n_0 m \frac{(2\pi\beta m)^{1/2}}{k} \frac{\omega^3}{k^3} k_{mij} e^{-\frac{1}{2}\beta m \omega^2/k^2}$$

$$- 3\sqrt{\pi}\frac{n_0 m}{k}\frac{\omega^2}{k^2}[k_m\kappa_{ij} + k_i\kappa_{jm} + k_j\kappa_{im}]e^{\frac{1}{2}\beta m \omega^2/k^2}, \quad (6\text{-}175)$$

$$K^0_{q^0_m T_{ij}}(\mathbf{k},\omega)_{\mathrm{B}} = n_0 \frac{(2\pi\beta m)^{1/2}}{\beta k}\frac{\omega^3}{k^3} k_{mij}\left[1 + \tfrac{1}{2}\beta m \frac{\omega^2}{k^2}\right] e^{-\frac{1}{2}\beta m \omega^2/k^2}$$

$$- 6\sqrt{\pi}\frac{n_0 m}{\beta k}\left[1 + \tfrac{1}{4}\beta m \frac{\omega^2}{k^2}\right]\frac{\omega^2}{k^2}[k_m\kappa_{[ij} + k_i\kappa_{jm} + k_j\kappa_{im}]$$

$$\times e^{-\frac{1}{2}\beta m \omega^2/k^2}. \qquad (6\text{-}176)$$

This completes evaluation of all the free-particle contributions to covariance functions involving the five basic microscopic field operators for a simple fluid. Note that in the Boltzmann limit all contain the characteristic factors

$$\left(\frac{\omega}{k}\right)^n e^{-\frac{1}{2}\beta m \omega^2/k^2}, \qquad (6\text{-}177)$$

for some n, thereby vanishing in an extremely strong fashion at $k = 0$. As a consequence these free-particle quantities can not contribute to the hydrodynamic limit, a result which is not too surprising in view of the observation that the HL is supposedly a collision-dominated regime. The inverse Fourier transforms will possess a characteristic dependence on the function $\exp(-\beta m r^2/2\tau^2)$.

SUMMARY

Calculation of covariance functions, or any correlation function for that matter, is seen to present formidable difficulties. The detailed discussion of perturbation calculations in Appendix B serves to reinforce this conclusion, despite the fact that this is the most straightforward approach one can take. Thus, a number of specialized devices have been employed in order to book some kind of progress. As mentioned above, some techniques developed specifically for hydrodynamic applications will be discussed further in Chapter 8.

It is not only the theoretical understanding of correlation functions that is weak, for the experimental situation is almost the same—with one exception. In Volume I we saw that the radial distribution function $g(r)$, which is related to density fluctuations in the equilibrium system, has been measured extensively in numerous fluids. Actually, its Fourier transform $S(k)$, the static structure factor, is the usual object of study, because it is related to scattering parameters. Similarly, the dynamic structure factor $S(\mathbf{k},\omega)$ has also been studied through light- and neutron-scattering experiments. (See, e.g., review discussions in Boon and Yip, 1980.) This function is

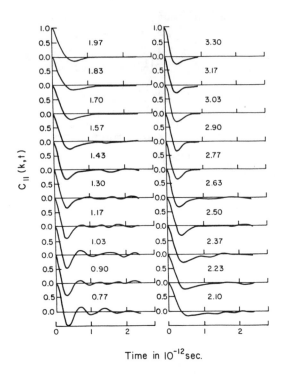

Time in 10^{-12} sec.

Fig. 6–2. Computer simulation of the longitudinal current-current correlation function for various values of k. [*Reproduced by permission from Ailawadi, et al (1971)*.]

effectively $S_{nn}(\mathbf{k}, \omega)$ in Eq.(6–69), and so provides information on $K_{nn}(\mathbf{k}, \omega)$ as well in the limit $\beta\hbar\omega \ll 1$. But this is essentially the only correlation function on which one possesses extensive data, and at that the ranges of wavelength and frequency are limited.

Owing to both theoretical and calculational difficulties, then, a strong interest has developed in computer simulations in the hope that they can provide further insight into the structure of correlation functions. These computer 'experiments' supply, in fact, almost our entire knowledge regarding current-current correlation functions, and not even they go much beyond K_{jj}. The results of computer calculations by Rahman have been discussed in conjunction with hydrodynamic models by Ailawadi, *et al* (1971), and in Figures 6–2 and 6–3 we exhibit the time dependence of the longitudinal and transverse current-current correlation functions, respectively. These calculations refer to $S_{nn}(k, t)$, rather than $K_{nn}(k, t)$, and pertain to a system of 500 particles interacting via Lennard-Jones (6-12) potentials. The corresponding frequency spectra are shown in Figures (6–4) and (6–5). A more extensive discussion of the computer simulations is also provided by Boon and Yip (1980).

There is little doubt that one of the more pressing problems of statistical me-

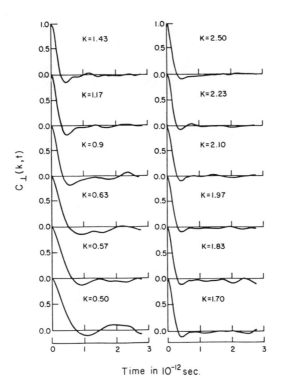

Fig. 6–3. Computer simulation of the transverse current-current correlation function for various values of k. [*Reproduced by permission from Ailawadi, et al (1971)*.]

chanics today remains that of calculating covariance, or correlation functions, and the following chapters will do nothing if not emphasize that point further.

Problems

6.1 Establish the following commutation relation between the system Hamiltonian $\hat{H}$ and the Fock-space field operator $\hat{\psi}(\mathbf{x})$:

$$[\hat{H}, \hat{\psi}(\mathbf{x})] = -\left[-\frac{\hbar^2}{2m}\nabla^2 + V_1(\mathbf{x}) + \int d^3r\, \hat{\psi}^\dagger(\mathbf{r})V_2(|\mathbf{x}-\mathbf{r}|)\hat{\psi}(\mathbf{r})\right]\hat{\psi}(\mathbf{x}).$$

Employ this and the complex-conjugate expression to verify the commutators of Eq.(6–31).

6.2 Evaluate the commutators on the right-hand sides of Eqs.(6–23) and (6–24) and verify the right-hand sides of Eqs.(6–26) and (6–27) in terms of Eqs.(6–33) and (6–34), respectively.

6.3 Recall the identity (6–32), which is developed in Appendix A.

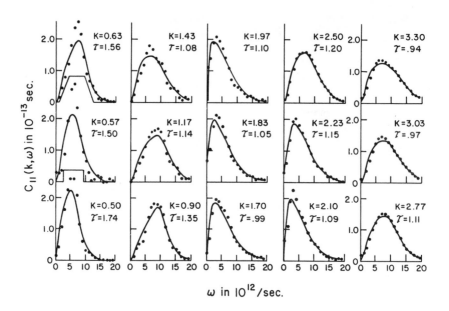

Fig. 6–4. Power spectra of the computer-simulated longitudinal current-current correlation function for several values of k (the dots), compared with calculations using a Gaussian memory function (see Chapter 8). [*Reproduced by permission from Ailawadi, et al (1971).*]

(a) Employ this identity to prove that in the *canonical* ensemble

$$\langle[\hat{A}(\mathbf{r},\tau),\hat{B}(\mathbf{0},0)]\rangle_0 = \overline{\langle[\hat{A}(\mathbf{r},\tau),\hat{H}]\hat{B}(\mathbf{0},0)\rangle_0},$$

in terms of the Kubo transform of the commutator. Hence, define an operator $\hat{C}(\mathbf{r},\tau) \equiv [\hat{A}(\mathbf{r},\tau),\hat{H}]$ to show that

$$K_{BC}(\mathbf{r},\tau) = \langle[\hat{A}(\mathbf{r},\tau),\hat{B}(\mathbf{0},0)]\rangle_0.$$

(b) Let $\hat{D}$ be a locally-conserved density and $\hat{\mathbf{J}}$ the corresponding current. If $\hat{A}$ is such a density, so that it commutes with $\hat{N}$, show that the first result in part (a) can be written

$$\langle[\hat{D}(\mathbf{r},\tau),\hat{B}(\mathbf{0},0)]\rangle_0 = -i\hbar\nabla_i K_{BJ_i}(\mathbf{r},\tau),\qquad(6\text{–}178)$$

where the summation convention is implied.

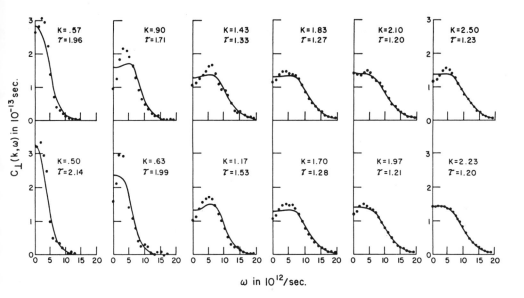

Fig. 6–5. Power spectra of the computer-simulated transverse current-current correlation function for several values of k (the dots), compared with calculations using a Gaussian memory function (see Chapter 8). [*Reproduced by permission from Ailawadi, et al (1971).*]

6.4 In a simple fluid we denote the ith component of the vector $\mathbf{k}$ as $(\mathbf{k})_i$ and then define the unit vector in the ith direction as $k_i \equiv (\mathbf{k})_i/k$. Products of unit vectors are written

$$\frac{(\mathbf{k})_i(\mathbf{k})_j}{k^2} = k_i k_j \equiv k_{ij}, \quad k_i k_j k_m \equiv k_{ijm},$$

etc., and the transverse basis tensor is $\kappa_{ij} \equiv \delta_{ij} - k_{ij}$. One can then decompose the stress-tensor operator into longitudinal and transverse parts as follows:

$$\hat{T}_{ij}(\mathbf{k}, \omega) = k_{ij}\hat{T}_\ell(k, \omega) + \kappa_{ij}\hat{T}_{tr}(k, \omega),$$

where it is noted explicitly that the operators depend only on the magnitude of $\mathbf{k}$. In order to account for all possible combinations of indices consistent with known symmetry properties, it is useful to define five scalar functions K_{TT}^i by writing

$$K_{T_{ij}T_{mn}}(\mathbf{k}, \omega) = k_{ijmn}K_{TT}^1(k, \omega) + [k_{ij}\kappa_{mn} + k_{mn}\kappa_{ij}]K_{TT}^2(k, \omega)$$

$$+ [k_{in}\kappa_{jm} + k_{jm}\kappa_{in} + k_{im}\kappa_{jn} + k_{jn}\kappa_{im}]K_{TT}^3(k,\omega)$$
$$+ \kappa_{ij}\kappa_{mn}K_{TT}^4(k,\omega) + [\kappa_{im}\kappa_{jn} + \kappa_{in}\kappa_{jm}]K_{TT}^5(k,\omega),$$

where we have intentionally isolated terms containing the pairs (i,j) and (m,n). This scheme is convenient because when $i = j$, or $m = n$, tensor traces are involved.

(a) Combine this last expression with the identity (6–63) to generate the following sequence of identities:

$$\frac{k}{\omega}k_i K_{T_{ij}T_{mn}}(\mathbf{k},\omega) = K_{j_j T_{mn}}(\mathbf{k},\omega)$$

$$= \frac{k}{\omega}k_{jmn}K_{TT}^1 + \frac{k}{\omega}k_j\kappa_{mn}K_{TT}^2$$

$$+ [k_m\kappa_{jn} + k_n\kappa_{jm}]\frac{k}{\omega}K_{TT}^3,$$

$$\frac{k^2}{\omega^2}k_{ij} K_{T_{ij}T_{mn}}(\mathbf{k},\omega) = K_{n T_{mn}}(\mathbf{k},\omega)$$

$$= \frac{k^2}{\omega^2}k_{mn}K_{TT}^1 + \frac{k^2}{\omega^2}\kappa_{mn}K_{TT}^2,$$

$$\frac{k^3}{\omega^3}k_{ijm} K_{T_{ij}T_{mn}}(\mathbf{k},\omega) = K_{n j_n}(\mathbf{k},\omega)$$

$$= \frac{k^3}{\omega^3}k_n K_{TT}^1,$$

$$\frac{k^4}{\omega^4}k_{ijmn} K_{T_{ij}T_{mn}}(\mathbf{k},\omega) = K_{nn}(\mathbf{k},\omega)$$

$$= \frac{k^4}{\omega^4}K_{TT}^1.$$

(b) Observe that a similar set of identities can be generated for each lower-rank-tensor covariance function by decomposing it in a way analogous to that above. By comparing all these identities verify the entries in Table 6–2.

6.5 Verify Eqs.(6–79) and (6–81), and by using the result of Problem 6.3(b) verify Eqs.(6–82) and (6–83).

6.6 Verify Eq.(6–90).

REFERENCES

Ailawadi, N.K., A. Rahman, and R. Zwanzig: 1971, 'Generalized Hydrodynamics and Analysis of Current Correlation Functions', *Phys. Rev. A* **4**, 1616.

Berne, B.J., and G.D. Harp: 1971, 'On the Calculation of Time Correlation Functions', *Adv. Chem. Phys.* **17**, 63.

Boon, J.P., and S. Yip: 1980, *Molecular Hydrodynamics*, McGraw-Hill, New York.

Copley, J.R.D., and S.W. Lovesey: 1975, 'The Dynamic Properties of Monatomic Liquids', *Repts. Prog. Phys.* **38**, 461.

Fetter, A.L., and J.D. Walecka: 1971, *Quantum Theory of Many-Particle Systems*, McGraw-Hill, New York.

Forster, D.: 1975, *Hydrodynamic Fluctuations, Broken Symmetry, and Correlation Functions*, Benjamin, Reading, MA.

Kadanoff, L.P., and P.C. Martin: 1963, 'Hydrodynamic Equations and Correlation Functions', *Ann. Phys. (N.Y.)* **24**, 419.

Maxwell, J.C.: 1867, 'On the Dynamical Theory of Gases', *Phil. Trans. Roy. Soc. London* **157**, 49.

Poisson, S.-D.: 1831, 'Mémoire sur les équations générales de l'équilibre et du mouvement des corps solides élastiques et des fluides', *J. École Poly.* **13**, 1.

Puff, R.D., and N.S. Gillis: 1968, 'Fluctuations and Transport Properties of Many-Particle Systems', *Ann. Phys. (N.Y.)* **6**, 364.

Zwanzig, R., and R.D. Mountain: 1965, 'High-Frequency Elastic Moduli of Simple Fluids', *J. Chem. Phys.* **43**, 4464.

Chapter 7

Thermally Driven Systems

The linear theory developed in the preceding chapters was deliberately kept rather general, in that Eqs.(5–61) and (5–62) refer to an unspecified information-gathering interval. In addition to the fact that $R(\mathbf{x}, t)$ is completely arbitrary, the entire attitude has been somewhat passive. One merely presumes that information about the system has been made available. But both the theoretician and experimentalist can exercise a certain amount of macroscopic dynamical control over the system of interest, in the case of thermal processes as well as for dynamical perturbations. Thus, a special case of the general formalism is the response to an external thermal driving mechanism. By the latter we mean any source which is *not* attributable to the addition of a clearly defined dynamic term to the Hamiltonian. In many ways, however, the ensuing theory of linear thermal response is analogous in form to that for dynamic response.

As a convenience let us collect together the pertinent equations of the prediction algorithm, and for simplicity temporarily omit the spatial dependence of variables. Suppose we have data on the average energy of the system in equilibrium, as well as on the expectation value of some other Heisenberg operator $\hat{F}(t)$ during some time interval $t \in R$. That is, from the data

$$\langle \hat{H} \rangle_0 = \text{Tr}[\hat{\rho}_0\, \hat{H}], \qquad \langle \hat{F}(t) \rangle = \text{Tr}[\hat{\rho}\hat{F}(t)], \quad t \in R, \tag{7-1}$$

we construct the new initial statistical operator

$$\hat{\rho} = \frac{1}{Z} \exp\left[-\beta\,\hat{H} + \int_R \lambda(t')\hat{F}(t')\,dt'\right], \tag{7-2}$$

where $\hat{H} \neq \hat{H}(t)$ is the equilibrium Hamiltonian and the Lagrange multipliers are determined from the constraints

$$\langle \hat{H} \rangle_0 = -\frac{\partial}{\partial \beta} \ln Z_0(\beta)\,,$$

$$\langle \hat{F}(t) \rangle = \frac{\delta}{\delta\lambda(t)} \ln Z[\beta, \lambda(t)]\,, \quad t \in R. \tag{7-3}$$

Note that this scenario envisions no new information in R regarding the energy, so that any changes in $\langle \hat{H} \rangle$ are determined from the new information on $\langle \hat{F}(t) \rangle$. Subscripts 0 denote expectation values in the equilibrium ensemble $\hat{\rho}_0$ which is relevant for t prior to R—in this ensemble we necessarily have $\lambda(t') = 0$.

In linear approximation the expectation value of some other operator $\hat{C}$ at time τ is

$$\langle \hat{C}(\tau) \rangle \simeq \langle \hat{C} \rangle_0 + \int_R \lambda(t') K_{CF}(t', \tau) \, dt' . \tag{7-4}$$

Recall that when $\tau > R$ this last expression provides the predicted future of $\langle \hat{C}(\tau) \rangle$ based on the initial information, whereas for $\tau < R$ we can obtain the retrodicted past in some cases. If we set $\hat{C} = \hat{F}$ and take $\tau \in R$, then we have a linear Fredholm integral equation of the first kind determining the Lagrange-multiplier function $\lambda(t)$. We presume β to have been identified in the equilibrium ensemble.

Let us now be a bit more specific about the information-gathering interval and take it to be $R : (0, t)$. Equation (7–4) then becomes

$$\langle \hat{C}(\tau) \rangle \simeq \langle \hat{C} \rangle_0 + \int_0^t \lambda(t') K_{CF}(t', \tau) \, dt' , \tag{7-5}$$

with the same comments as above, and various scenarios now suggest themselves. The first of these is that measurement has provided continuous information on $\langle \hat{F}(t) \rangle$ throughout the interval $(0, t)$, so that Eq.(7–5) yields the predicted value of $\langle \hat{C}(\tau) \rangle$ for any τ, based on this information. If we take $\hat{C} = \hat{F}$ we have an integral equation for $\lambda(t')$ when $0 \le \tau \le t$. In particular, we could be interested in the value $\tau = t$, and then Eq.(7–5) takes on the *appearance* of a Volterra integral equation. This can be misleading, though, because there is no indication that t is a continuous variable; in general it is just a number.

But it is possible that there may be available, in addition to quantitative data, further information of a qualitative kind. One may know that there is an external source driving the system, and yet remain somewhat ignorant of its details. It can be anything from a Bunsen burner to a jackhammer, for eventually a Lagrange-multiplier function will provide all the relevant details we need to know. The existence of the source can be inferred from a continuous development of $\langle \hat{F}(t) \rangle$, which is monitored. Knowledge of this quantity at any given moment allows prediction of any other quantity at that time.

A. Theory of Thermal Driving

Consider now an experimental arrangement in which, for a system initially in thermal equilibrium, the expectation value $\langle \hat{F}(t) \rangle$ is actually specified throughout the interval $[0, t]$, in the sense that its value is carefully controlled. In other words, the variable described by $\hat{F}$ is *thermally driven* in a prescribed way by an external source, a statement that can now be considered a definite part of the initial information defining the problem. This is similar to the view expressed in connection with the steady-state situation. Although a steady-state process must be continuously driven, we did not there presume any information to be known about the nature of the source or how the driving was effected. Indeed, no knowledge was available as to whether the driving would continue or whether the system would relax. If it did relax, from our point of view that was due to a lack of information, rather than to definite knowledge of the experimental situation. When thermal driving is specified we know exactly what is happening to the system, and when to look for relaxation. One presumes that t is under complete experimental control and that primary interest lies with the value of $\langle \hat{C}(\tau) \rangle$ at $\tau = t$. Because the external source is driving the system, expectation values at times $t > \tau$ will depend on whether the driving is continued, or whether the external source is disconnected and the system allowed to thermally relax. In any event, the scenario is much akin to that of the mechanically-driven linear response arrangement, except that here no claims are made regarding knowledge of the complete Hamiltonian encompassing both system and source.

With these thoughts in mind, reconsider Eq.(7–5) for the case $\hat{C} = \hat{F}$, so that in linear approximation

$$F(t) \equiv \langle \hat{F}(t) \rangle - \langle \hat{F} \rangle_0 = \int_0^t \lambda(t') K_{FF}(t-t')\, dt' , \qquad (7\text{–}6)$$

where $F(t)$ is presumed known on the interval $(0,t)$ by the presumption of thermal driving. Because we have complete control of the driving, the expectation value is now a continuous function of t. In this case Eq.(7–6) is indeed a Volterra integral equation of the first kind determining $\lambda(t)$. Note, however, that $\lambda(t)$ is only *defined* on the interval $(0,t)$, and not beyond.

Although this prescription is equivalent to that of Eq.(7–3), we now have the entire theory of Volterra integral equations at our disposal (e.g., Tricomi, 1957). For example, if the equal-time covariance function $K_{FF}(t,t)$ vanishes nowhere in the allowed range of t, and if the first derivatives of $F(t)$ and $K_{FF}(t',t)$ exist and are continuous, then Eq.(7–6) can always be reduced to a Volterra equation of the second kind. When $\hat{\mathsf{H}}$ is time independent the kernel takes the form $K_{FF}(t-t')$ and further simplification occurs. Suppose the thermal driving began in the infinite past, so that the basic interval is $(-\infty, t)$. Then the integral transform on the right-hand side of Eq.(7–6) carries any periodic function $\lambda(t')$ with period T into another periodic function with the same period if and only if $K_{FF}(t',t) = K_{FF}(t-t')$. Also, this form of the kernel permits one to use the convolution theorem for Laplace transforms to immediately effect the solution to Eq.(7–6), under the appropriate conditions.

In a more general and purely formal way, and without recourse to any linear approximation, the PME provides the statistical operator for thermal driving:

$$\hat{\rho}_t = \frac{1}{Z_t} \exp\left[-\beta\,\hat{\mathsf{H}} + \int_0^t \lambda(t')\hat{F}(t')\, dt' \right],$$

$$Z_t[\beta, \lambda(t)] = \mathrm{Tr}\,\exp\left[-\beta\,\hat{\mathsf{H}} + \int_0^t \lambda(t')\hat{F}(t')\, dt' \right]. \qquad (7\text{–}7)$$

Actually, rather than the PME, we have employed here the principle of minimum cross-entropy [e.g., Volume I; or, Shore and Johnson, (1980)] in order to include an additional criterion for thermal driving. This criterion specifies that $\hat{F}(t)$ is driven without any definite statement being made about $\hat{H}$, so that any subsequent changes in $\langle\hat{H}\rangle$ arise only as a consequence of those in $\hat{F}(t)$. Thus, we minimize

$$H \equiv K\,\text{Tr}[\hat{\rho}(\ln\hat{\rho} - \ln\hat{\rho}_0)]\,, \quad K > 0\,, \tag{7-8}$$

subject to the constraints $\text{Tr}\,\hat{\rho} = \text{Tr}\,\hat{\rho}_0 = 1$, and knowledge of $\langle\hat{F}(t)\rangle$ over the interval $[0,t]$. The prior probabilities in the present case are just the eigenvalues of $\hat{\rho}_0$. In Problem 7.1 the reader is asked to carry out the proper derivation in detail of Eqs.(7–7).

One could employ a more general equilibrium operator $\hat{A}$ in place of $-\beta\,\hat{H}$, as well as include spatial dependence and a number of operators $\hat{F}_i$, but Eq.(7–7) is sufficient for subsequent discussion. Clearly, $\hat{\rho}_t$ can now be considered a function of the continuous variable t, and its time derivative is just

$$\partial_t\hat{\rho}_t = \hat{\rho}_t\lambda(t)\big[\overline{\hat{F}(t)} - \langle\hat{F}(t)\rangle_t\big]\,, \tag{7-9}$$

where we have employed the following operator identity from Appendix A:

$$\partial_\lambda e^{\hat{A}(\lambda)} = e^{\hat{A}(\lambda)}\int_0^1 e^{-x\hat{A}(\lambda)}\,(\partial_\lambda\hat{A})e^{x\hat{A}(\lambda)}\,dx = e^{\hat{A}(\lambda)}\,\overline{(\partial_\lambda\hat{A})}\,. \tag{7-10}$$

The expectation value at time t and the Kubo transform both refer to the ensemble $\hat{\rho}_t$, and *not* to the equilibrium ensemble $\hat{\rho}_0$. The expectation value of any other operator at time t is then

$$\langle\hat{C}\rangle_t = \text{Tr}[\hat{\rho}_t\hat{C}(t)]\,. \tag{7-11}$$

These are completely nonlinear expressions, and if one makes the choice $\hat{C} = \hat{F}$ along with a linear approximation, then Eq.(7–6) is regained.

Although Eq.(7–9) appears to be a differential equation for $\hat{\rho}_t$, it is really $\lambda(t)$ which is the essential unknown. The latter function is defined only on the interval $0 \le t' \le t$, and so we interpret Eq.(7–9) as merely specifying the rate at which $\hat{\rho}_t$ has been changing in that interval, but at time t. For time-independent $\hat{H}$, and in the Heisenberg picture, $\hat{\rho}_t$ does not evolve unitarily but explicitly, as in Eq.(7–9). In turn this implies that the theoretical entropy

$$S_t = -\kappa\,\text{Tr}[\hat{\rho}_t\ln\hat{\rho}_t] \tag{7-12}$$

is now unambiguously and explicitly time dependent.

Equation (7–9) now permits calculation of the time derivative of the expectation value of any other operator at time t. Direct differentiation in Eq.(7–11) yields

$$\partial_t\langle\hat{C}(t)\rangle_t = \text{Tr}[\hat{C}(t)\partial_t\hat{\rho}_t + \hat{\rho}_t\partial_t\hat{C}(t)]$$
$$= \langle\dot{\hat{C}}(t)\rangle_t + \lambda(t)K_{CF}^t(t,t)\,, \tag{7-13}$$

where the superposed dot denotes partial differentiation and

$$K_{CF}^t(t', t) \equiv \overline{\langle \hat{F}(t')\hat{C}(t) \rangle}_t - \langle \hat{F}(t') \rangle_t \langle \hat{C}(t) \rangle_t . \qquad (7\text{--}14)$$

Additional convenience of notation is achieved in Eq.(7–13) by writing

$$\dot{\sigma}_C(t) \equiv \partial_t \langle \hat{C}(t) \rangle_t - \langle \dot{\hat{C}}(t) \rangle_t = \lambda(t) K_{CF}^t(t, t) , \qquad (7\text{--}15)$$

and the equal-time function $K_{CF}^t(t, t)$ usually vanishes when $\hat{C} \neq \hat{F}$ is not a driven variable. These are completely nonlinear expressions, because expectation values, Kubo transforms, and covariance functions are all written in terms of the full operator $\hat{\rho}_t$—there are no approximations of any kind. Although Eq.(7–15) has the appearance of a differential, or kinetic equation for $\langle \hat{C}(t) \rangle_t$, it is redundant in that sense owing to Eq.(7–11) and the practical need to make a linear approximation eventually. In any event, one must first determine $\lambda(t)$.

Equation (7–15), however, leads to an illuminating interpretation if the choice $\hat{C} = \hat{F}$ is made. Then,

$$\dot{\sigma}_F(t) = \partial_t \langle \hat{F}(t) \rangle_t - \langle \dot{\hat{F}}(t) \rangle_t = \lambda(t) K_{FF}^t(t, t) . \qquad (7\text{--}16)$$

A relation of this type was first noticed by Mitchell (1967), and mentioned in passing elsewhere (Robertson and Mitchell, 1971). Mitchell has provided an important piece of insight to be derived from it. Owing to the specification of thermal driving, $\partial_t \langle \hat{F}(t) \rangle_t$ is the time rate-of-change of $\langle \hat{F}(t) \rangle$ in the system at time t, whereas $\langle \dot{\hat{F}}(t) \rangle_t$ is the negative rate of change produced by internal relaxation. Hence, $\dot{\sigma}_F(t)$ must be the rate at which $\hat{F}$ is driven by the external source, and is often what is directly measured or controlled experimentally. One need know nothing else about the details of the source, because its total effect on the system is expressed by the second equality in Eq.(7–16). The effect of the source on other variables is described by Eq.(7–15). Notice that Eq.(7–16) is a macroscopic conservation law having the appearance of a generalized first law of thermodynamics. It is also worth noticing that this is a nonlinear equation determining $\lambda(t')$ in the interval $0 \leq t' \leq t$. In linear approximation it gives the immediate solution, as well as the physical interpretation of λ as the fluctuation-normalized source strength. Because of symmetry arguments, the equal-time covariance functions often vanish identically, but this is not the case when $\hat{C} = \hat{F}$.

MODE COUPLING

Mitchell (1967) has also suggested how the thermal driving scenario can be extended to the case in which $\hat{F}$ is driven subject to the constraint that some other quantity $\hat{G}(t)$ is explicitly *not* driven. That is, although $\langle \hat{G}(t) \rangle_t$ may have a definite value, this results only from a possible coupling of $\hat{G}$ to the driven variable $\hat{F}$. The appropriate statistical operator is then

$$\hat{\rho}_t = \frac{1}{Z_t} \exp\left[-\beta \hat{H} + \int_0^t \lambda(t')\hat{F}(t')\, dt' + \int_0^t \gamma(t')\hat{G}(t')\, dt' \right] , \qquad (7\text{--}17)$$

under the constraints

$$\dot{\sigma}_F(t) = \lambda(t) K_{FF}^t(t,t) + \gamma(t) K_{FG}^t(t,t), \tag{7-18a}$$

$$0 = \lambda(t) K_{GF}^t(t,t) + \gamma(t) K_{GG}^t(t,t). \tag{7-18b}$$

Therefore, $\gamma(t)$ is actually determined by $\lambda(t)$, so that

$$\dot{\sigma}_F(t) = \lambda(t) K_{FF}^t(t,t) \left[1 - \frac{[K_{FG}^t(t,t)]^2}{K_{FF}^t(t,t) K_{GG}^t(t,t)} \right]. \tag{7-19}$$

It is both revealing and convenient to define an operator $\hat{F}'$ as the linear combination

$$\hat{F}'(t) \equiv \hat{F}(t) - \frac{K_{GF}^t(t,t)}{K_{GG}^t(t,t)} \hat{G}(t), \tag{7-20}$$

in terms of which Eq.(7-19) becomes

$$\dot{\sigma}_F(t) = \lambda(t) K_{F'F'}^t(t,t). \tag{7-21}$$

In the linear case, at least, the constraint merely alters the normalization of the source strength. More important, though, one uncovers a completely natural mechanism through which two or more dynamical variables may be coupled in the system, a phenomenon discussed extensively, for example, by Kawasaki (1970).

The preceding development provides a thermal analogy to the nonlinear theory of dynamical response. But thermal driving is far richer, and at this point admits an additional interpretation induced by the physical identification of the source term $\dot{\sigma}_F(t)$. We began by envisioning the expectation value $\langle \hat{F}(t) \rangle$ to be specified throughout the interval $0 \le t' \le t$, and added the observation that $\hat{F}$ was being driven. Rather, thermal driving was specified, though specific behavior of the source itself was not given. But with identification of the quantity $\dot{\sigma}_F$ the equations allow us to turn the situation around slightly and suppose it is the actual source term which is controlled. That is, we can specify $\dot{\sigma}_F$ and consider $\langle \hat{F}(t) \rangle$ to be predicted from a knowledge of the rate of driving provided by the source. Yet another scenario emerges, and is the one often realized experimentally.

As an example of this version of thermal driving let us consider the effects of space- and time-dependent heating on an ideal simple fluid. Generalization of the preceding equations to include spatial variation is essentially trivial and, although the region R can be arbitrary, we take it here to be the entire system volume V. If the energy density $\hat{h}(\mathbf{x},t)$ is the driven variable and the equilibrium system is uniform, then in linear approximation Eq.(7-16) becomes

$$\dot{\sigma}_h(\mathbf{x},t) = \int \lambda_h(\mathbf{x}',t') K_{hh}(\mathbf{x}-\mathbf{x}',t{=}0)\, d^3x', \tag{7-22}$$

and the linear response of any other variable $\hat{F}$ is

$$\langle \hat{F}(\mathbf{x},t) \rangle_t = \langle \hat{F} \rangle_0 + \int_0^t dt' \int d^3x'\, \lambda_h(\mathbf{x}',t') K_{Fh}(\mathbf{x}-\mathbf{x}',t-t'). \tag{7-23}$$

It will be convenient to employ the notation $F(\mathbf{x},t) \equiv \langle \hat{F}(\mathbf{x},t)\rangle_t - \langle \hat{F}\rangle_0$ and to eliminate the volume integral by Fourier transformation of the spatial variable. In the present problem interest lies primarily with choosing $\hat{F}$ as either the number density $\hat{n}(\mathbf{x},t)$ or the energy density $\hat{h}(\mathbf{x},t)$, and we shall presume the Boltzmann limit to provide an adequate description. Thus, one obtains from Eqs.(6–148), (6–157), and (6–173) the relevant ideal gas covariance functions:

$$K^0_{nn}(\mathbf{k},t) = ne^{-kt^2/2\beta m}, \tag{7–24a}$$

$$K^0_{hh}(\mathbf{k},t) = \frac{1}{4\beta^2}\left[15 - 10\frac{k^2t^2}{\beta m} + \frac{k^4t^4}{(\beta m)^2}\right]K^0_{nn}(\mathbf{k},t), \tag{7–24b}$$

$$K^0_{hn}(\mathbf{k},t) = \frac{1}{\beta}\left[\frac{3}{2} - \frac{k^2t^2}{2\beta m}\right]K^0_{nn}(\mathbf{k},t), \tag{7–24c}$$

where $n = N/V$ is the equilibrium number density.

Pure heating implies that only the energy density is driven, and therfore under this constraint Eqs.(7–17)-(7–21) are appropriate. Thus, in the Fourier-transformed representation, and in linear approximation, we define the operator

$$\hat{h}'(\mathbf{k},t) \equiv \hat{h}(\mathbf{k},t) - \frac{K_{hn}(\mathbf{k},0)}{K_{nn}(\mathbf{k},0)}\hat{n}(\mathbf{k},t) = \hat{h}(\mathbf{k},t) - \tfrac{3}{2\beta}\hat{n}(\mathbf{k},t). \tag{7–25}$$

The heating is now fully described by the following set of equations:

$$\dot{\sigma}_h(\mathbf{k},t) = \lambda_h(\mathbf{k},t)K_{h'h'}(\mathbf{k},t{=}0), \tag{7–26}$$

$$h'(\mathbf{k},t) = \int_0^t \lambda_h(\mathbf{k},t')K_{h'h'}(\mathbf{k},t-t')\,dt', \tag{7–27}$$

$$n(\mathbf{k},t) = \int_0^t \lambda_h(\mathbf{k},t')K_{nh'}(\mathbf{k},t-t')\,dt', \tag{7–28}$$

in the notation introduced following Eq.(7–23).

A simple model of the driving mechanism, also considered by Mitchell (1967), consists of the instantaneous addition of an amount of thermal energy $\Delta h(\mathbf{k})$ to mode $\mathbf{k}$. This model describes a pulse source $\dot{\sigma}_h(\mathbf{k},t) = \delta(t)\Delta(\mathbf{k})$, and we find

$$h'(\mathbf{k},t) = \Delta(\mathbf{k})\left[1 - \frac{2}{3}\frac{k^2t^2}{\beta m} + \frac{1}{6}\frac{k^4t^4}{(\beta m)^2}\right]e^{-k^2t^2/2\beta m} \tag{7–29}$$

for the energy density, and

$$n(\mathbf{k},t) = -\frac{1}{3}\frac{k^2t^2}{m}\Delta h(\mathbf{k})e^{-k^2t^2/2\beta m} \tag{7–30}$$

for the number density. Hence, the number density builds up 180° out of phase with the initial heat pulse. A second example is discussed in Problem 7.2.

ENTROPY

An explicit expression for the theoretical entropy is obtained by substitution of Eq.(7–7) into Eq.(7–12):

$$\frac{1}{\kappa} S_t = \ln Z_t + \beta \langle \hat{H} \rangle_t - \int_0^t \lambda(t') \langle \hat{F}(t') \rangle_t \, dt' \, . \tag{7–31}$$

The time rate-of-change of entropy is then found by direct calculation,

$$\frac{1}{\kappa} \partial_t S_t = \beta \lambda(t) K_{HF}^t(t,0) - \lambda(t) \int \lambda(t') K_{FF}^t(t,t') \, dt' \, , \tag{7–32}$$

which contains no approximations. If we take $\hat{F} = \hat{H} \neq \hat{H}(t)$, for example, then

$$\partial_t S_t = \kappa \beta(t) \frac{d \langle \hat{H} \rangle_t}{dt} \, , \tag{7–33a}$$

with

$$\beta(t) \equiv \beta - \int_0^t \lambda(t') \, dt' \, . \tag{7–33b}$$

In the event that $\hat{H}$ *is* time dependent, a linear approximation yields

$$\frac{1}{\kappa} \partial_t S_t \simeq \beta \lambda(t) K_{HF} = \beta \frac{d}{dt} \langle \hat{H} \rangle_t \, , \tag{7–34}$$

so that in either case we obtain the *form* $dS_t = dQ/T$.

But what is the exact physical significance of these results? That is, so far we have been able to define entropy and temperature only for equilibrium states, and that is certainly not the reference for the functions S_t and $\beta(t)$. In fact, S_t is just the time-dependent version of the information-theoretic entropy S_I of Volume I, and we have only been able to relate this to the physical experimental entropy S_e at thermal equilibrium. Although Eq.(7–32) describes a rate of 'entropy production', it is not at all clear that this is a physically meaningful quantity—let alone observable.

In Eq.(1–56) we considered a quantity σ defined as a rate of entropy production in the phenomenological theories, and subsequently found it rather difficult to extend the definitions of equilibrium quantities to nonequilibrium states in a precise way. If entropy-production theories are to be of any utility whatsoever, a way must be found to eliminate the ambiguities inherent in attempts to define temperature and entropy in nonequilibrium states. One such approach was initiated twenty years ago by Tykodi (1967), who employed the analogy of the Joule-Thomson porous-plug experiment to examine processes whose initial and final states could be described in terms of equilibrium parameters. Then, for steady-state processes at least, the rate of entropy production can be defined in terms of stationary fluxes and well-defined thermodynamic potentials. Whether this approach can be adequately generalized to a complete nonequilibrium thermodynamics remains an open question, but it does demonstrate that there exist some situations in which the above ambiguities can be removed.

A second scenario by means of which the ambiguities associated with nonequilibrium states can be avoided deals simply with processes which actually begin and end in states of thermal equilibrium. The discussion is most readily carried out by means of a model based on the example given above. Let a homogeneous system in thermal equilibrium be described by the grand canonical ensemble, and suppose an external source to be coupled to the energy density such that the heating drives the system from equilibrium. Presume that only the source is controlled, so that in linear approximation the relevant equations of thermal driving are

$$\langle \hat{h}(\mathbf{x},t)\rangle - \langle \hat{h}\rangle_0 = \int_V d^3x' \int_0^t dt'\, \lambda(\mathbf{x}',t') K_{hh}(\mathbf{x}-\mathbf{x}',t-t')\,, \qquad (7\text{–}35a)$$

$$\dot{\sigma}_h(\mathbf{x},t) = \int_V \lambda(\mathbf{x}',t) K_{hh}(\mathbf{x}-\mathbf{x}',t{=}0)\, d^3x'\,. \qquad (7\text{–}35b)$$

After a well-defined period of driving the source is turned off at $t = t_1$, the system is again isolated, and we expect equilibrium to ensue. It is supposed that this is a reproducible experiment. We shall consider general relaxation processes in more detail in the following section, but the simple example being discussed here will serve several purposes.

Let us integrate over the system volume in Eqs.(7–35), which results in a conversion of the densities into total Hamiltonians. Moreover, owing to uniformity and cyclic invariance of the traces, the covariance function in Eq.(7–35a) is then independent of the time. Combination of the two equations for $t > t_1$, when the source has been removed, now yields

$$\langle \hat{H}(t)\rangle - \langle \hat{H}\rangle_0 = \int_0^{t_1} \dot{\sigma}_H(t')\, dt'\,, \quad t > t_1\,, \qquad (7\text{–}36)$$

which is actually independent of time. That is,

$$\langle \hat{H}\rangle = \langle \hat{H}\rangle_0 + \Delta E\,, \qquad (7\text{–}37)$$

and the total energy of the isolated system is now known.

Once again we are dealing with constants of the motion—to the best of our knowledge—so that we expect an equilibrium situation to ensue, with total energy given by Eq.(7–37). Maximization of the entropy subject to this *predicted* constraint for $t \gg t_1$ should yield a correct prediction for the new Lagrange multiplier β_1 in the new equilibrium state. The theoretical expression for the entropy is given by Eq.(7–31), which in the present case is

$$S = \kappa \ln Z + \kappa \beta_0 \langle \hat{H}\rangle$$
$$- \kappa \int_V d^3x' \int_0^{t_1} \lambda(\mathbf{x}',t') \langle \hat{h}(\mathbf{x}',t')\rangle\, dt'\,, \qquad (7\text{–}38)$$

and subscripts 0 will denote quantities referring to the original unperturbed system.
Now make a linear approximation, so that through $O(\lambda)$

$$S \simeq \kappa \ln Z_0 + \kappa \beta_0 \langle \hat{H} \rangle_0$$
$$+ \kappa \beta_0 \int d^3 x' \int_0^{t_1} \lambda(\mathbf{x}', t') K_{Hh}(\mathbf{x} - \mathbf{x}', t - t') \, dt' . \qquad (7\text{-}39)$$

But the covariance function contains the total Hamiltonian as one of its operators, so that K_{Hh} is actually independent of time (and space). Consequently, the change in theoretical entropy is

$$S - S_0 \simeq \kappa \beta_0 K_{Hh} \int_V d^3 x' \int_0^{t_1} \lambda(\mathbf{x}', t') \, dt' . \qquad (7\text{-}40)$$

Note, however, that S can not be related to the experimental (i.e., physical) entropy unless the final state actually refers to an equilibrium state. If it does, then differentiation of the entropy in this state with respect to the new energy yields the predicted temperature for the final equilibrium state:

$$\beta_1 \simeq \beta_0 \left[1 - \langle \hat{h} \rangle_0 \int_V d^3 x' \int_0^{t_1} \lambda(\mathbf{x}', t') \, dt' \right] . \qquad (7\text{-}41)$$

Whether or not the final state is indeed one of equilibrium can be confirmed only from measurement, or from a rigorous prediction of relaxation. We return to this question presently.

RELATION TO DYNAMICAL DRIVING

An important feature of the thermal driving formalism is that the actual details of the driving source are irrelevant. Only the *rates* at which system variables are driven by the external forces enter the equations. As a consequence, it should make no difference whether the driving is thermal or, in fact, mechanical.

In Chapter 5 we considered a problem in which the two approaches led to equivalent results, but the comparison was essentially restricted to conditions of relaxation. Envision now the same problem, but let the driving be continuous under the separate scenarios of thermal and mechanical driving, and employ the same notation. If the two approaches are to yield equivalent predictions, then we must have in linear approximation

$$J(t) = \int_0^t \phi_{JF}(t - t') v(t') \, dt' = \int_0^t \lambda(t') K_{JF}(t - t') \, dt' . \qquad (7\text{-}42)$$

The identity (5–70) leads to the conclusion that

$$\lambda(t') = \beta \left[\delta(t - t') - \frac{d}{dt'} \right] v(t') = \frac{\dot{\sigma}_F(t)}{K_{FF}(0)} , \qquad (7\text{-}43)$$

which also identifies the source term for thermal driving. If the driving is suddenly removed in step-function fashion at $t = t_0$, so that $v(t')$ is replaced by $v(t')\theta(t_0 - t')$, then Eq.(7–43) introduces the anomalous factor of 2 discussed in connection with Eq.(5–78).

This discussion demonstrates that one can always find a thermal driving term which will produce the same ensemble as a dynamical driving term. The converse, however, is not true, for the statistical operator $\hat{\rho}(t)$ of dynamic response theory evolves unitarily, whereas $\hat{\rho}_t$ does not. It is seen from Eq.(7–43) that an external field has the same *effect* as a source term, and so one can always identify a pseudo-Hamiltonian and make the two descriptions *look* completely equivalent. But without Eq.(7–43) there is no unambiguous way to choose such a pseudo-Hamiltonian, and one must generally adopt a very specific model in order to make sense of the resulting theory. Moreover, if a number of variables are driven the identification involves complicated linear combinations of Heisenberg operators, a complexity which arises owing to the greater degree of *microscopic* control presumed by dynamical driving. As might be expected, *most* of this complexity disappears in linear approximation.

B. Relaxation Processes

In the relaxation model described above one gains a strong feeling that the final state for $t > t_1$ is indeed one of equilibrium. Barring actual measurement, however, this view can only be supported by means of a detailed theory demonstrating that all relevant quantities are constants of the motion in the final state. For example, a prediction of relaxation would include expressions asserting that the number density is essentially uniform and that all currents have ceased in that state. But these observations follow from the linear prediction equations of that model:

$$\langle \hat{n}(\mathbf{x},t) \rangle - \langle \hat{n} \rangle_0 = \int_V d^3x' \int_0^{t_1} \lambda(\mathbf{x}',t') K_{nh}(\mathbf{x}-\mathbf{x}',t-t')\, dt', \qquad (7\text{–}44\text{a})$$

$$\langle \hat{\mathbf{q}}(\mathbf{x},t) \rangle = \int_V d^3x' \int_0^{t_1} \lambda(\mathbf{x}',t') K_{qh}(\mathbf{x}-\mathbf{x}',t-t')\, dt', \qquad (7\text{–}44\text{b})$$

valid for $t > t_1$. Relaxation is completely dependent on the behavior of the covariance functions, whose structure will also yield expressions for the relaxation times and correlation lengths. Although many correlation functions should decay to zero, such as those in Eqs.(7–44), some will only approach a constant. An example of the latter behavior is evident in Eq.(7–39), where the covariance function is actually independent of time. This behavior, of course, is clearly consistent with the physical interpretation of that equation, but nevertheless serves to warn against dogmatic stances regarding decays of correlations.

Let us now generalize this scenario in a formal way so as to construct the rudiments of a theory of relaxation. We consider a simple fluid, as usual, and drive the system explicitly from equilibrium to a well-defined initial nonequilibrium state. Let the driving interval be $[-\infty, 0]$, so that the driving mechanism is removed at $t = 0$ and we expect the system to relax. Essentially the same problem has already been considered in a more general form in Chapter 5, where the driving was not specified. For simplicity we shall focus on a single locally-conserved density $\hat{e}(\mathbf{x}, t)$ and consider both dynamical and thermal driving, beginning with the former.

As discussed earlier, dynamical response theory has been employed to simulate a thermal source by constructing a pseudo-Hamiltonian through addition to $\hat{H}_0$ of a term of the form

$$\int \hat{e}(\mathbf{x}', t) v(\mathbf{x}', t) \, d^3x', \tag{7-45}$$

despite its intrinsic ambiguity. The system is taken to be in equilibrium in the infinite past when the force is turned on adiabatically in the form

$$v(\mathbf{x}, t) = v_0(\mathbf{x}) e^{\epsilon t}, \quad \epsilon > 0, \tag{7-46}$$

such that $\epsilon \to 0^+$ at the end of the calculation. The linear response for $t \geq 0$ is then

$$e(\mathbf{x}, t) \equiv \langle \hat{e}(\mathbf{x}, t) \rangle - \langle \hat{e} \rangle_0$$
$$= \int d^3x' \int_{-\infty}^{0} \phi(\mathbf{x}-\mathbf{x}', t-t') v(\mathbf{x}', t') \, dt', \tag{7-47}$$

where here there is no ambiguity in omitting subscripts on the linear response function $\phi_{ee} \equiv \phi$.

At $t=0$ the initial nonequilibrium state can be described by

$$e(\mathbf{x}, t=0) = \int d^3x' \, v_0(\mathbf{x}') \int_0^{\infty} \phi(\mathbf{x}-\mathbf{x}', t') e^{-\epsilon t'} \, dt', \tag{7-48}$$

which is presumed reproducible. Let us carry out a spatial Fourier transform, and also introduce the notation of Eq.(3–35), or (6–64), for the Fourier-Laplace time transform:

$$e(\mathbf{k}, t=0) = v_0(\mathbf{k}) \int_0^{\infty} dt' \, \phi(\mathbf{k}, t') e^{-\epsilon t'}$$
$$= v_0(\mathbf{k}) \tilde{\phi}(\mathbf{k}), \tag{7-49a}$$

where

$$\tilde{\phi}(\mathbf{k}) \equiv \lim_{z \to 0} \tilde{\phi}(\mathbf{k}, z)\big|_{z=i\epsilon} = \int_{-\infty}^{\infty} \frac{d\omega}{2\pi} \frac{\chi''(\mathbf{k}, \omega)}{\omega}, \tag{7-49b}$$

and χ'' was defined in Eq.(3–31)—$i\chi''$ is the imaginary part of the full Fourier transform of the linear response function, and is real. For $t > 0$ we can also write [e.g., Eq.(6–68)]

$$\tilde{e}(\mathbf{k}, z) = \int_0^{\infty} dt \, e^{izt} \int_{-\infty}^{0} d\tau \, \phi(\mathbf{k}, t-\tau) e^{\epsilon\tau} \, v_0(\mathbf{k})$$
$$= \int_{-\infty}^{\infty} \frac{d\omega}{i\pi} \frac{\chi''(\mathbf{k}, \omega)}{\omega(\omega - z)} \, v_0(\mathbf{k})$$
$$= \frac{1}{iz} [\tilde{\phi}(\mathbf{k}, z) - \tilde{\phi}(\mathbf{k}, i0)] v_0(\mathbf{k})$$
$$= \beta \tilde{K}_{ee}(\mathbf{k}, z) v_0(\mathbf{k}). \tag{7-50}$$

By combining Eqs.(7–49a) and (7–50) we have succeeded in formulating the relaxation model as an initial-value problem:

$$\tilde{e}(\mathbf{k}, z) = \frac{\tilde{K}_{ee}(\mathbf{k}, z)}{\beta^{-1}\tilde{\phi}(\mathbf{k})} e(\mathbf{k}, t=0) , \qquad (7\text{–}51)$$

and in the hydrodynamic limit the denominator becomes a thermodynamic derivative.

The usual argument made at this point is that the last expression is correct regardless of how the initial state was constructed. That is, it is independent of the character of the driving force, and therefore describes relaxation from any nonequilibrium state. By reformulating the problem in terms of thermal driving we are able to demonstrate that this argument is essentially correct, at least in linear approximation, despite its conceptual weakness.

Thermal driving describes the preceding scenario in linear approximation by means of the equations

$$C(\mathbf{x}, t) = \int d^3x' \int_{-\infty}^{t} dt' \, \lambda(\mathbf{x}', t') K_{Ce}(\mathbf{x}-\mathbf{x}', t-t') , \qquad (7\text{–}52a)$$

$$\dot{\sigma}_e(\mathbf{x}, t) \equiv \partial_t \langle \hat{e}(\mathbf{x}, t) \rangle_t - \langle \partial_t \hat{e}(\mathbf{x}, t) \rangle_t$$

$$= \int d^3x' \, \lambda(\mathbf{x}', t) K_{ee}(\mathbf{x}-\mathbf{x}', t=0) , \qquad (7\text{–}52b)$$

where $\hat{e}$ is driven and $C(\mathbf{x}, t) \equiv \langle \hat{C}(\mathbf{x}, t) \rangle_t - \langle \hat{C} \rangle_0$ is the thermal response of any other variable. The view is that we specify the source term up to the time $t = 0$, and then remove it. Although there is no mention of changing a Hamiltonian in this description, we know the system is again closed for $t > 0$ and is then indeed described completely by a Hamiltonian. Note that the expectation values in the first line of Eq.(7–52b) refer to the statistical operator for thermal driving,

$$\hat{\rho}_t = \frac{1}{Z_t} \exp\left[-\beta \hat{\mathrm{H}} + \int d^3x' \int_{-\infty}^{t} \lambda(\mathbf{x}', t') \hat{e}(\mathbf{x}', t') \, dt'\right] , \qquad (7\text{–}53)$$

whereas the second line is a linear approximation. Moreover, the upper limit of integration in this last expression *becomes and remains* zero for $t \geq 0$, which therefore defines the relaxation mechanism. Hence, the notation $\hat{\rho}_0$ does *not* refer here to an equilibrium state, but to the initial nonequilibrium ensemble. This point is so important that it warrants further digression.

There is a subtle and persistent conceptual difficulty in nonequilibrium statistical mechanics which is only rarely addressed explicitly (e.g., Jaynes, 1982). Given a Hamiltonian system in an initial nonequilibrium state described by $\hat{\rho}_0$, we believe on the one hand that if it is left to itself the statistical operator will evolve unitarily to

$$\hat{\rho}(t) = e^{-i\hat{\mathrm{H}}\,t/\hbar}\,\hat{\rho}(0)e^{i\hat{\mathrm{H}}\,t/\hbar} .$$

On the other hand, because there are no external forces, we also believe that the system will relax to an equilibrium state described by the canonical statistical operator $\hat{\rho}_{can} \propto \exp(-\beta\,\hat{H})$, where now β is a parameter of this final equilibrium state. But under unitary transformation each eigenvalue of $\hat{\rho}(t)$ is a constant of the motion, so that unless the eigenvalues of $\hat{\rho}(0)$ are identical to those of $\hat{\rho}_{can}$ there is no mathematical way that $\hat{\rho}(0)$ can evolve into $\hat{\rho}_{can}$. Rather, the best we can hope to achieve is a demonstration that the physical predictions of macroscopic quantities made by $\hat{\rho}_0$ in our model will eventually be the same as those predicted by the equilibrium ensemble describing the final state to which the system relaxes. As we shall see, even this goal remains a difficult one.

Because $\hat{e}(\mathbf{x},t)$ is a locally conserved density it satisfies a local continuity equation:

$$\partial_t \hat{e}(\mathbf{x},t) + \nabla \cdot \hat{\mathbf{J}}(\mathbf{x},t) = 0, \tag{7-54}$$

irrespective of the state of the system. When this identity is substituted into the right-hand side of the first line of Eq.(7–52b), and we note from Eq.(7–53) that the entire $\mathbf{x}$-dependence of the expectation value is just that of the operator $\hat{e}(\mathbf{x},t)$, we obtain immediately the *macroscopic* conservation law:

$$\partial_t \langle \hat{e}(\mathbf{x},t) \rangle_t + \nabla \cdot \langle \hat{\mathbf{J}}(\mathbf{x},t) \rangle_t = \dot{\sigma}(\mathbf{x},t), \tag{7-55}$$

which is completely nonlinear. If the driving is removed at $t=0$ one can *not* simply set the right-hand side to zero in order to obtain the corresponding equation for the relaxation mode, for a review of the thermal-driving scenario indicates that Eq.(7–52b) is no longer valid at this point. But the microscopic equation (7–54) is still valid during relaxation, and the relevant statistical operator is $\hat{\rho}_0$, describing the initial nonequilibrium state from which the system is relaxing. This is obtained from Eq.(7–53), again independent of $\mathbf{x}$, and so by taking expectation values of Eq.(7–54) we find the conservation law valid during relaxation:

$$\partial_t \langle \hat{e}(\mathbf{x},t) \rangle_0 + \nabla \cdot \langle \hat{\mathbf{J}}(\mathbf{x},t) \rangle_0 = 0. \tag{7-56}$$

Again note that the subscripts do *not* refer to equilibrium, but to the ensemble $\hat{\rho}_{t=0}$ given by Eq.(7–53).

The conservation laws are, in fact, the equations of motion of the system. In the following chapter we shall return to these in order to obtain the relevant equations of hydrodynamics.

Let us now return to the relaxation scenario, in which the Fourier-transformed predictions for $\hat{e}$ and $\hat{\mathbf{J}}$ for $t > 0$ follow from Eq.(7–52a):

$$e(\mathbf{k},t) = \int_t^\infty K_{ee}(\mathbf{k},t')\lambda(\mathbf{k},t-t')\,dt', \tag{7-57}$$

$$\mathbf{J}(\mathbf{k},t) = \int_t^\infty K_{Je}(\mathbf{k},t')\lambda(\mathbf{k},t-t')\,dt'. \tag{7-58}$$

Both these equations, as well as the conservation law (7–56), are now continued into the complex plane by means of the Fourier-Laplace transform and the convolution theorem. Thus,

$$\tilde{e}(\mathbf{k}, t=0) + iz\tilde{e}(\mathbf{k}, z) = i\mathbf{k} \cdot \tilde{\mathbf{J}}(\mathbf{k}, z), \qquad (7\text{--}59)$$

$$\tilde{e}(\mathbf{k}, z) = \tilde{K}_{ee}(\mathbf{k}, z)\tilde{\lambda}(\mathbf{k}, -z), \qquad (7\text{--}60)$$

$$\tilde{\mathbf{J}}(\mathbf{k}, z) = \tilde{K}_{\mathbf{J}e}(\mathbf{k}, z)\tilde{\lambda}(\mathbf{k}, -z). \qquad (7\text{--}61)$$

The Lagrange multiplier is then eliminated between Eqs.(7–60) and (7–61) and substitution into Eq.(7–59) for $\tilde{\mathbf{J}}$ yields

$$\tilde{e}(\mathbf{k}, z) = \tilde{e}(\mathbf{k}, t=0)\frac{\tilde{K}_{ee}(\mathbf{k}, z)}{i\mathbf{k} \cdot \tilde{K}_{\mathbf{J}e}(\mathbf{k}, z) - iz\tilde{K}_{ee}(\mathbf{k}, z)}. \qquad (7\text{--}62)$$

As in Eq.(6–62), the local conservation law implies that

$$\partial_t K_{eB}(\mathbf{x}, t) = -\nabla \cdot K_{\mathbf{J}B}(\mathbf{x}, t), \qquad (7\text{--}63a)$$

for arbitrary $\hat{B}$, and this can be transformed into

$$K_{eB}(\mathbf{k}, t=0) + iz\tilde{K}_{eB}(\mathbf{k}, z) = i\mathbf{k} \cdot \tilde{K}_{\mathbf{J}B}(\mathbf{k}, z). \qquad (7\text{--}63b)$$

Hence, in the present case,

$$i\mathbf{k} \cdot K_{\mathbf{J}e}(\mathbf{k}, z) = iz\tilde{K}_{ee}(\mathbf{k}, z) + K_{ee}(\mathbf{k}, t=0), \qquad (7\text{--}64)$$

so that Eq.(7–62) can now be rewritten as

$$\tilde{e}(\mathbf{k}, z) = \frac{\tilde{K}_{ee}(\mathbf{k}, z)}{K_{ee}(\mathbf{k}, t=0)}e(\mathbf{k}, t=0). \qquad (7\text{--}65)$$

(The appropriate results in the case that we had desired to predict a different density—e_1, say—are studied in Problem 7.3.)

Equation (7–65) is the general relaxation equation for this scenario, and implies nothing about the specific form of driving mechanism. It is essentially the same result as found in Eq.(7–51), thereby verifying the latter. In particular, we note the relation

$$\begin{aligned} K_{ee}(\mathbf{k}, t=0) &= \int_{-\infty}^{\infty} \frac{d\omega}{2\pi} K_{ee}(\mathbf{k}, \omega) \\ &= \beta^{-1} \int_{-\infty}^{\infty} \frac{d\omega}{2\pi} \frac{\chi_{ee}''(\mathbf{k}, \omega)}{\omega} \\ &= \beta^{-1}\phi_{ee}(\mathbf{k}). \end{aligned} \qquad (7\text{--}66)$$

This model scenario must usually be expanded somewhat for application to actual processes, of course, in order to include several densities.

Finally, an inverse time transformation in Eq.(7–65) yields the more transparent form

$$e(\mathbf{k}, t) = \frac{K_{ee}(\mathbf{k}, t)}{K_{ee}(\mathbf{k}, 0)} e(\mathbf{k}, 0) . \tag{7–67}$$

Once more we observe a 'Markovian' flavor to this process, in that the linear response depends only on the parameters of the initial nonequilibrium state. The reason for this is evident, because the past history has been eliminated by eliminating λ from the final prediction.

TIME-DECAY OF COVARIANCE FUNCTIONS

Although the foregoing discussion provides a framework for a rigorous theory of relaxation, it is by no means complete. Eventually one must be able to describe the time decay in detail, and once again the need for a complete understanding of the structure of covariance functions emerges. Indeed, even to construct the above framework it was necessary to presume nontrivial asymptotic properties for the functions involved so as to carry out the integral transforms. In lieu of a comprehensive theory of covariance functions, which would reveal precisely how fast they decay, it is important to establish arguments suggesting that they do in fact decay under appropriate conditions. Once this is established it is then possible to demonstrate that the left-hand sides of expressions such as those in Eqs.(7–44) also decay to zero. For physical reasons $\lambda(t)$ must be bounded along the entire real axis—in fact, it vanishes outside the interval $[0, t_1]$. This assertion is probably true even in the vicinity of critical points. One can then appeal to Tauberian theorems and use the free-particle covariance functions as comparison kernels to demonstrate the expected decay of the nonequilibrium expectation values (e.g., Wiener, 1932). But the arguments depend critically on the asymptotic behavior of the covariance functions themselves, which in turn depends on their structure.

It is perhaps useful to summarize briefly the general situation regarding the asymptotic time behavior of correlation functions. For convenience we shall consider only the autocorrelation of an operator $\hat{A}$ in the form

$$K(t) \equiv K_{AA}(t) = \langle \overline{\delta\hat{A}} \, \delta\hat{A}(t) \rangle_0 , \tag{7–68}$$

where $\delta\hat{A} \equiv \hat{A} - \langle\hat{A}\rangle_0$, as usual, and spatial dependence is suppressed throughout the present discussion. (Subscripts 0 *will* refer to the equilibrium ensemble throughout the remainder of this chapter.) The Laplace transform will be denoted by

$$\overline{K}(s) \equiv \int_0^\infty e^{-st} K(t) \, dt , \tag{7–69}$$

which necessarily vanishes as the real variable $s \to \infty$. A well-known final-value theorem (e.g., Doetsch, 1974) states that

$$s\overline{K}(s) \xrightarrow[s \to 0]{} K(\infty) , \tag{7–70}$$

implying that $K(\infty)$ vanishes *unless* $\overline{K}(s)$ has a pole at $s = 0$. That is, if

$$\overline{K}(0) = \int_0^\infty K(t)\,dt < \infty\,, \tag{7-71}$$

we expect the covariance function to decay to zero for large t, whereas if $K(t)$ approaches a constant as $t \to \infty$ it is likely that $\overline{K}(s)$ possesses a singularity at the origin. A general proof , or disproof, of Eq.(7-71) does not seem to be available at this time. Various possible decay modes based on behavior of the Laplace transform are investigated further in Problem 7.4.

Consider now a finite system and rewrite Eq.(7-68) explicitly in a representation in which $\hat{H}$, and therefore $\hat{\rho}_0$, are diagonal. One finds that

$$K(t) = \sum_{m,n} \frac{1 - e^{-\beta\hbar\omega_{mn}}}{\beta\hbar\omega_{mn}} \langle n|\hat{\rho}_0|n\rangle\langle n|\delta\hat{A}|m\rangle$$

$$\times\, \langle m|\delta\hat{A}|n\rangle e^{i\omega_{mn}t}\,, \tag{7-72}$$

where $\omega_{mn} \equiv (E_m - E_n)/\hbar$, and E_n is an eigenvalue of $\hat{H}$. A more useful and obvious notation yields

$$K(t) = \sum_{m,n} F_{mn} e^{i\omega_{mn}t} \le K(0)\,, \tag{7-73}$$

where $F_{mn} \ge 0$. (Other quantum numbers needed to specify the states are included implicitly.) Equation (7-73) provides a Fourier-series representation for the finite system in terms of its discrete spectrum.

From this last expression one is able to see qualitatively how the decay of $K(t)$ arises. The function initially decays from its upper bound at $t = 0$ owing to a dynamical dephasing of the matrix elements under the unitary transformation. If the level spacing is small and there are many nonzero off-diagonal matrix elements in the sum, then the decay is fairly rapid—but not necessarily to zero. For finite systems this destructive interference can again become constructive after sufficiently long times, and the correlation function can return arbitrarily close to its initial value. This almost-periodic behavior of $K(t)$ is just the Poincaré recurrence of Chapter 1 (Poincaré, 1890).

One can observe this behavior explicitly by taking the Laplace transform in Eq.(7-73):

$$\overline{K}(s) = \sum_{m,n} \frac{F_{mn}}{s - i\omega_{mn}}\,. \tag{7-74}$$

Unless $F_{mn} = 0$ precisely where $\omega_{mn} \ne 0$, there is a series of poles along the imaginary axis, implying almost-periodic behavior. In the unlikely event that the above condition were met, $K(t)$ would equal $K(0)$ for all time.

If the spectrum of $K(t)$ were continuous, Eq.(7-73) would be replaced by the Fourier integral

$$K(t) = \int_{-\infty}^\infty e^{i\omega t} K(\omega)\,d\omega\,, \tag{7-75}$$

and the analysis changes considerably (Andrews, 1965). One need only show that $K(\omega)$ is absolutely integrable in order to invoke the Riemann-Lebesgue lemma and conclude that $K(t)$ vanishes as $t \to \infty$. But Andrews' argument that the spectrum for a finite system is always continuous, except in certain unphysical cases, is not entirely convincing. The spectrum is almost continuous for large systems, and it is certainly so in the infinite-volume limit.

A simple model illustrating these comments was developed by Schrödinger (1914), which consists of a ring of N equal mass points coupled by identical springs. Scalapino (1961) considered a similar system consisting of N exchange-coupled spins. In either case the covariance function takes the form $K(t) = |f(t)|^2$, and for finite N it is shown that $K(t)$ is almost-periodic. In the limit of a large system, however,

$$\lim_{N \to \infty} f(t) = J_0(\alpha t), \qquad (7\text{–}76)$$

where α is a constant, so that $K(t)$ decays as t^{-1}. The time of Poincaré recurrence is proportional to N^2, and at least this much time must elapse before $f(t)$ departs appreciably from its limiting value in Eq.(7–76). In a large system Poincaré recurrence is simply unimportant with respect to macroscopic quantities.

Ever since Maxwell's initial theory of relaxation it has been a common belief that correlations generally decay exponentially (Maxwell, 1867). Yet, a large number, if not a majority, of physical relaxation processes are observed to follow other forms of decay laws—a varied list of references is given in Ngai, *et al* (1983). There has never been developed a rigorous microscopic theory of relaxation rates, and we can now see why: such a theory requires detailed knowledge of the structure of covariance functions, and this is very difficult to come by analytically.

Because of these difficulties attention turned to computer simulations over 25 years ago, and a major surprise ensued when Alder and Wainwright (1967) discovered that in a molecular dynamics system of hard spheres the velocity autocorrelation function $\langle v_x(0)v_x(t) \rangle$ decayed as $t^{-3/2}$ in three dimensions. We have earlier illustrated the results obtained when using a more realistic Lennard-Jones potential in Figures 6–2 and 6–3. Although there seems to be essentially no experimental evidence for these persistent correlations, or 'long time tails', all computer 'experiments' continue to predict them, so that the need for a fundamental theoretical analysis is evident. Despite a number of claims to the contrary, such an analysis remains to be seen. Although a bit dated, the review of this situation by Pomeau and Résibois (1975) remains rather valuable.

Nevertheless, there are numerous theoretical arguments indicating that exponential decay of correlation functions is perhaps an exception rather than the rule. An immediate piece of evidence arises in dilute systems, where the dominant behavior might be thought to be given by the free-particle covariance functions. For example, from Eq.(6–148) in the Boltzmann limit we calculate

$$K_{nn}(\tau) \equiv \int \frac{d^3k}{(2\pi)^3} \int_{-\infty}^{\infty} \frac{d\omega}{2\pi} e^{-i\omega\tau} K_{nn}^0(\mathbf{k}, \omega)_{\mathrm{B}} = \frac{n}{4} \left(\frac{2\beta m}{\pi} \right)^{3/2} \frac{1}{\tau^3}. \qquad (7\text{–}77)$$

In Appendix B one notes that this behavior tends to persist through the leading-order terms of perturbation theory. More generally, on physical grounds there is little reason to believe that K_{nn} falls off very rapidly in either space or time. As we shall see in Chapter 9, the propagation of intelligible sounds depends on persistent long-range correlation of density fluctuations. Although this may not necessarily allow us to infer similar behavior for other covariance functions, it does alert us to the need for exercising great care in presuming any particular decay mode.

Ironically, claims to theoretical understanding of persistent correlations are very narrowly rooted in calculations on transport processes, involving the hydrodynamic limit (HL). Yet we shall see in the following chapter that a resolution of precisely how correlations decay may eventually emerge from a sharper definition and a deeper understanding of the HL itself.

Problems

7.1 Employ the principle of minimum cross-entropy, Eq.(7–8), to derive the statistical operator for thermal driving, Eq.(7–7).

7.2 In the pure heating model of Eqs.(7–22)-(7–28) suppose the driving mechanism to be that of steady continuous heating: $\dot{\sigma}_h(\mathbf{k}, t) = \Delta h(\mathbf{k})$. Calculate the resulting $h'(\mathbf{k}, t)$ and $n(\mathbf{k}, t)$ analogous to Eqs.(7–29) and (7–30).

7.3 Recall the thermal-driving/relaxation scenario of Eqs.(7–57)-(7–65), and suppose one had wished to predict the decay of another density e_1, rather than the driven e. Derive an expression for $\tilde{e}(\mathbf{k}, z)$ similar to that of Eq.(7–65).

7.4 By studying the behavior of the Laplace transform of the autocovariance function $K(t)$, Eq.(7–69), elaborate the possible decay modes for $K(t)$. In particular, as $s \to 0$ investigate the consequences if $\overline{K}(s)$ behaves as: (a) s^n; (b) B/s, B a constant; (c) $B\Gamma(\lambda + 1)/s^{\lambda+1}$, $\mathrm{Re}\lambda > -1$; (d) $B\Gamma(\lambda+1)/(s - s_0)^{\lambda+1}$, $\mathrm{Re}\lambda > -1$; (e) s^λ, λ a rational fraction. Examine the possibility that $K(t)$ has the asymptotic form t^{-n}, n a positive integer, by investigating the Laplace transform of $(t + \alpha)^{-n}$, $\alpha > 0$.

REFERENCES

Alder, B.J., and T.E. Wainwright: 1967, 'Velocity Autocorrelations for Hard Spheres', *Phys. Rev. Letters* **18**, 988.

Andrews, F.C.: 1965, 'On the Nature of the Spectrum of the Time Correlation Function Matrix for Stochastic Processes', *Proc. Natl. Acad. Sci. (U.S.A.)* **53**, 1284.

Doetsch, G.: 1974, *Introduction to the Theory and Application of the Laplace Transformation*, Springer-Verlag, New York.

Jaynes, E.T.: 1982, 'Book Review', *Physics Today*, August, p.57.

Maxwell, J.C.: 1867, 'On the Dynamical Theory of Gases', *Phil. Trans. Roy. Soc. London* **157**, 49.

Mitchell, W.C.: 1967, 'Statistical Mechanics of Thermally Driven Systems', *Ph.D. thesis*, Washington Univ. (unpublished).

Ngai, K.L., A.K. Rajagopal, R.W. Rendell, and S. Teitler: 1983, 'Paley-Wiener Criterion for Relaxation Functions', *Phys. Rev. B* **28**, 6073.

Poincaré, H.: 1890, 'Sur le problème des trois corps et les équations de la dynamique', *Acta math.* **13**, 1.

Pomeau, Y., and P. Résibois: 1975, 'Time Dependent Correlation Functions and Mode-Mode Coupling Theories', *Phys. Repts.* **19**, 63.

Robertson, B., and W.C. Mitchell: 1971, 'Equations of Motion in Nonequilibrium Statistical Mechanics. III.Open Systems', *J. Math. Phys.* **12**, 563.

Scalapino, D.J.: 1961, 'Irreversible Statistical Mechanics and the Principle of Maximum Entropy', *Ph.D. thesis*, Stanford Univ. (unpublished).

Schrödinger, E.: 1914, 'Zur Dynamik elastisch gekoppeter Punktsysteme', *Ann. Phys.* **44**, 916.

Shore, J.E., and R.W Johnson: 1980, 'Axiomatic Derivation of the Principle of Maximum Entropy and the Principle of Minimum Cross-Entropy', *IEEE Trans. Inf. Th.* **IT-26**, 26.

Tricomi, F.G.: 1957, *Integral Equations*, Interscience, New York.

Tykodi, R.J.: 1967, *Thermodynamics of Steady States*, Macmillan, New York.

Wiener, N.: 1932, 'Tauberian Theorems', *Ann. Math.* **33**, 1.

Chapter 8

Transport Processes
and
Hydrodynamics

In this chapter and the next we provide a detailed application of the linear approximation to simple fluids, a choice anticipated by the context in which covariance functions were examined earlier. Although these models have been studied extensively, the aim of the present development is to introduce some needed clarity and rigor, as well as to provide some new insights. Perhaps the most common example of near-equilibrium phenomena is provided by linear hydrodynamics, which is taken to be valid in the so-called hydrodynamic limit discussed earlier. This model has the considerable merit of possessing a well-developed macroscopic theory, which facilitates identification of quantities generated by the microscopic theory with those observed experimentally. In other words, one can readily make a physical identification of the Lagrange multipliers in this case and thus treat them as the independent variables of the model. Thus, we shall recall briefly, and then continue, the analysis of the conventional approach to these problems set out in Chapter 1.

The study of most physical problems, both experimentally and theoretically, consists of perturbing some characteristic quantity of the system and observing the response. If the perturbation is small enough, the response is usually linear (but not always). As noted in Chapter 1, a long history of such studies in the case of simple fluids has provided a number of phenomenological dissipation laws. For example, a small perturbation of the local number density produces a particle current, a response known as Fick's law of diffusion; when the local energy density is disturbed there results a heat flow and the ensuing relationship is known as Fourier's law of thermal conduction. Perturbation of the momentum density leads to a somewhat more complicated phenomenological law describing, among other things, internal friction in the medium. The essential feature of all these laws is that they relate the local currents to gradients of macroscopic thermodynamic quantities, and the proportionality factors are called transport coefficients.

In a simple fluid the characteristic densities are those of number, energy, and momentum. Let us denote any one of these generically by $e(\mathbf{x}, t)$, and its associated current by $\mathbf{J}(\mathbf{x}, t)$. The phenomenological transport laws, or *constitutive relations* all take the form

$$\mathbf{J} = -D(\nabla \lambda) \,, \tag{8–1}$$

relating a flux to a macroscopic gradient by means of a phenomenological transport coefficient D. In turn, the density and current are related through equations of motion, or continuity equations,

$$\partial_t e(\mathbf{x}, t) + \nabla \cdot \mathbf{J}(\mathbf{x}, t) = 0, \tag{8-2}$$

which are just reflections of conservation laws. Combination of Eqs.(8–1) and (8–2) for all the relevant densities, along with some physical arguments involving 'local equilibrium' thermodynamics, yields the conventional coupled equations of linear hydrodynamics (e.g., Landau and Lifshitz, 1959). Depending on the complexity of the processes under study, these are usually referred to as the Euler or Navier-Stokes equations.

The role of statistical mechanics, presumably, is to derive the above macroscopic equations of motion and constitutive relations from microscopic physics. In addition, one would like to express the phenomenological transport coefficients in terms of the physical behavior of the microscopic constituents of the system. The pursuit of these goals has been underway for over 100 years.

The key to understanding the preceding phenomenological picture can be elucidated by first noting that when a many-body system is perturbed from thermal equilibrium the resulting internal situation is one of considerable chaos. In a fluid containing a very large number of particles there is a corresponding large number of degrees of freedom. If the system is allowed to relax, most of these degrees of freedom return rather quickly to their equilibrium values in ways determined by the microscopic characteristics of the system. One observes, however, that this relaxation can be described macroscopically by only a few long-lived modes which decay relatively slowly; in fact, they decay with relaxation times proportional to some power of their wavelengths. Because mechanical systems usually have as many normal modes of oscillation as they have degrees of freedom, the puzzle as to why only a few of these modes actually are manifest in a many-body system has sometimes been called *Hilbert's paradox*.

Some reflection leads to the realization that these long-lived modes are related to the existence of only a few locally-conserved quantities in the macroscopic system, primarily because the number of microscopic conservation laws is limited. Because they satisfy local conservation laws, local excesses of these quantities can disappear neither locally nor quickly, but must relax by spreading out over the entire system. Classical hydrodynamics describes these long-lived modes by exploiting the fact that they vary slowly in space and time, so that each small portion of the system can be considered almost in thermal equilibrium—a concept found difficult to render precise.

In a simple fluid the only locally-conserved quantities are those of Eqs.(6–25)-(6–27): the number density, energy density, and three components of momentum density, resulting in five hydrodynamic modes. If we again represent any one of these densities by the generic symbol $\hat{e}(\mathbf{x}, t)$, the five local conservation laws all have the form

$$\partial_t \hat{e}(\mathbf{x}, t) + \nabla \cdot \hat{\mathbf{J}}((\mathbf{x}, t) = 0, \tag{8-3}$$

which are now *microscopic operator* equations. In order to maintain some degree of simplicity and focus on the essential arguments, we shall carry out much of the ensuing discussion in terms of these generic quantities.

The conventional development now proceeds by means of a number of presumptions, which we outline in sequence. The purpose is to isolate these essential features as clearly as possible.

1. A presumption of *local equilibrium* is made which, although consisting of several physical ideas, amounts to the inference that Eq.(8–3) is also valid as a macroscopic law. That is,

$$\partial_t \langle \hat{e}(\mathbf{x}, t) \rangle = -\nabla \cdot \langle \hat{\mathbf{J}}(\mathbf{x}, t) \rangle, \tag{8–4}$$

where the microscopic operators are replaced by (generally not-clearly-defined) nonequilibrium expectation values representing observable quantities. This is sometimes referred to as a 'regression of fluctuations', and implied is a restriction to small wavenumbers and low frequencies.

2. Macroscopic *gradient expansions* of the currents are presumed to exist, in analogy with the phenomenological laws of Eq.(8–1). In linear approximation,

$$\langle \hat{\mathbf{J}}(\mathbf{x}, t) \rangle = -D\nabla \langle \hat{e}(\mathbf{x}, t) \rangle, \tag{8–5}$$

and the right-hand side contains a further use of the local equilibrium idea. One now combines Eqs.(8–4) and (8–5) to obtain a closed equation of motion:

$$\partial_t \langle \hat{e}(\mathbf{x}, t) \rangle = D\nabla^2 \langle \hat{e}(\mathbf{x}, t) \rangle. \tag{8–6}$$

At this point it is recognized that the theory is not quite broad enough to describe the full range of hydrodynamic response. Therefore, one replaces D with a nonlocal, memory-retaining function $D(\mathbf{x} - \mathbf{x}', t - t')$, in an *ad hoc* way, and integrates the right-hand side of Eq.(8–6) over the primed variables. The solution to this equation is then effected by means of the Fourier-Laplace transform techniques discussed previously. With the notation $e \equiv \langle \hat{e} \rangle - \langle \hat{e} \rangle_0$, the deviation from the equilibrium value, we find the dispersion law

$$\tilde{e}(\mathbf{k}, z) = \frac{ie(\mathbf{k}, t=0)}{z + ik^2 \tilde{D}(\mathbf{k}, z)}. \tag{8–7}$$

Owing to the presumption of local equilibrium, this equation is valid for small k and z, and incorporates a belief that the disturbance must disappear at long distances and after long times. Note that the hydrodynamic mode is here made manifest by the pole in Eq.(8–7). Application of the inverse frequency transformation yields

$$e(\mathbf{k}, t) = e^{-k^2 Dt} e(\mathbf{k}, 0), \tag{8–8}$$

exhibiting the ubiquitous exponential attenuation. One infers a lifetime for this mode of $\tau(k) = (k^2 D)^{-1}$.

3. The scenario of *linear dynamic response* is introduced in order to provide the mechanism which drives the system from equilibrium. In the past this step has traditionally been taken in terms of the Boltzmann equation, or a master equation, but the more modern technique is to employ a pseudo-Hamiltonian in order to simulate the perturbation. As already discussed in the preceding chapter, in linear approximation it matters little from a practical point of view, because the external force does not appear in the final result. One finds for the response, from Eq.(7–51),

$$\tilde{e}(\mathbf{k}, z) = \frac{\tilde{K}_{ee}(\mathbf{k}, z)}{K_{ee}(\mathbf{k}, t=0)} \, e(\mathbf{k}, t=0) \,, \tag{8–9}$$

conveniently written here in terms of density-density covariance functions. This equation is valid for *all* k and z.

Comparison of Eqs. (8–7) and (8–9) allows one to eliminate the densities and obtain a relation between $\tilde{K}_{ee}$ and $\tilde{D}$. Through a somewhat ingenious coupling of a number of ideas one has succeeded in relating the dissipative transport coefficients to functions describing the absorptive microscopic processes in the system. By carrying out the appropriate mathematics on the real axis and employing the hydrodynamic limit one readily extracts the phenomenological transport coefficient:

$$D = \lim_{\omega \to 0}\left[\lim_{k \to 0} \frac{\omega}{k^2}\chi''(\mathbf{k}, \omega)\right], \tag{8–10}$$

in terms of the dissipative part of the linear response function. Note that the order of the limiting processes can *not* be interchanged. That is, one must first take the hydrodynamic limit in order to eliminate all the rapidly-decaying modes, a procedure to which we shall return presently for further discussion. By use of conservation laws and the fluctuation-dissipation theorem one can rewrite the above expression for D in terms of current-current correlation functions. These yield the so-called Kubo formulas, and extrapolations away from the hydrodynamic limit can be made through a more careful study of the function $\tilde{D}(\mathbf{k}, z)$. Finally, the complete description is obtained by considering the coupled equations for all five conserved densities.

The particular approach to hydrodynamics outlined here was pioneered by Kadanoff and Martin (1963), then extended and summarized in some detail by Puff and Gillis (1968), and Forster (1975). Although the preceding discussion was designed to expose the conceptual weaknesses in this approach, it must be admitted that the theory leads to definite expressions for transport coefficients and a physical description of the dissipative hydrodynamic processes. The five microscopic local conservation laws predict five macroscopic hydrodynamic modes. In the hydrodynamic limit two of the solutions are propagating sound modes with dispersion relations of the form $\omega = \pm ck - ik^2\Gamma/2$, where time-reversal invariance guarantees that propagating modes occur in pairs. The remaining three solutions are of the form $\omega = -ik^2 D$, describing a thermal diffusion mode and two nonpropagating transverse shear waves. Although these are the only hydrodynamic modes in a simple fluid, others appear in more complicated systems when additional local conservation laws arise. For example, a molecular fluid with weak noncentral two-body interactions exhibits a long-lived collective mode associated with angular momentum fluctuations (Ailawadi, *et al*, 1971). The scope of hydrodynamics is further extended when it is realized that in some systems hydrodynamic modes can arise which are *not* related to continuity equations. In ordered systems a long-range order exists owing to a spontaneously broken symmetry, in which the stable state of the system does not possess the symmetry of the Hamiltonian. In these cases the so-called Goldstone modes emerge (Goldstone, 1961), and the resulting hydrodynamics is thus far richer than that of the simple fluid. For example, in superfluid helium one finds *two pairs* of propagating modes, corresponding to first and second sound.

In summary, the conventional approach to transport processes and hydrodynamics consists of a number of somewhat disparate theoretical contributions and *ad hoc* procedures. Moreover, actual equations of motion are simply taken to be those of the classical macroscopic theory described in Chapter 1. We now turn to the task of providing a unified and coherent microscopic foundation for the macroscopic theory.

A. Linear Transport Processes

When constructing a theory of transport processes one must realize that there exist various possible experimental arrangements, and a specific choice can affect some aspects of the theory. It may be convenient to define fluxes differently from one experimental situation to another, for example, owing to the freedom of choice as to what is measured as a density. Moreover, the experimenter must also choose a reference frame in which to measure currents, and that choice can again affect what are to be identified as transport coefficients in particular processes. In principle, each type of experiment defines a different transport coefficient, and for some experimental arrangements it may be neither sensible nor possible to identify an isolated coefficient. We shall see that, in fact, it makes more sense in general to emphasize the covariance functions themselves rather than transport coefficients.

It will be useful to examine two general procedures for the identification and study of transport coefficients. The first is based on the steady-state scenario of Section 5-A, because determinations of static transport coefficients are often carried out by observing stationary processes. Recall that if information is specified over a region $R(x)$ in the form of a time-independent expectation value $\langle \hat{F}(\mathbf{x}) \rangle$, then in linear approximation we have for any other operator $\hat{C}$

$$\langle \hat{C}(\mathbf{x}) \rangle - \langle \hat{C} \rangle_0 \simeq \int_R \lambda(\mathbf{x}') K_{CF}(\mathbf{x}-\mathbf{x}')\, d^3x'$$

$$- \lim_{\epsilon \to 0^+} \int_R d^3x' \int_0^\infty e^{-\epsilon t}\, \lambda(\mathbf{x}') K_{C\dot{F}}(\mathbf{x}-\mathbf{x}',t)\, dt\,, \quad (8\text{–}11)$$

with the help of Eqs.(5–17) and (5–23). The unperturbed system is presumed uniform in space and time.

For present purposes we are interested in the case that $\hat{F} = \hat{e}$, one of the locally-conserved densities in the system. Because the covariance function refers to the equilibrium ensemble, the microscopic conservation law of Eq.(8–3) can be employed and we immediately replace $K_{C\dot{e}}$ in Eq.(8–11) with $-\nabla' \cdot K_{C\mathbf{J}}$. An integration by parts yields

$$\langle \hat{C}(\mathbf{x}) \rangle - \langle \hat{C} \rangle_0 = \int_R \lambda(\mathbf{x}') K_{Ce}(\mathbf{x} - \mathbf{x}')\, d^3 x'$$

$$- \lim_{\epsilon \to 0^+} \int_0^\infty e^{-\epsilon t}\, dt \int_R \nabla'\lambda(\mathbf{x}') \cdot K_{C\mathbf{J}}(\mathbf{x} - \mathbf{x}', t)\, d^3 x'$$

$$+\, (\text{ surface term}). \qquad (8\text{–}12)$$

Although it can be calculated in principle, the surface term may or may not be of interest—in the study of the hydrodynamic behavior of a large system it usually is not. Therefore, *for the first time*, we introduce the infinite-volume limit so as to omit rigorously the surface term. This is merely a mathematically precise statement of just what is meant by a large system. Consistency now demands that we set $R = V$, the system volume, so that the initial data are presumed specified over all of V.

The vast majority of transport processes with which we experiment take place in the hydrodynamic region, where one presumes that macroscopic quantities vary slowly in space and time. Therefore, in the present context we *define* the hydrodynamic limit (HL) by asserting that, away from critical points, the short range of the correlation function in space and time and the slow variation of the gradient permit the latter to be extracted from the integrals and the static transport coefficient identified. Thus, for long-wavelength and low-frequency disturbances the HL allows us to rewrite Eq.(8–12) as the approximation

$$\langle \hat{C}(\mathbf{x}) \rangle - \langle \hat{C} \rangle_0 \simeq \int_V \lambda(\mathbf{x}') K_{Ce}(\mathbf{x} - \mathbf{x}')\, d^3 x'$$

$$- \nabla\lambda \cdot \lim_{\epsilon \to 0^+} \int_0^\infty e^{-\epsilon t}\, dt \int_V K_{C\mathbf{J}}(\mathbf{x} - \mathbf{x}', t)\, d^3 x'. \qquad (8\text{–}13)$$

The quantity multiplying the gradient of the Lagrange multiplier is said to be a *linear transport coefficient*, here identified by means of a series of very definite approximations.

Two features of this last expression are to be emphasized. First, the left-hand side of Eq.(8–13) can in no way depend on the time, for the physical process is stationary. Presently this will emerge as a rather useful observation. Second, one observes that the linear approximation has automatically provided the first term in a nonlocal *gradient expansion*—we shall see in Section C below an example of how second-order nonlinear terms are included. In the present context the Lagrange-multiplier function can be identified physically as an independent variable of the theory, although it could be determined rigorously in principle from the integral equation provided by Eq.(8–11) if we set $\hat{C} = \hat{F}$. Hence, several *ad hoc* features of the conventional approach are eliminated immediately.

Quite often, and more generally, one does not observe a steady density. Rather, a time variation $\partial \hat{e} / \partial t$ is inferred from measurements over a space-time information-gathering region $R(\mathbf{x}, t)$, and in linear approximation the expectation value of another operator $\hat{C}(\mathbf{x}, t)$ is predicted to be

$$\langle \hat{C}(\mathbf{x}, t) \rangle - \langle \hat{C} \rangle_0 = \int_R \lambda(\mathbf{x}', t') K_{C\hat{e}}(\mathbf{x}, t; \mathbf{x}', t') \, d^3 x' \, dt' . \tag{8–14}$$

One proceeds exactly as above in order to identify a transport coefficient, and in the HL we find

$$\langle \hat{C}(\mathbf{x}, t) \rangle - \langle \hat{C} \rangle_0 \simeq \nabla \lambda \cdot \int_V K_{C\mathbf{J}}(\mathbf{x} - \mathbf{x}'; t - t') \, d^3 x' \, dt' . \tag{8–15}$$

Although the coefficient multiplying the gradient is formally space and time dependent, homogeneity usually removes these dependencies.

Let us hasten to point out that one need *not* take the step leading to Eqs.(8–13) and (8–15), the HL. Equation (8–12), for example, provides a completely rigorous description of the transport process in linear approximation and is surely perfectly well behaved. Further approximation is carried out at possible risk to rigor, but we shall continue to invoke the HL because it is the conventional limit of hydrodynamics. We make the point here, and return to it below, because it is not clear that the HL is mathematically well defined in general.

The next order of business is to apply these expressions to some familiar transport scenarios. For the remainder of this section we shall focus primarily on stationary processes, so that Eq.(8–13) is relevant. In addition, we shall presume that all currents are measured with respect to a fixed laboratory frame.

PARTICLE DIFFUSION

When density gradients exist across macroscopic regions of a system one expects on intuitive grounds that; in the absence of special constraints, particle currents will ensue. Generally the resulting process is an interdiffusion of one species through others in at least a two-component system. In this simplest case a diffusion current appears for each species and two equations are necessary for a complete description. But elementary conservation laws imply that the currents are equal and opposite, so that the hydrodynamic diffusion coefficient is the same for each and one need only consider a single equation. This situation is often regarded as a model for self diffusion, which is not observable in the strict sense except in the limit of infinite dilution of one species. The concept of self diffusion is useful, however, and allows one to employ the Nernst-Einstein relation in the form of Eq.(1–21) in general discussions, for example. It is possible to perform self-diffusion experiments when the diffusion coefficient is independent of density and the mass ratio of the two species is close to unity.

In gases at normal temperature and pressure D is quite generally on the order of 0.1–$1.0 \, \mathrm{cm}^2/\mathrm{sec}$ and is proportional to η/mn, where η is the shear viscosity. For liquids D is smaller by a factor of 10^{-4}, and is inversely proportional to the viscosity.

Consider a system not far from an equilibrium state described by a Hamiltonian $\hat{H}$, temperature T, and number density n for a single species. Presume that data are available over a spatial region R in terms of a time-independent density gradient characterized by the deviation $\delta\hat{n}(\mathbf{x}) \equiv \hat{n}(\mathbf{x}) - \langle\hat{n}\rangle_0\hat{1}$, where $\langle\hat{n}\rangle_0 = n$. Then, from Eq.(8–13), the expectation value of the particle current density is predicted to be

$$\langle\hat{\jmath}(\mathbf{x})\rangle = \int_R \lambda(\mathbf{x}') \langle \overline{\hat{n}(\mathbf{x}')} \, \hat{\jmath}(\mathbf{x}) \rangle_0 \, d^3x'$$
$$- \nabla\lambda \cdot \int_0^\infty e^{-\epsilon t} \, dt \int_R \langle \overline{\hat{\jmath}(\mathbf{x}', -t)} \, \hat{\jmath}(\mathbf{x}) \rangle_0 \, d^3x' , \qquad (8\text{–}16)$$

where we have noted explicitly that the equilibrium current vanishes.

Note now that the first term on the right-hand side of Eq.(8–16) must vanish in a stationary process, because it is odd under time reversal. Also, reference to standard discussions of the phenomenological theory reveals that the most general form of the steady-state diffusion equation with constant coefficient is (e.g., Jost, 1960)

$$\mathbf{J}(\mathbf{x}) = -\beta D n(\mathbf{x}) \nabla\mu(\mathbf{x}) , \qquad (8\text{–}17)$$

where $\mu(\mathbf{x})$ is a generalized chemical potential. In linear approximation we take $n(\mathbf{x}) \simeq n$ and employ dimensional considerations to identify the Lagrange-multiplier function as

$$\lambda(\mathbf{x}) = \beta\mu(\mathbf{x}) . \qquad (8\text{–}18)$$

In this hydrodynamic approximation we are then able to extract the diffusion coefficient from Eq.(8–16):

$$D_{ij}(\mathbf{x}) = \frac{1}{n} \lim_{\epsilon \to 0^+} \int_0^\infty e^{-\epsilon t} \, dt \int_R K_{j_i j_j}(\mathbf{x}', -t; \mathbf{x}) \, d^3x'$$
$$= D_{ji}(\mathbf{x}) . \qquad (8\text{–}19)$$

This expression is equivalent to those obtained by other methods (e.g., Nakajima, 1958), and translational invariance usually renders it independent of $\mathbf{x}$. One can obtain a diffusion coefficient in the more general situation in which the process is not stationary by considering initial data in the form of an expectation value of $\partial\hat{n}(\mathbf{x}, t)/\partial t$. In this case $D_{ij}(\mathbf{x}, t)$ follows immediately from Eq.(8–15). The generalization to multicomponent systems is also straightforward and we find a number of diffusion coefficients involving the different species. Common interest is in a self-diffusion coefficient, however, which is usually independent of $\mathbf{x}$, and then one can derive a rather important relation.

If we refer back to Chapter 4 and the discussion of electrical conductivity, it is seen that the conductivity tensor per unit volume is also a covariance function of two current-density operators. One can generalize the theory of linear response to a nonlocal description in a way suggested by Eq.(7–45), and introduce an adiabatic switching factor as discussed earlier. When the resulting expression, obtained in Problem 8.1, is compared with Eq.(8–19) we find that

$$\sigma_{ij} = \beta n e^2 D_{ij} \,, \tag{8-20}$$

which is the Nernst-Einstein relation (Nernst, 1888; Einstein, 1905). As Nakajima (1958) has pointed out, if a magnetic field is present the relation is only valid for the symmetric parts of the tensors. This is because the argument on time reversal concerning the first term on the right-hand side of Eq.(8–16) is no longer valid, and that term contributes to the antisymmetric components. Finally, it is interesting to note that Eq.(8–20) is probably the oldest theoretical expression of a fluctuation-dissipation theorem. It has been verified in a number of different model systems (Ferrari, *et al*, 1985).

How does the diffusion coefficient depend on the viscosity, say? Clearly, this is not a question for the microscopic theory, for viscosity is a macroscopic transport coefficient just like the diffusion coefficient. Interdependences of this kind can only be determined by examining the full set of hydrodynamic equations. We can express the transport coefficients in terms of covariance functions and then, through the macroscopic equations of motion, invert the process and find the covariance functions in terms of the transport coefficients. Depending on the degree of approximation one employs, this may not be a simple problem, and we shall not pursue it in any detail here.

THERMAL CONDUCTIVITY

In almost complete analogy with the preceding discussion one can formulate expressions for currents produced in a system by the presence of thermal gradients. We again focus on steady-state processes, because it appears that most heat-flow experiments are performed in this way. Presume, therefore, that information is available over a spatial region R in terms of a time-independent deviation in the energy density: $\delta \hat{h}(\mathbf{x}) \equiv \hat{h}(\mathbf{x}) - \langle \hat{h} \rangle_0 \hat{1}$. The same kind of arguments as used above then lead to an expectation value for the energy current density in the hydrodynamic limit:

$$\langle \hat{\boldsymbol{q}}(\mathbf{x}) \rangle \simeq \nabla \lambda \cdot \int_0^\infty e^{-\epsilon t} \, dt \int_R \langle \overline{\hat{\boldsymbol{q}}(\mathbf{x}', -t)} \, \hat{\boldsymbol{q}}(\mathbf{x}) \rangle_0 \, d^3 x' \,, \tag{8-21}$$

where once more the equilibrium current vanishes.

Equation (8–21) appears to be a generalized form of Fourier's law of heat conduction, and so we interpret $\lambda(\mathbf{x})$ as a local 'temperature' function:

$$\lambda(\mathbf{x}) \equiv \beta(\mathbf{x}) \equiv [\kappa T(\mathbf{x})]^{-1} \,. \tag{8-22}$$

This interpretation is in keeping with both the dimensions and meaning of the Lagrange multiplier associated with $\hat{h}(\mathbf{x})$. Hence, in leading order,

$$\nabla \beta(\mathbf{x}) = -\beta(\mathbf{x}) \frac{\nabla T(\mathbf{x})}{T(\mathbf{x})} \simeq -\frac{1}{\kappa T^2} \nabla T(\mathbf{x}) \,, \tag{8-23}$$

and the steady-state thermal conductivity tensor is identified as

$$\lambda_{ij}(\mathbf{x}) = -\frac{1}{\kappa T^2} \lim_{\epsilon \to 0^+} \int_0^\infty e^{-\epsilon t}\, dt \int_R K_{q_i q_j}(\mathbf{x}', -t; \mathbf{x})\, d^3 x'$$
$$= \lambda_{ji}(\mathbf{x}). \qquad (8\text{-}24)$$

We emphasize that $T(\mathbf{x})$ can not really be identified in this nonequilibrium state with the Kelvin temperature, but it is *replaced* in this last expression by the absolute temperature of the perturbed equilibrium state.

Again, the preceding calculation can be generalized to non-stationary processes by measuring the time variation of $\hat{h}(\mathbf{x}, t)$ over a space-time region $R(\mathbf{x}, t)$. It is also a simple matter to extend the discussion to multicomponent systems, and in the usual applications λ_{ij} is considered independent of $\mathbf{x}$. Thermal conductivities of gases and liquids are of the order 10^{-4} cal/sec·cm·°C, whereas those for solids are roughly 1000 times greater.

Experimentally it is found that the off-diagonal elements of λ_{ij} vanish in the absence of a magnetic field (Soret, 1893; Voight, 1903). When such an external field is introduced these elements are nonzero and give rise to the Righi-Leduc effect, and reciprocity relations of the kind indicated in Eq.(8–24) hold only when the field itself is also reversed. We shall discuss these effects further below.

THERMAL DIFFUSION

Recall for a moment the phenomenological equation (8–17), in which the generalized chemical potential can also be considered a function of number density and temperature. In a 'local-equilibrium' approximation one might then expect the total derivative to take the form

$$\nabla \mu = \left(\frac{\partial \mu}{\partial n}\right)_T \nabla n + \left(\frac{\partial \mu}{\partial T}\right)_n \nabla T. \qquad (8\text{-}25)$$

Thus, the particle current can have components arising from *both* density and temperature gradients. Similarly, if the function $\beta(\mathbf{x})$ is also a function of n and T we can write

$$\nabla \beta = \left(\frac{\partial \beta}{\partial T}\right)_n \nabla T + \left(\frac{\partial \beta}{\partial n}\right)_n \nabla n, \qquad (8\text{-}26)$$

and heat flow arises from both types of gradient. The partial derivatives are just ordinary thermodynamic derivatives. Although these are heuristic considerations, they indicate the possibility for observing such coupled effects— which, in fact, *are* observed. It does not seem possible to make these predictions from elementary kinetic theory, as was the case with the laws of Fick and Fourier.

Consider now a two-component system in which all measurements are made on a single component. Once again it appears that most measurements are made in the steady state, so we presume information available on the expectation values of both $\hat{n}(\mathbf{x})$ and $\hat{h}(\mathbf{x})$ over a spatial region R. Equation (8–13) is now employed for the sum of the two operators and we wish first to predict the expectation value of the particle current density. The familiar arguments on time-reversal invariance and surface terms are valid, and in the hydrodynamic limit we find that

$$
\langle \hat{\jmath}(\mathbf{x}) \rangle \simeq \nabla \lambda_1 \cdot \int_0^\infty e^{-\epsilon t}\, dt \int_R \langle \overline{\hat{\jmath}(\mathbf{x}', -t)}\, \hat{\jmath}(\mathbf{x}) \rangle_0\, d^3 x'
$$

$$
- \nabla \lambda_2 \cdot \int_0^\infty e^{-\epsilon t}\, dt \int_R \langle \overline{\hat{q}(\mathbf{x}', -t)}\, \hat{\jmath}(\mathbf{x}) \rangle_0\, d^3 x', \qquad (8\text{-}27)
$$

in the limit $\epsilon \to 0^+$.

The identity of the Lagrange-multiplier functions proceeds physically, as before. Although we shall *not* adopt the heuristic considerations of Eqs.(8–25) and (8–26), it *is* true that $\nabla \mu = (\partial \mu / \partial n) \nabla n$. The partial derivative is evaluated in Boltzmann statistics, and we find that

$$
\beta \nabla \mu = \tfrac{1}{n} \nabla n. \qquad (8\text{-}28)
$$

Equation (8–27) now takes the form

$$
\langle \hat{\jmath}_k(\mathbf{x}) \rangle = -[\nabla n(\mathbf{x})]_j\, D_{jk}(\mathbf{x}) + [\nabla T(\mathbf{x})]_j\, D_{jk}^T(\mathbf{x}), \qquad (8\text{-}29)
$$

where D_{jk} is the ordinary diffusion coefficient of Eq.(8–19), and we have defined the *thermal diffusion coefficient*

$$
D_{jk}^T(\mathbf{x}) \equiv \frac{1}{\kappa T^2} \lim_{\epsilon \to 0^+} \int_0^\infty e^{-\epsilon t}\, dt \int_R K_{j_j q_k}(\mathbf{x}', -t; \mathbf{x})\, d^3 x'. \qquad (8\text{-}30)
$$

Note that we have used the summation convention in Eq.(8–29) and that the x-dependence in these coefficients is usually omitted in standard experimental arrangements.

This demixing of a system—called *thermal diffusion*—was first suggested by Feddersen (1873), although it had been noticed earlier in liquids by Ludwig (1856), and was rediscovered by Soret (1879). It is well documented experimentally (e.g., Mason, *et al*, 1966).

Suppose that instead of predicting the expectation value of $\hat{\jmath}(\mathbf{x})$ we had chosen to study $\hat{q}(\mathbf{x})$. Then in place of Eq.(8–29) we would have obtained

$$
\langle \hat{q}_k(\mathbf{x}) \rangle = -[\nabla T(\mathbf{x})]_j\, \lambda_{jk}(\mathbf{x}) - [\nabla n(\mathbf{x})]_j\, D_{jk}^D(\mathbf{x}), \qquad (8\text{-}31)
$$

where λ_{jk} is the coefficient of thermal conductivity, Eq.(8–24). We have defined here the *Dufour coefficient*

$$
D_{jk}^D(\mathbf{x}) \equiv \frac{1}{n} \lim_{\epsilon \to 0^+} \int_0^\infty e^{-\epsilon t}\, dt \int_R K_{q_j jk}(\mathbf{x}', -t; \mathbf{x})\, d^3 x', \qquad (8\text{-}32)
$$

and Eq.(8–31) predicts a heat flow owing to a gradient in number density. This is called the diffusion-thermo effect, and was first observed by Dufour (1873).

Although these coefficients are usually considered to be independent of $\mathbf{x}$, they are also subject to several generalizations. For example, if the processes are not stationary one finds the definitions

$$D_{ij}^T(\mathbf{x},t) = -\frac{1}{\kappa T^2} \int_R \langle \overline{\hat{q}_j(\mathbf{x}',t')} \, j_i(\mathbf{x},t) \rangle_0 \, d^3x' \, dt' \,, \tag{8-33a}$$

$$D_{ij}^D(\mathbf{x},t) = -\frac{1}{n} \int_R \langle \overline{\hat{j}_j(\mathbf{x}',t')} \, \hat{q}_i(\mathbf{x},t) \rangle_0 \, d^3x' \, dt' \,. \tag{8-33b}$$

The expressions (8–29) and (8–31) comprise the first example of a number of coupled transport processes. By making a simple renormalization of the constants and appealing to the symmetry of the covariance functions, one obtains the Onsager-type reciprocity relations

$$D_{ij}^T = D_{ji}^D \,. \tag{8-34}$$

The quantity usually considered in discussing thermal diffusion is the thermal-diffusion ratio

$$k_T \equiv \frac{D_{ij}^T(\mathbf{x})}{D_{ij}^D(\mathbf{x})} \,, \tag{8-35}$$

although only the diagonal elements are normally calculated and the spatial dependence is omitted. Experimentally it is found that k_T is proportional to a product of the concentration parameters appropriate to the specific experiment involving two species. It is then convenient to measure the thermal-diffusion factor α_T, which is just k_T divided by this product of concentrations. Thus, a specific expression for α_T depends on the choice of a definite experimental arrangement.

Viscoelasticity

In the preceding discussion we have seen that direct observation of a gradient in a density $\hat{e}$ leads to a straightforward description of dissipative effects in any system. If the corresponding approximations are valid we can obtain equally direct expressions for the transport coefficients in terms of covariance functions. There remains one more density to study in a simple fluid, that for linear momentum, and to do so we consider a general viscoelastic medium which is isotropic and homogeneous when in thermal equilibrium, and described by a time-independent Hamiltonian. We shall continue to presume a steady-state situation and suppose that information is available over a spatial region R in the form of a steady current $\langle \hat{\jmath}(\mathbf{x}) \rangle$. From Eq(8–13) we then have for any other operator $\hat{C}$, in the hydrodynamic approximation,

$$\delta C(\mathbf{x}) \equiv \langle \hat{C}(\mathbf{x}) \rangle - \langle \hat{C} \rangle_0$$

$$\simeq \int_R \lambda_i(\mathbf{x}') \langle \overline{\hat{\jmath}^i(\mathbf{x}')} \, \hat{C}(\mathbf{x}) \rangle_0 \, d^3x'$$

$$- (\nabla_j \lambda_i) \int_0^\infty e^{-\epsilon t} \, dt \int_R \langle \overline{\hat{T}^{ij}(\mathbf{x}',-t)} \, \hat{C}(\mathbf{x}) \rangle_0 \, d^3x' \,. \tag{8-36}$$

We again employ the summation convention, and note that the equilibrium expectation value of the current density vanishes.

Equation (8–36) contains a great deal of information about the medium, which must be extracted in a careful sequence of steps. First let us consider the case where $\hat{C}$ is the total momentum operator $\hat{P}$, which is essentially the volume-integrated current density. From our work in Chapter 6 we see that $\langle \hat{j}(x') \hat{P}(x) \rangle_0 = \delta_{ij} n_0 / \beta$, from Eq.(6–82), and time-reversal invariance in the steady state implies that $\langle \hat{T}^{ij}(x', -t) \hat{P}(x) \rangle_0 = 0$. Hence, from Eq.(8–36),

$$\int \langle m \hat{j}_i(\mathbf{x}) \rangle \, d^3x = \frac{n_0}{\beta} \int_R \lambda_i(\mathbf{x}') \, d^3x' . \tag{8–37}$$

If we presume that the region R is just the entire volume, it follows that the Lagrange-multiplier function has the form $\lambda_i(\mathbf{x}) = \beta m v_i(\mathbf{x})$, where $v_i(\mathbf{x})$ is a fluid velocity. That is,

$$\langle m \hat{j}_i(\mathbf{x}) \rangle = m n v_i(\mathbf{x}) , \tag{8–38}$$

and we have unambiguously identified $\boldsymbol{\lambda}(\mathbf{x})$.

In the same manner we can make the choice that $\hat{C}$ is the volume-integrated energy current and employ the identity (6–83). Then,

$$\langle \hat{q}_i \rangle = (h_0 + P_0) v_i , \tag{8–39}$$

where h_0 and P_0 are the equilibrium energy density and pressure, respectively.

The preceding observations are consistent with earlier discussion on Galilean invariance, of course, but they also contain a more subtle point. Equations (8–38) and (8–39) merely indicate what the expectation values of these currents are in the laboratory frame when the available information indicates that the medium is convected with respect to that frame. These values are completely determined by the Lagrange-multiplier function associated with the actually measured quantity. It is clear that these values are reactive and not dissipative, because one needs actual data on the corresponding density in order to infer that a current is dissipative. If, for example, we also measured a steady-state heat flow, then a term of the form (8–21) would be added to the right-hand side of Eq.(8–39). In the present case this information is not available.

Dissipative effects *are* manifest if we ask for the expectation value of the stress tensor, $\hat{T}^{mn}(\mathbf{x})$, which is the current associated with the density whose gradient we have just measured (or specified). Equation (8–36), along with the usual argument on time-reversal invariance, then yields

$$\delta T^{mn}(\mathbf{x}) \equiv \langle \hat{T}^{mn}(\mathbf{x}) \rangle - \langle \hat{T}^{mn} \rangle_0$$
$$= -\nabla_j \lambda_i \int_0^\infty e^{-\epsilon t} \, dt \int_R \overline{\langle \hat{T}^{ij}(\mathbf{x}', -t) \, \hat{T}^{mn}(\mathbf{x}) \rangle_0} \, d^3x' . \tag{8–40}$$

This expression is precisely equivalent to the phenomenological equation (1–7) where, because we are working in the laboratory frame and have made the linear approximation, the kinetic-energy tensor t^{ij} does not appear. Therefore, this is actually of the form (1–3) and we can interpret $\langle \hat{T}^{mn} \rangle$ as S^{mn}. Moreover, our earlier work shows that the equilibrium expectation value of $\hat{T}^{mn}$ is just the hydrostatic pressure, and thus the second line of Eq.(8–40) represents the shear tensor, p^{mn}.

In the discussion associated with Table 6–2 we noted that the space-integrated covariance functions involving stress tensors possess a limited number of nonzero components in the isotropic medium. Along with Eq.(6–38), this implies that we can rewrite Eq.(8–40) as

$$\delta T^{mn}(\mathbf{x}) = -\beta m \frac{\partial v_i}{\partial x^i} \int_0^\infty e^{-\epsilon t}\, dt \int_V \langle \tfrac{1}{3} \overline{\hat{T}^j}_{\,j}(\mathbf{x}', -t)\, \hat{T}^{mn}(\mathbf{x}) \rangle_0\, \delta_{mn}\, d^3 x'$$
$$- \beta m \left(\frac{\partial v_m}{\partial x^n} + \frac{\partial v_n}{\partial x^m} \right) \int_0^\infty e^{-\epsilon t}\, dt \int_V \langle \overline{\hat{T}^{mn}(\mathbf{x}', -t)}\, \hat{T}^{mn}(\mathbf{x}) \rangle_0$$
$$\times (1 - \delta_{mn})\, d^3 x', \qquad (8\text{–}41)$$

where we have introduced explicitly the identification of the Lagrange-multiplier functions, and taken the region R to be the entire system volume. One can now identify the transport coefficients.

Let us focus first on a prediction of the shear components $\hat{T}^{ij}$, $i \neq j$. Equation (8–41) then gives for the shear viscosity the expression

$$\eta = \frac{\beta m}{V} \lim_{\epsilon \to 0^+} \int_0^\infty e^{-\epsilon t} \langle \overline{\hat{T}^{ij}(-t)}\, \hat{T}^{ij} \rangle_0\, dt, \qquad i \neq j, \qquad (8\text{–}42)$$

and now we are dealing only with the space-integrated operators. Although any values of $i \neq j$ can be used here in the isotropic fluid, they are usually chosen in accord with the initially-measured current component. If a diagonal component is predicted only traces are involved, and the bulk viscosity is given by

$$\varsigma = \frac{1}{9} \frac{\beta m}{V} \lim_{\epsilon \to 0^+} \int_0^\infty e^{-\epsilon t} \langle \overline{\hat{T}(-t)}\, \hat{T} \rangle_0\, dt, \qquad (8\text{–}43)$$

where $\hat{T}$ is the space-integrated tensor trace.

It is possible to obtain more general expressions than these, of course, if the initial data do not imply a steady state. Indeed, this will be exactly the situation in Section C below. Also, a dependence of η on the shear rate γ, say, can only arise from consideration of the entire set of macroscopic hydrodynamic equations, as pointed out above. We have by no means extracted all the information contained in Eq.(8–40), but the immediate point here is simply to illustrate that viscosity coefficients are indeed contained in that equation.

Another example of generalization occurs in the high-frequency regime characterizing the elastic limit of an ideal *solid* [rather than the liquid considered in Eqs.(6–94) and (6–95)]. That is, the stresses do not relax unless the external perturbation is removed, but then the strains relax with the stresses and we are therefore interested in the short-memory limit. If the stress tensor is driven, either thermally or mechanically, we obtain

$$\delta T^{mn}(\mathbf{x},t) \equiv \langle \hat{T}^{mn}(\mathbf{x},t) \rangle - \langle \hat{T}^{mn} \rangle_0$$
$$= \int_0^t dt' \int \nabla_j' \lambda_i(\mathbf{x}',t') K_{T^{mn}T^{ij}}(\mathbf{x}-\mathbf{x}',t-t') \, d^3x'. \qquad (8\text{-}44)$$

In the short-memory limit the covariance function is effectively independent of the time coordinate and so contributes nothing to the time integration. The Lagrange-multiplier function has been identified above as proportional to the fluid velocity. But the latter is also the time derivative of the displacement vector, which is the more appropriate parameter in the present case. Thus, in the short-memory limit Eq.(8-44) reduces to

$$\delta T^{mn}(\mathbf{x},t) \simeq \beta m \int_0^t \frac{\partial}{\partial t'} \sigma_{i,j}(t') \, dt' \int K_{T^{mn}T^{ij}}(\mathbf{x}-\mathbf{x}') \, d^3x'$$
$$= \beta m \frac{\partial \sigma_i}{\partial x^j} \int K_{T^{mn}T^{ij}}(\mathbf{x}-\mathbf{x}') \, d^3x', \qquad (8\text{-}45)$$

in the independent limits $k \to 0$, $\omega \to \infty$. This result can now be compared with the Cauchy-Hooke equations and the elastic moduli identified. The resulting expressions are precisely those found in Eqs.(6-94) and (6-95).

In summary, merely by observing that a process is taking place in a system we find that the linear approximation to the equilibrium theory immediately yields a gradient expansion, and in the hydrodynamic limit provides explicit expressions for the appropriate transport coefficients. These results are automatic when the initial data refer to conserved local densities. In order to evaluate the transport coefficients, of course, one must eventually calculate covariance functions and perform the indicated integrations; this is not a trivial problem, as we have seen. Moreover, in order to complete the picture one must actually drive several densities so as to obtain the familiar coupled processes of hydrodynamics, and we shall address this problem in Section C below.

With the specific choices we have made for the conserved densities, expressions for transport coefficients much like those given above have been obtained in many different ways by Mori (1956, 1958), Kadanoff and Martin (1963), Luttinger (1964), Puff and Gillis (1968), Zubarev (1974), and Forster (1975). It is reasonable to ask why all approaches seem to lead to similar results in these cases. Apparently the linear transport coefficients are very insensitive to the theory used to obtain them, most likely because of the extreme approximation represented by the limit of local hydrodynamics. Almost any theory will work in this limit. This is a familiar situation in statistical mechanics, as evidenced by studies of the second virial coefficient in dilute gases. The data can be fit by calculating $B(T)$ using almost any potential function containing a repulsive core, an attractive well, and a few adjustable parameters, as illustrated long ago by Hirschfelder, *et al* (1954).

At very low densities the phenomena depend primarily on the *existence* of particle interactions, and not so much on specific details of the potential. It appears that linear hydrodynamics plays a similar role for nonequilibrium phenomena, although here the problems are more vicious and one suspects that considerable differences will appear in nonlinear and intrinsically nonlocal processes.

We shall return to a discussion of this and other peculiarities of the hydrodynamic limit presently, in Section D. But here we must emphasize that obtaining *forms* of the transport coefficients which are somewhat familiar is only a small point. Of equal, or more, importance is the *way* one arrives there, and the above derivations are certainly free of *ad hoc*, or otherwise questionable presumptions. In the following chapter we shall see that other applications of the theory have little to do in general with the HL, even when discussing related
matters of hydrodynamics.

B. Thermomechanical Processes

Thermal diffusion and the Dufour effect represent just two of the many coupled processes which can occur in a nonequilibrium many-body system. Irreversible processes of this kind are among the most interesting to study, so that it is of value to catalog some others of similar character. Of particular interest are those which possess some elements of both dynamical and thermal driving. Although we shall restrict the discussion to the linear approximation and hydrodynamic limit, the phenomena described below are primarily studied in solids, so that what is meant by the HL must be interpreted carefully in application.

While studying ordinary diffusion above we envisioned information obtained regarding a space-time variation in number density, say, and this led (implicitly) to an initial statistical operator

$$\hat{\rho} = \frac{1}{Z} \exp\left\{-\beta\,\hat{H}_0 + \int_R \beta\mu(\mathbf{x}',t')\hat{n}(\mathbf{x}',t')\,d^3x'\,dt'\right\}, \qquad (8\text{--}46)$$

in which $\mu(\mathbf{x}',t')$ is identified as a general chemical potential. If, in fact, the number density turns out to be uniform throughout the system, then the operator in the exponential is just the effective Hamiltonian of the grand canonical ensemble:

$$\hat{K} \equiv \hat{H}_0 - \mu \int \hat{n}(\mathbf{x})'\,d^3x'$$
$$= \hat{H}_0 - \mu\hat{N}, \qquad (8\text{--}47)$$

where $\hat{H}_0$ is the Hamiltonian of the equilibrium system.

Similarly, a dynamical perturbation in the form of an external nonlocal scalar potential $\phi(\mathbf{x},t)$ can couple to the system through the number-density operator, resulting in the addition to $\hat{H}_0$ of a well-defined term of the form

$$\int \hat{n}(\mathbf{x},t)\phi(\mathbf{x},t)\,d^3x. \qquad (8\text{--}48)$$

For example, such a term could describe the imposition upon the system of an external electric field derivable from a scalar potential. Or, ϕ could describe the gravitational potential present in every earthbound laboratory. In this way one can develop a simple modification of the theory encompassing both mechanical and thermal disturbances. The only *caveat* to be observed here is that when one employs potentials, as in Eq.(8–48), attention must be given to questions of gauge invariance. That such invariance is always maintained has been argued positively elsewhere (Adu-Gyamfi, 1986).

THERMOELECTRIC PHENOMENA

Consider now a system of charged particles each with charge e and under the influence of a static external electric field derived from a potential per unit charge $\phi(\mathbf{x}, t)$. The *electrochemical potential* is defined as

$$\psi(\mathbf{x}, t) \equiv \mu(\mathbf{x}, t) + e\phi(\mathbf{x}, t), \tag{8–49}$$

and we suppose that additional information consists of knowledge of the number and energy densities over a region R. Presume this information implies a steady-state process. The expectation value of any other operator $\hat{B}(\mathbf{x})$ will then take the form of Eq.(8–27), and the Lagrange-multiplier functions have already been identified in the hydrodynamic limit. In particular, that associated with $\hat{h}(\mathbf{x})$ is approximated by the equilibrium quantity β, as in Eq.(8–24).

Specific application to thermoelectric processes follows from the successive choices for prediction of $\langle \hat{q}(\mathbf{x}) \rangle$ and $\langle \hat{\jmath}(\mathbf{x}) \rangle$, where $\hat{\jmath}$ is the *electric*-current operator. One then obtains the set of coupled equations

$$\langle \hat{\jmath}(\mathbf{x}) \rangle = -\frac{1}{e\kappa T} \nabla \psi \cdot \mathbf{L}_{11}(\mathbf{x}) + \frac{1}{\kappa T^2} \nabla T \cdot \mathbf{L}_{12}(\mathbf{x}), \tag{8–50}$$

$$\langle \hat{q}(\mathbf{x}) \rangle = -\frac{1}{e\kappa T} \nabla \psi \cdot \mathbf{L}_{21}(\mathbf{x}) + \frac{1}{\kappa T^2} \nabla T \cdot \mathbf{L}_{22}(\mathbf{x}), \tag{8–51}$$

where we have defined the second-rank tensors

$$\mathbf{L}_{11}(\mathbf{x}) \equiv \int_0^\infty e^{-\epsilon t}\, dt \int_R \langle \overline{\hat{\jmath}(\mathbf{x}')}\, \hat{\jmath}(\mathbf{x}, t) \rangle_0\, d^3 x', \tag{8–52}$$

$$\mathbf{L}_{12}(\mathbf{x}) \equiv \int_0^\infty e^{-\epsilon t}\, dt \int_R \langle \overline{\hat{q}(\mathbf{x}')}\, \hat{\jmath}(\mathbf{x}, t) \rangle_0\, d^3 x', \tag{8–53}$$

$$\mathbf{L}_{21}(\mathbf{x}) \equiv \int_0^\infty e^{-\epsilon t}\, dt \int_R \langle \overline{\hat{\jmath}(\mathbf{x}')}\, \hat{q}(\mathbf{x}, t) \rangle_0\, d^3 x', \tag{8–54}$$

$$\mathbf{L}_{22}(\mathbf{x}) \equiv \int_0^\infty e^{-\epsilon t}\, dt \int_R \langle \overline{\hat{q}(\mathbf{x}')}\, \hat{q}(\mathbf{x}, t) \rangle_0\, d^3 x', \tag{8–55}$$

which tacitly include the limits $\epsilon \to 0^+$. Note that we have employed cyclic invariance of the trace to relocate the variable t in the integrands. These tensors are the *thermoelectric coefficients*, and in most experimental arrangements are actually independent of $\mathbf{x}$. In an isotropic medium the symmetry properties of the covariance functions imply the following Onsager-like reciprocity relations:

$$\mathbf{L}_{12}(\mathbf{x}) = \mathbf{L}_{21}(\mathbf{x}). \tag{8-56}$$

One can, of course, readily generalize these to time-dependent functions in the event that the process is not stationary.

There is a great deal of information contained in Eqs.(8–50)-(8–55), much of which has been known phenomenologically for over 160 years, and some of which has already been discussed in Chapter 1. For example, if two different metal wires are connected in parallel and the junctions maintained at different temperatures, then a thermoelectric emf will develop in the circuit. Equation (8–50) predicts this phenomenon by illustrating that an electric current will be produced by *both* an electric potential *and* a temperature gradient. The coefficient $\mathbf{L}_{11}$ is just the electrical conductivity, whereas $\mathbf{L}_{12}$ is called the *Seebeck coefficient*. This effect was discovered in 1822 (Seebeck, 1826). The fact that two different substances are involved is manifested by a space-dependent Hamiltonian—presumably $\mathbf{L}_{12}$ will vanish if this space dependence disappears.

In a similar manner, Eq.(8–51) predicts a heat flow arising from an electric potential, even in the absence of a temperature gradient. (The coefficient $\mathbf{L}_{22}$ is the thermal conductivity.) This is the *Peltier effect*, and can be observed by connecting two dissimilar metals in series, passing an electric current through the conductors, and noting a heating or cooling at the junction (Peltier, 1834). One refers to $\mathbf{L}_{21}$ as the *Peltier coefficient*—unlike Joule heating, this effect is seen to be linear in the electric current.

Actually, the prediction of Eq.(8–51) remains valid in a homogeneous material if an electric potential *and* a temperature gradient are both present. The quantity $\langle \hat{q}(\mathbf{x}) \rangle$ is then known as the *Thomson heat*, representing a contribution over and above the Joule heat (Thomson, 1854). Indeed, Thomson already recognized that all three of the thermoelectric effects are clearly related to one another. A rather detailed review of these effects in metals and alloys has been provided by Huebner (1972).

THERMOMAGNETIC EFFECTS

Other thermomechanical effects arise when, instead of an electric field, the system is exposed to an external magnetic field in addition to any other processes that may be taking place. Recall the equations of the canonical ensemble describing thermal equilibrium in the absence of a field:

$$\hat{\rho}_0 = \frac{1}{Z_0} e^{-\beta \hat{H}_0}, \qquad Z_0 = \text{Tr}\left[e^{-\beta \hat{H}_0} \right], \tag{8-57}$$

which refer to time-independent $\hat{H}_0$. Following imposition of a static and uniform magnetic field $\mathbf{H}$ the statistical operator has the same form,

$$\hat{\rho}_1 = \frac{1}{Z_1} e^{-\beta \hat{H}_1}, \qquad Z_1 = \text{Tr}\left[e^{-\beta \hat{H}_1} \right], \tag{8-58}$$

but now the time-independent Hamiltonian is $\hat{H}_1 = \hat{H}_0 - \hat{M} \cdot H$, and $\hat{M}$ is the total magnetic-moment operator for the system. We refer to the description of Eq.(8–58) as the *magnetic equilibrium ensemble*.

Now suppose that information is obtained in the form of an expectation value of some Heisenberg operator $\hat{F}(\mathbf{x}, t)$ over some space-time region R. In linear approximation the expectation value of any other operator $\hat{B}(\mathbf{x}, t)$ is then

$$
\begin{aligned}
\langle \hat{B}(\mathbf{x}, t) \rangle = {} & \langle \hat{B}(\mathbf{x}, t) \rangle_1 \\
& + \int_R \lambda(\mathbf{x}', t') \big[\overline{\langle \hat{F}(\mathbf{x}', t') } \, \hat{B}(\mathbf{x}, t) \rangle_1 \\
& \qquad - \langle \hat{F}(\mathbf{x}') \rangle_1 \langle \hat{B}(\mathbf{x}) \rangle_1 \big] \, d^3 x' .
\end{aligned}
\tag{8–59}
$$

The quantity in square brackets is just the covariance function defined in terms of the equilibrium ensemble of Eq.(8–58).

As a simple example of the utility of this ensemble let us study the steady-state thermal conductivity. A now-familiar calculation yields, for an otherwise isotropic system, the field-dependent quantity

$$
\lambda_{ij}(\mathbf{x}, H) = -\frac{1}{\kappa T^2} \lim_{\epsilon \to 0^+} \int_0^\infty e^{-\epsilon t} \, dt \int \langle \overline{\hat{q}_j(\mathbf{x}', -t)} \, \hat{j}_i(\mathbf{x}) \rangle_1 \, d^3 x' ,
\tag{8–60}
$$

which is usually independent of $\mathbf{x}$. If we notice that the magnetic-moment operator is odd under PT, a detailed examination of the right-hand side of this last expression yields the Onsager-like reciprocity relation

$$
\lambda_{ij}(H) = \lambda_{ji}(-H) .
\tag{8–61}
$$

Recall that the experimental results indicate that the off-diagonal elements of λ_{ij} vanish in zero field. In the presence of a field, however, these elements are nonzero and, in linear approximation, proportional to H. Thus a heat flow can be established transverse to the original temperature gradient when an external field is applied, and this is known as the Righi-Leduc effect (Righi, 1883; Leduc, 1887). It is the thermal analog of the Hall effect.

Next consider the construction of Eqs.(8–50) and (8–51), and carry out the calculation anew when an external magnetic field is present. The initial equilibrium ensemble is that of Eq.(8–58), and the expressions (8–52)-(8–55) retain the same form except that now we write $L_{ij}(\mathbf{x}, H)$ and replace the expectation values $\langle \cdots \rangle_0$ with $\langle \cdots \rangle_1$. These are now *thermomagnetic coefficients*, and Eq.(8–56) is replaced by

$$
L_{12}(\mathbf{x}, H) = L_{21}(\mathbf{x}, -H) .
\tag{8–62}
$$

With these coefficients one obtains essentially three new effects. The coefficient $L_{21}(\mathbf{x}, H)$ describes the production of a temperature gradient perpendicular to the plane of the magnetic field and the electric current when the two are not collinear. This transverse phenomenon is called the *Ettinghausen effect* (Ettinghausen, 1887).

In addition to effects discussed previously, the coefficient $L_{12}(\mathbf{x}, \mathbf{H})$ now refers to production of a gradient in electrochemical potential owing to a temperature gradient in the presence of a magnetic field. The longitudinal result is due to Nernst (1888), and the transverse phenomenon is called the *Ettinghausen-Nernst effect* (Ettinghausen and Nernst, 1887). Further specific details of these phenomena can be found in standard references (e.g, Harman and Honig, 1967).

C. The Equations of Hydrodynamics

In Chapter 7 we were able to uncover quite generally the natural way in which macroscopic conservation laws arise from the corresponding microscopic equations. Whether the driving is done thermally or mechanically is irrelevant to the results, so that we shall tend not to be specific about the nature of sources here.

Let us recall that the particle current $\hat{\mathbf{J}}$ serves also as the momentum density in the simple fluid. The corresponding current is then the stress tensor $\hat{T}_{ij}$, and the two are related through the microscopic conservation law

$$m\partial_t \hat{\jmath}_i(\mathbf{x}, t) + \nabla_k \hat{T}_{ki}(\mathbf{x}, t) = 0. \tag{8-63}$$

Thus, suppose the momentum density (or particle current) to be driven uniformly by an outside force. Then the microscopic expression (8–63) leads directly to the nonlinear macroscopic law of Eq.(7–55), which in the present case gives us

$$\partial_t \langle m\hat{\jmath}_i(\mathbf{x}, t)\rangle + \nabla_k \langle \hat{T}_{ki}(\mathbf{x}, t)\rangle = \dot{\sigma}_i(\mathbf{x}, t), \tag{8-64}$$

where $\dot{\sigma}_i$ is the rate at which the source drives the component $m\hat{\jmath}_i$. Equation (8–64) exhibits the macroscopic equations of motion, but a more explicit representation can only emerge from a study of the expectation value of the stress tensor.

In order to evaluate $\langle \hat{T}_{ki}(\mathbf{x}, t)\rangle$, let us first recall the earlier work involving the steady-state scenario, in which $\langle \hat{\jmath}\rangle$ was time independent and the Lagrange-multiplier function was identified in Eqs.(8–37) and (8–38). But, although the current can now vary in time, the physical interpretation of the Lagrange-multiplier function must remain unchanged. Hence, for the general non-stationary case of thermal driving we observe a variation $\langle m\hat{\jmath}(\mathbf{x}, t)\rangle$ and predict the linear approximation

$$\langle \hat{T}_{ki}(\mathbf{x}, t)\rangle \simeq \langle \hat{T}_{ki}(\mathbf{x})\rangle_0$$
$$+ \int_V d^3x' \int_0^t \nabla_j \lambda_\ell(\mathbf{x}', t') \overline{\langle \hat{T}_{j\ell}(\mathbf{x}', t')} \hat{T}_{ki}(\mathbf{x}, t)\rangle_0 \, dt', \tag{8-65}$$

after an integration by parts. We have employed the scenario of Eq.(8–14), as well as the summation convention.

Equation (8–65) is as far as one can proceed in general without additional approximation. Although we shall study this general expression further in the next section, here we recall that classical hydrodynamics refers to long wavelengths and low frequencies, so that the hydrodynamic limit (HL) is appropriate. With the identification

$$\boldsymbol{\lambda}(\mathbf{x}, t) \equiv \beta m \mathbf{v}(\mathbf{x}, t)$$

$$= \frac{\beta}{n_0} \langle m \hat{\mathbf{j}}(\mathbf{x}, t) \rangle, \qquad (8\text{-}66)$$

where β is the temperature parameter of the equilibrium system, we need only carry out calculations similar to those of Eqs.(8–41)-(8–43) with this generalization in order to complete the evaluation of $\langle \hat{T}_{ki}(\mathbf{x}, t) \rangle$.

In terms of the space-integrated tensors one now makes the identifications

$$\eta(t) \equiv -\frac{\beta m}{V} \int_0^t \overline{\langle \hat{T}_{ki}(t') \, \hat{T}_{ki}(t) \rangle_0} \, dt', \quad k \neq i, \qquad (8\text{-}67)$$

$$\varsigma(t) \equiv -\frac{1}{9} \frac{\beta m}{V} \int_0^t \overline{\langle \hat{T}(t') \, \hat{T}(t) \rangle_0} \, dt', \qquad (8\text{-}68)$$

the time-dependent shear and bulk viscosities, respectively. Actually, the correlations relax rapidly enough that these quantities are effectively time independent on a macroscopic scale—as will be seen explicitly in the next section. We have introduced the notation $\hat{T}$ for the tensor trace of $\hat{T}_{ij}$, and noted that in the isotropic fluid one can employ *any* shear components in Eq.(8–67).

Further identification is achieved by noting from Eq.(6–43) that $\langle \hat{T}_{ij}(\mathbf{x}, t) \rangle_0 = P_0 \delta_{ij}$, the equilibrium hydrostatic pressure, and reference to the phenomenological macroscopic theory (e.g., Currie, 1974). In linear approximation $\langle \hat{T}_{ij}(\mathbf{x}, t) \rangle$ is taken to be the shear tensor, and minus one-third its trace is defined as the mechanical pressure in the system. In turn, though, the mechanical pressure in the phenomenological theory is equal to the thermodynamic pressure P minus the quantity $\varsigma (\nabla \cdot \mathbf{v})$ (see, e.g., Problem 1.2). In accordance with the notation of Chapter 1, we define the shear tensor as

$$P_{ki} \equiv \eta \left(\frac{\partial v_k}{\partial x^i} + \frac{\partial v_i}{\partial x^k} - \frac{2}{3} \delta_{ki} \frac{\partial v_\ell}{\partial x^\ell} \right) + \varsigma \frac{\partial v_\ell}{\partial x^\ell} \delta_{ki}, \qquad (8\text{-}69a)$$

where η and ς are given by Eqs.(8–67) and (8–68), respectively. The stress tensor of Eq.(8–65) is then

$$\langle \hat{T}_{ki}(\mathbf{x}, t) \rangle \simeq P_0 \delta_{ki} + P \delta_{ki} - P_{ki}. \qquad (8\text{-}69b)$$

If this result is now substituted into Eq.(8–64), and it is noted that Eq.(8–66) is also to be employed here, then one obtains the macroscopic equations of motion corresponding to what are conventionally called the Stokes equations:

$$m n_0 \partial_t v_i + \nabla_k (P \delta_{ki} - P_{ki}) = \dot{\sigma}_i, \qquad (8\text{-}70)$$

and one *could* write the source term as $\dot{\sigma}_i = n_0 F_i$, in terms of an external force. In this case the linear approximation corresponds to a restriction to low Reynolds number. Equations (8–70) are sometimes referred to as the linearized Navier-Stokes equations, but that is slightly inaccurate—the Navier-Stokes equations refer to an incompressible fluid, which requires $\nabla \cdot \mathbf{v} = 0$. Coupled to these equations, of course, is that for conservation of energy. Because only the particle current is considered here, this is just the linear approximation

$$\partial_t \langle \hat{h}(\mathbf{x}, t) \rangle + \nabla \cdot \langle \hat{q}(\mathbf{x}, t) \rangle = 0. \tag{8–71}$$

These equations arise from consideration of the microscopic motions in, and refer to, the laboratory reference frame. It is conventional instead to refer the equations to an observer convected with the fluid, in which case the time derivatives must be replaced by the so-called *hydrodynamic derivative*:

$$\frac{D}{Dt} \equiv \frac{\partial}{\partial t} + (\mathbf{v} \cdot \nabla). \tag{8–72}$$

Replacement of $\partial/\partial t$ by D/Dt in Eqs.(8–70) and (8–71) then introduces nonlinear convective terms into the equations of motion, and hence leads to the full Navier-Stokes equations (Navier, 1821).

In addition to nonlinearities arising from convection—which are essentially kinetic in origin—there are true nonlinear corrections arising from higher-order approximations to the expectation values $\langle \hat{T}_{ij}(\mathbf{x}, t) \rangle$ and $\langle \hat{q}(\mathbf{x}, t) \rangle$. For example, in the present scenario the expansion of expectation values through second order as expressed in Eq.(5–114) takes the form

$$\langle \hat{C}(\mathbf{x}, t) \rangle = \langle \hat{C}(\mathbf{x}) \rangle_0 + \int_0^t dt' \int \nabla_j \lambda_\ell \, K_{CT_{j\ell}} \, d^3 x'$$
$$+ \int_0^t dt_1 \int_0^t dt_2 \int d^3 x_1 \int d^3 x_2 \, (\nabla_m \lambda_\ell)(\nabla_n \lambda_k) K_{CT_{m\ell} T_{nk}} \tag{8–73}$$

where

$$K_{CT_{m\ell} T_{nk}}(\mathbf{x}_1, \mathbf{x}_2, \mathbf{x}; t_1, t_2, t) =$$
$$\langle \hat{T}_{nk}(\mathbf{x}_1, t_1) \hat{T}_{m\ell}(\mathbf{x}_2, t_2) \, \hat{C}(\mathbf{x}, t) \rangle_0$$
$$- \langle \hat{T}_{nk}(\mathbf{x}_1, t_1) \hat{T}_{m\ell}(\mathbf{x}_2, t_2) \rangle_0 \langle \hat{C}(\mathbf{x}) \rangle_0, \tag{8–74}$$

with double Kubo transforms. We have, as usual, integrated by parts and dropped the surface term, thereby illustrating the structure of the gradient expansion in Eq.(8–73).

In linear approximation the expectation value $\langle \hat{q}(\mathbf{x}, t) \rangle$ in Eq.(8–71) has the general form (8–21), proportional to a temperature gradient by means of the thermal conductivity. A second-order contribution arises in this hydrodynamic scenario if we set $\hat{C} = \hat{q}_i$ in Eq.(8–73), in which case the second term on the right-hand side of

Eq.(8–74) vanishes. In this way one accounts for the term $S^{jk}v_{k,j}$ in Eq.(1–15b), describing dissipation of heat owing to internal viscous forces. To spell this out in complete detail is not trivial, of course, for it requires analysis of a three-point correlation function, $K_{q_i T_{m\ell} T_{nk}}$. Nevertheless, it is now clear that the macroscopic equations of motion are contained within the present theory.

A final point should be made here. We could have specified that all three locally-conserved densities are driven in the process. Rather than Eq.(8–65) being the operative expression, then, we would have obtained something of the form

$$\langle \hat{C}(\mathbf{x},t) \rangle - \langle \hat{C}(\mathbf{x}) \rangle_0 \simeq \int \nabla_j \lambda_\ell K_{CT_{j\ell}}$$

$$+ \int \nabla \Lambda \cdot K_{C\mathbf{q}}$$

$$+ \int \nabla \mu \cdot K_{C\mathbf{j}}.$$

But in the hydrodynamic limit the symmetry properties of the covariance functions would have rendered a number of terms on the right-hand side zero, depending on the choice of $\hat{C}$. In linear approximation we obtain exactly the same results as above, but this may not be true in higher order.

D. The Hydrodynamic Limit

Although it can be argued that we have a coherent and comprehensive derivation of the equations of hydrodynamics, it is clear that further exposition and development are necessary. As we have emphasized continually, much depends on a detailed understanding of the structure of covariance functions, including not only their time-decay properties, but also their role in defining a 'hydrodynamic limit'. It is not at all clear at this point that this limit has been defined in a mathematically correct way—rather, it serves as a stopgap in place of a complete understanding of covariance functions.

Presumably, if this understanding were improved considerably, we would uncover a structure applicable to simple fluids which would lead quite directly to the phenomenological classical hydrodynamic modes under the appropriate conditions. For example, with m the particle mass and n_0 the equilibrium density, we define the thermal diffusion constant as

$$D_T \equiv \frac{\lambda}{mn_0 C_p}, \tag{8–75a}$$

in terms of the thermal conductivity λ, and the longitudinal diffusion constant as

$$D_\ell \equiv \frac{1}{mn_0}\left(\tfrac{4}{3}\eta + \varsigma\right), \tag{8–75b}$$

in terms of the viscosities. Define further the quantity

$$\Gamma \equiv D_\ell + D_T\left(\frac{C_p}{C_V} - 1\right), \tag{8–76}$$

where C_x will represent heat capacities at constant x, and c denotes the speed of sound in the medium. Then, in the coupled-mode form of hydrodynamics, one would expect the relevant covariance functions to have a structure something like the following:

$$\frac{\tilde{K}_{ij}(\mathbf{k}, z)}{K_{ij}(\mathbf{k}, t=0)} = i\left[\frac{Z_{ij}^{(+)}(k)}{z - ck + \frac{i}{2}k^2\Gamma} + \frac{Z_{ij}^{(-)}(k)}{z + ck - \frac{i}{2}k^2\Gamma} + \frac{Z_{ij}^{(T)}(k)}{z + ik^2 D_T}\right],$$

$$(8\text{--}77)$$

where the $Z_{ij}^{(\alpha)}(k)$ are the residues at the indicated poles. Such a structure provides a close connection to experiment—alas, it has yet to be derived from first principles. Note that this is a small-k form, but we have *not* taken the limit $k \to 0$.

If we focus on density-density correlations in the limit $\beta\hbar\omega \ll 1$ and take twice the real part of the ratio (8–77) evaluated at $z = \omega + i0$, the resulting *structure factor* can be plotted as in Figure 8–1. The central peak is known as the Rayleigh component, whereas those located symmetrically about $\omega = \pm ck$ comprise the Brillouin doublet. All three stem from Lorentzian functions, so that the linewidths are readily identifiable. The Rayleigh linewidth is

$$\Gamma_R = D_T k^2, \qquad (8\text{--}78)$$

and that of each Brillouin component is

$$\Gamma_B = \tfrac{1}{2}\Gamma k^2. \qquad (8\text{--}79)$$

The integrated intensity of the Rayleigh component is given by

$$I_R = 1 - \frac{C_V}{C_P}, \qquad (8\text{--}80)$$

whereas that of the doublet is

$$2I_B = \frac{C_V}{C_P}, \qquad (8\text{--}81)$$

and hence $I_R + 2I_B = 1$. Thus, a great deal of experimental information about the hydrodynamic parameters can be obtained from light scattering experiments—information that must already be contained in K_{nn} itself.

TRANSPORT COEFFICIENTS

In order to underscore further the difficulties with the HL, let us examine more closely the identification of a general transport coefficient. Recall the steady-state form given in Eq. (8–13), and suppose we are interested in a current $\hat{\mathbf{J}}$:

$$\langle\hat{\mathbf{J}}(\mathbf{x})\rangle \simeq -\lim_{\epsilon \to 0^+}\int_0^\infty e^{-\epsilon t}\,dt\int \nabla'\lambda(\mathbf{x}') \cdot K_{\mathbf{J}\mathbf{J}}(\mathbf{x} - \mathbf{x}', t)\,d^3x'. \qquad (8\text{--}82)$$

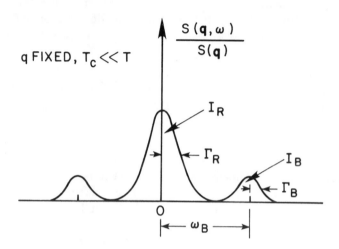

Fig. 8–1. The structure factor corresponding to the ratio in Eq.(8–77) as a function of frequency, for fixed $\mathbf{k}=\mathbf{q}$ and $\beta\hbar\omega \ll 1$.

We assert that in the HL we can extract the gradient from the integral, and what remains is a transport coefficient, D. Indeed, this is really the *definition* of the HL. With $\hat{\mathbf{k}}$ a unit vector, this generic coefficient is just

$$
\begin{aligned}
D &= \lim_{\epsilon\to 0^+}\lim_{k\to 0}\int_0^\infty e^{-\epsilon t}\,\hat{\mathbf{k}}\cdot K_{\mathbf{JJ}}(\mathbf{k},t)\cdot\hat{\mathbf{k}}\,dt \\
&= \lim_{z\to 0}\lim_{k\to 0}\big[\hat{\mathbf{k}}\cdot\tilde{K}_{\mathbf{JJ}}(\mathbf{k},z)\cdot\hat{\mathbf{k}}\big]_{z=i\epsilon},
\end{aligned}
\tag{8–83}
$$

which is essentially the standard form found in *any* correlation-function theory (e.g., Forster, 1975). Strangely, the order of the limits here is important, which one would not expect in a well-defined formulation. We shall examine this further presently, after first illustrating the situation with a simple model.

Consider a steady-state process in which n is measured and $\mathbf{J}$ predicted, which should yield the diffusion coefficient in the HL. Only the xx-component will be considered, and in the Boltzmann limit. Then, in first-order perturbation theory,

$$
\begin{aligned}
K_{J_x J_x}(\mathbf{k},\omega)_{\mathrm{B}} &= \frac{\omega^2}{k^2}\hat{k}_x\hat{k}_x K_{nn}(\mathbf{k},\omega)_{\mathrm{B}} \\
&\simeq \frac{\omega^2}{k^2}\hat{k}^2\frac{n_0(2\pi\beta m)^{1/2}}{k}[1-\beta n_0 v(\mathbf{k})]e^{-\frac{1}{2}\beta m\omega^2/k^2}.
\end{aligned}
\tag{8–84}
$$

A short calculation—confirmed in Problem 8.3—yields

$$
\begin{aligned}
\int_0^\infty e^{-\epsilon t}\,K_{J_x J_x}(\mathbf{k},t)_{\mathrm{B}}\,dt &\simeq -2\pi^2 n_0\hat{k}^2[1-\beta n_0 v(\mathbf{k})]\frac{(\beta m)^{1/2}}{k}e^{\epsilon^2\beta m/4k^2} \\
&\quad\times\big[2U(\tfrac{5}{2},\epsilon\sqrt{\beta m}/k)-U(\tfrac{1}{2},\epsilon\sqrt{\beta m}/k)\big],
\end{aligned}
\tag{8–85}
$$

where $U(a, x)$ is the parabolic cylinder function. For a moderate the following limiting values are appropriate (e.g., Abramowitz and Stegun, 1964):

$$U(a, x) \xrightarrow[x \to 0]{} \frac{\sqrt{\pi}}{2^{a/2 + 1/4} \, \Gamma\left(\frac{3}{4} + \frac{1}{2}a\right)}$$

$$\xrightarrow[x \to \infty]{} \frac{e^{-\frac{1}{4}x^2}}{x^{a+1/2}}. \tag{8-86}$$

With these limiting forms we find that, indeed, the expression (8–83) for D depends on which limit is performed first:

$$D \xrightarrow[k \to 0]{} \epsilon^{-1},$$

$$\xrightarrow[\epsilon \to 0]{} 0. \tag{8-87}$$

There are several observations about this simple calculation which should be made at this point, the first being that the covariance function itself is completely well behaved in this order, as is evident from Eq.(8–84). One takes ω or k, in any order, to zero or infinity and obtains the same vanishing result. The Laplace transform, however, has a pole at the origin if the HL is taken first, indicating that the $(k \to 0)$-limit of $K_{J_z J_z}(\mathbf{k}, t)_\mathrm{B}$ may *not* vanish as $t \to \infty$.

Most important, our experience with the general perturbation expansion, as exhibited in Appendices B and C, indicates that the frequency dependence of the covariance functions in the Boltzmann limit remains identical with that displayed in Eq.(8–84) in all orders. Anomalies arise only if the HL is invoked (see Appendix C). Hence, it appears that any kind of straightforward perturbation theory will be totally inadequate for discussing transport coefficients, because the essential time dependence in the hydrodynamic region is not obtained in this way. In turn, this means that a naive density expansion of transport coefficients obtained in the HL may not be well behaved either—such density expansions may exist, but they will have to account in some detail for the essential hydrodynamic structure of the covariance (or correlation) functions. The development in Appendix C has not been carried out far enough to determine if any logarithmic terms appear in second order, although they are not ruled out. To this writer's knowledge there is not a shred of experimental evidence requiring such terms, whereas there is ample reason to believe that they may be simply an artifact of the hydrodynamic limit.

One would think, in any event, that a density expansion of transport coefficients would be only marginally useful. If the density of a system is high enough so that collisions are very important and the normal hydrodynamic modes are supported, then a reasonably large number of terms must be calculated. Otherwise, the system must be somewhat dilute, and we are faced with the possibility that linear departures from equilibrium are inadequate for describing the ensuing transport processes. We shall return to a more specific discussion of this 'diluteness' problem in the following chapter.

STRUCTURE OF COVARIANCE FUNCTIONS

Time and again one is forced back to the problem of ascertaining the analytic structure of covariance functions in order to make progress. As we have also noted repeatedly, detailed calculation of these functions is a decidedly nontrivial problem at this time, so that it would be very useful to obtain equations satisfied by the covariance function from which its general properties might be readily deduced.

One such method, which has proved to be of some utility, begins with the observation made in connection with Eq.(6–56) that K_{AB} satisfies all the requirements for the scalar product of a linear vector space. The operators themselves form the state vectors of this 'superspace' and, without fear of confusion in the remainder of this section, we denote the states by $|\hat{A}\rangle$. Hence, the scalar product reads

$$\langle \hat{B}|\hat{A}\rangle = K_{AB}, \qquad (8\text{–}88)$$

at time $t=0$.

But the Heisenberg operators evolve from $t=0$ in a well-defined manner, which we restate as

$$\partial_t \hat{A} = \frac{1}{i\hbar}[\hat{A}, \hat{H}] \equiv -iL\hat{A}, \qquad (8\text{–}89)$$

in terms of the so-called Liouville operator that operates on operators in the prescribed way. (We here define L with a difference in sign from what is customary, so that our definition of scalar product will be in harmony with others.) Time evolution of states in the superspace is then described by

$$|\hat{A}(t)\rangle = e^{-iLt}|\hat{A}(0)\rangle. \qquad (8\text{–}90)$$

As long as we deal with only with Hermitian operators in the Hilbert space, L is Hermitian in the superspace and Eq.(8–90) describes a unitary transformation which rotates the vector $|\hat{A}\rangle$. Thus, the scalar product between vectors $|\hat{A}(t)\rangle$ and $|\hat{B}\rangle$ can be written

$$K_{AB}(t) = \langle \hat{B}|\hat{A}(t)\rangle = \langle \hat{B}|e^{-iLt}|\hat{A}\rangle, \qquad (8\text{–}91)$$

which is just the regular time-dependent covariance function. Spatial dependence is momentarily suppressed because it has no bearing on the following development.

One can now define a projection operator onto the state $|\hat{A}\rangle$ in this superspace, as follows:

$$P \equiv \frac{|\hat{A}\rangle\langle \hat{A}|}{\langle \hat{A}|\hat{A}\rangle} = 1 - Q, \qquad (8\text{–}92)$$

such that Q projects orthogonally to P. It is easy to demonstrate that P is idempotent and that P and Q commute. Below we shall also need the following operator identity for any two operators in the superspace:

$$\frac{1}{X+Y} = \frac{1}{X} - \frac{1}{X}Y\frac{1}{X+Y}. \qquad (8\text{–}93)$$

In order to analyze the matrix element in Eq.(8–91) it is useful to take the Fourier-Laplace transform with complex z and write formally

$$\tilde{K}_{AB}(z) = \langle\hat{B}|\frac{i}{z-L}|\hat{A}\rangle\,, \tag{8–94}$$

as in Eq.(6-64). With the observation that $L = LP + LQ$, and the identity (8–93), some algebra leads to the expression

$$z\tilde{K}_{AB}(z) = iK_{AB}(0) + \frac{1}{K_{AA}(0)}\langle\hat{B}|L|\hat{A}\rangle\tilde{K}_{AB}(z)$$
$$+ \frac{1}{K_{AA}(0)}\langle\hat{B}|LQ\frac{1}{z-LQ}L|\hat{A}\rangle\tilde{K}_{AB}(z)\,, \tag{8–95}$$

where $K_{AB}(0) = K_{AB}(t{=}0)$. Define two quantities

$$\Omega_{AB} \equiv \frac{1}{K_{AA}(0)}\langle\hat{B}|L|\hat{A}\rangle = \frac{1}{K_{AA}(0)}\langle\hat{B}|\dot{\hat{A}}\rangle = \frac{K_{\dot{A}B}(0)}{K_{AA}(0)}\,, \tag{8–96a}$$

$$\tilde{\Sigma}_{AB} \equiv \frac{1}{K_{AA}(0)}\langle\hat{B}|LQ\frac{i}{z-LQ}L|\hat{A}\rangle$$
$$= \frac{1}{K_{AA}(0)}\langle\dot{\hat{B}}|Q\frac{i}{z-QLQ}Q|\dot{\hat{A}}\rangle\,, \tag{8–96b}$$

the last because Q is also idempotent. One can now rewrite Eq.(8–95) as

$$\tilde{K}_{AB}(z) = \frac{iK_{AB}(0)}{z - \Omega_{AB} + i\tilde{\Sigma}_{AB}(z)}\,, \tag{8–97}$$

which describes an initial-value problem. Owing to time-reversal symmetry, Ω_{AB} is often zero.

There is a definite advantage to being able to write the covariance function in this form, because it reveals some details of its structure, at least in the frequency plane. In the unlikely case that $\tilde{\Sigma}_{AB}(z)$ is actually independent of z, the covariance function would have a Lorentzian shape. The quantity $\tilde{\Sigma}_{AB}$ acts much like a self-energy and is itself a correlation function of projected time derivatives, wherein the slowly-decaying modes have been projected out. One might hope that it has a simpler behavior than the covariance function, so that the study of the latter is somewhat facilitated. For example, by means of a series of physical arguments as to the behavior of $\tilde{\Sigma}_{AB}$, one observes directly in Eq.(8–97) hydrodynamic mode structure, as we shall see.

If one applies the inverse Fourier-Laplace transformation to Eq.(8–97), there emerges an integro-differential equation determining the covariance function:

$$(\partial_t + i\Omega_{AB})K_{AB}(t) + \int_0^t \Sigma_{AB}(t{-}t')K_{AB}(t')\,dt' = 0\,. \tag{8–98}$$

The kernel of this equation is called the *memory function* and in another context this formalism has been used extensively in hydrodynamics, particularly in the derivation of generalized Langevin equations (Kubo, 1966). Projection operators were introduced in this context by Zwanzig (1961) and Robertson (1966), and their use extended considerably by Mori (1965a,b). The technique has also been used to study hydrodynamic modes arising from broken symmetries (e.g., Forster, 1875), in which the formal discussion is greatly facilitated. Unfortunately, the projection operators do not always have an obvious or simple construction, so that these expressions are as complicated as any others.

Although Eqs.(8–97) and (8–98) are usually taken to determine the correlation function, given a model for the memory function, it is also possible to turn the interpretation around. Because the covariance function can always be written in the *form* of Eq.(8–97), we take this equation to be an *ansatz*. For locally-conserved densities this takes the form

$$\tilde{K}_{ee}(\mathbf{k}, z) = \frac{i}{z + i\tilde{D}_e(\mathbf{k}, z)} K_{ee}(\mathbf{k}, t{=}0) , \qquad (8\text{–}99)$$

and its utility depends on our ability to solve the following integral equation for D_e, given K_{ee}:

$$\partial_t K_{ee}(t) = - \int_0^t K_{ee}(t') D_e(t{-}t') \, dt' . \qquad (8\text{–}100)$$

This is probably not of great value in general, but one can extract some useful information.

First of all, from Eq.(6–62) one can generate the identity

$$K_{eB}(\mathbf{k}, t{=}0) + iz\tilde{K}_{eB}(\mathbf{k}, z) = i\mathbf{k} \cdot \tilde{K}_{JB}(\mathbf{k}, z) , \qquad (8\text{–}101)$$

for any operator $\hat{B}$. This can be used in different forms to obtain the further identity

$$\tilde{K}_{ee}(\mathbf{k}, z) = \frac{1}{z^2}\mathbf{k} \cdot \tilde{K}_{JJ}(\mathbf{k}, z) \cdot \mathbf{k} + \frac{i}{z} K_{ee}(\mathbf{k}, t{=}0) . \qquad (8\text{–}102)$$

Comparison with Eq.(8–99) and some further algebra yields

$$\mathbf{k} \cdot \tilde{K}_{JJ}(\mathbf{k}, z) \cdot \mathbf{k} = \frac{z\tilde{D}_e(\mathbf{k}, z)}{z + i\tilde{D}_e(\mathbf{k}, z)} K_{ee}(\mathbf{k}, t{=}0) . \qquad (8\text{–}103)$$

Thus, if $\tilde{D}_e$ behaves as k^2 times a well-behaved D, then in the hydrodynamic limit D is identified as the transport coefficient, Eq.(8–83).

Secondly, we note that Eq.(8–100) always has a particular solution proportional to the derivative of a δ-function. In addition, the $z-k$ relation implied by the pole in Eq.(8–99) has the form

$$iz = \tilde{D}(\mathbf{k}, z) \xrightarrow[k \to 0]{} k^2 d(\mathbf{k}, z)$$

$$\simeq k^2 d(\mathbf{0}, 0) + O(k^4) . \qquad (8\text{–}104)$$

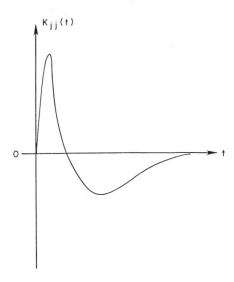

Fig. 8–2. Expected qualitative temporal behavior of the current-current covariance function.

Together, these observations suggest that $K_{JJ}(t)$ has the $(k \to 0)$-behavior

$$K_{JJ}(t) \propto \delta'(t) + k^2 A(t) . \qquad (8\text{–}105)$$

As $k \to 0$ the pole may actually become independent of z, so that in the hydrodynamic limit the $\delta'(t)$-term is replaced by $\delta(t)d(0,0)$. One then suspects that a correct treatment of $K_{JJ}(\mathbf{x},t)$ in the HL should yield two contributions: one proportional to $\delta(t)$ multiplied by the static transport coefficient, and another which does not contribute in the HL, but which may make a negative contribution to the integrands in Eqs.(8–82) and (8–83)—and may or may not drop off slowly.

This qualitative analysis actually clarifies a number of points, and we begin by recalling Figure 6–2, illustrating the results of computer simulation of the current-current correlation function. One gains the impression that the covariance function has the general behavior depicted in Figure 8–2. Thus, the fact that the total time integral of the covariance function may vanish—as in the example of Eqs.(8–85)-(8–87), and as once suggested by Lebowitz (1972) in the classical case—is readily understood.

Moreover, we can now see why possibly the order of the limiting operations in Eq.(8–83) might be important, and not necessarily a shortcoming of the theory. If the ϵ-limit is performed first the entire integral vanishes. But *all* of the normal modes are included in the integrand, including the rapidly decaying ones. Yet we are generally interested only in the long-lived modes $(k \rightarrow 0)$, for it is they which relate to the phenomenological transport coefficients. Hence, in order to identify them, these modes must first be isolated by means of the HL before performing the time integration, or they become lost in the total integral. These hydrodynamic modes may be long-lived, but the piece of the correlation function from which they arise need not be—indeed, this portion may simply be of considerably larger amplitude than the 'fast' modes, and of much larger spatial range. The isolation of these separate contributions is precisely what the projection-operator techniques mentioned above were designed to accomplish.

Although we gain some understanding, the problems are far from solved. This is clear from the perturbation model of Eq.(8–85), for which the 'correct' ordering of limits, according to Eq.(8–87), leads to a divergent result. The methods by which the hydrodynamic limit is invoked may be ill defined and poorly understood, thereby creating artificial mathematical difficulties. In order to understand and overcome many of these sore points it will be necessary to develop a considerably better ability than we have at present for determining the structure of covariance functions. Undoubtedly the linear transport processes are properly described by the function $\tilde{K}_{AB}(\mathbf{k}, z)$, but the structure of these functions must be much better known before rushing to take various limits. Until this problem is resolved we are not going to understand in much further detail irreversibility and relaxation, let alone nonlinear problems.

Problems

8.1 Generalize the theory of linear dynamical response to provide a nonlocal expression for electrical conductivity, compare the result with Eq.(8–19), and thus verify Eq.(8–20).

8.2 Carry out the indicated steps to verify the reciprocity relation (8–34).

8.3 Carry out the necessary integrations to verify the result of Eq.(8–85).

8.4 Derive the identities of Eqs.(8–101) and (8–102), and then carry out the algebra leading to Eq.(8–103).

REFERENCES

Abramowitz, M., and I.A. Stegun (eds.): 1964, *Handbook of Mathematical Functions*, AMS 55, Natl. Bur. Stds., Washington.

Adu-Gyamfi, D.: 1986, 'On the Gauge Invariance of Linear Response Theory', *J. Phys. A: Math. Gen.* **19**, 3443.

Ailawadi, N.K., B.J. Berne, and D. Forster: 1971, 'Hydrodynamics and Collective Angular-Momentum Fluctuations in Molecular Fluids', *Phys. Rev.* **A3**, 1462.

Currie, I.G.: 1974, *Fundamental Mechanics of Fluids*, McGraw-Hill, New York.

Dufour, L.: 1873, *Pogg. Ann.* **148**, 490.

Ettinghausen, A, ':1887', *Ueber eine polare Wirkung des Magnetismus auf die galvanische Wärme in gewissen Substanzen* **Ann. d. Phys.**, 31. 737.

Ettinghausen, A., and W. Nernst: 1887, *Akad. Sitzungsber.* **96** .

Feddersen, W.: 1873, *Pogg. Ann.* **148**, 308.

Ferrari, P.A., S. Goldstein, and J.L. Lebowitz: 1985, 'Diffusion, Mobility, and The Einstein Relation', in J. Fritz, A. Jaffe, and D. Szász (eds.), *Statistical Physics and Dynamical Systems*, Birkhäuser, Boston,p.405.

Forster, D.: 1975, *Hydrodynamic Fluctuations, Broken Symmetry, and Correlation Functions*, Benjamin, Reading, MA.

Goldstone, J.: 1961, 'Field Theories with "Superconductor" Solutions', *Nuovo Cimento* **19**, 154.

Harman, T.C., and J.M. Honig: 1967, *Thermoelectric and Thermomagnetic Effects and Applications*, McGraw-Hill, New York.

Hirschfelder, J.O., C.F. Curtiss, and R.B. Bird: 1954, *Molecular Theory of Gases and Liquids*, Wiley, New York.

Huebner, R.P.: 1972, 'Thermoelectricity in Metals and Alloys', in H. Ehrenreich, F. Seitz, and D. Turnbull (eds.), *Solid State Physics, Vol.27*, Academic Press, New York.

Jost, W.: 1960, *Diffusion*, Academic Press, New York.

Kadanoff, L.P., and P.C. Martin: 1963, 'Hydrodynamic Equations and Correlation Functions', *Ann. Phys. (N.Y.)* **24**, 419.

Kubo, R.: 1966, 'The Fluctuation-Dissipation Theorem', *Repts. Prog. Phys.* **29**, 255.

Landau, L.D., and E.M. Lifshitz: 1959, *Fluid Mechanics*, Addison-Wesley, Reading, MA.

Lebowitz, J.L.: 1972, 'Hamiltonian Flows and Rigorous Results in Nonequilibrium Statistical Mechanics', in S.A. Rice, K.F. Freed, and J.C. Light (eds.), *Statistical Mechanics: New Concepts, New Problems, New Applications*, Univ. of Chicago Press, Chicago.

Leduc, S.-A.: 1887, 'Nouvelle Méthode pour la Mesure des Champs Magnétiques', *J. de Phys.* **6**, 184.

Ludwig, C.: 1856, *Sitzungsber. Akad. Wiss. Wien, Math.-Naturw. Kl.* **20**, 539.

Luttinger, J.M.: 1964, 'Theory of Thermal Transport Coefficients', *Phys. Rev.* **135**, A1505.

Mason, E.A., R.J. Munn, and F.J. Smith: 1966, 'Thermal Diffusion in Gases', in D.R. Bates (ed.), *Advances in Atomic and Molecular Physics, Vol.2*, Academic Press, New York.

Mori, H.: 1956, 'A Quantum-Statistical Theory of Transport Processes', *J. Phys. Soc. Japan* **11**, 1029.

Mori, H.: 1958, 'Statistical-Mechanical Theory of Transport in Fluids', *Phys. Rev.* **112**, 1829.

Mori, H.: 1965a, 'Transport, Collective Motion, and Brownian Motion', *Prog. Theor. Phys.* **33**, 423.

Mori, H.: 1965b, 'A Continued-Fraction Representation of the Time-Correlation Functions', *Prog. Theor. Phys.* **34**, 399.

Nakajima, S.: 1958, 'On Quantum Theory of Transport Phenomena', *Prog. Theor. Phys.* **20**, 948.

Navier, C.-L.-M.-H.: 1821, 'Sur les lois du mouvement des fluides, en ayant égard à l'adhesion des molécules', *Ann. Chimie* **19**, 244.

Nernst, W.: 1888, *Z. Physik. Chem.* **2**, 613.

Peltier, J.C.A.: 1834, 'Nouvelles Expériences sur la Caloricité des courans électriques', *Ann. Chim. Phys.* **56**, 371.

Puff, R.D., and N.S. Gillis: 1968, 'Fluctuations and Transport Properties of Many-Particle Systems', *Ann. Phys. (N.Y.)* **6**, 364.

Righi, A.: 1883, *Acc. Sci. Mem.* **5**.

Robertson, B.: 1966, 'Equations of Motion in Nonequilibrium Statistical Mechanics', *Phys. Rev.* **144**, 151.

Seebeck, T.J.: 1826, *Pogg. Ann.* **6**, 133.

Soret, Ch.: 1879, *Arch. de Genève* **2**, 48.

Soret, Ch.: 1893, *Arch. Sci. Phys. nat. (Genève)* **29**, 4.

Thomson, W.: 1854, 'A Mechanical Theory of Thermoelectric Currents in Crystalline Solids', *Proc. Roy. Soc. Edinburgh* **3**, 255.

Voight, W.: 1903, *Gött. Nachr.* , 87.

Zubarev, D.N.: 1974, *Nonequilibrium Statistical Mechanics*, Consultants Bureau, New York.

Zwanzig, R.: 1961, 'Statistical Mechanics of Irreversibility', in W.E. Brittin, B.W. Downs, and J.Downs (eds.), *Lectures in Theoretical Physics, Vol.III*, Interscience, New York.

Chapter 9

Sound Propagation and Attenuation

Propagation of sound waves has traditionally been viewed as a hydrodynamic phenomenon. Indeed, the discussion in the preceding chapter intimated that two long-lived modes in a simple fluid are propagating sound modes. Unquestionably, long-wavelength, low-frequency sound propagation and attenuation are intimately related to the transport properties of the medium in the hydrodynamic limit. But the important physical properties of the acoustic field far transcend this limit, and a general theory of sound should be independent of it. In this chapter we illustrate how the present theory of irreversible processes provides a general description of sound propagation and attenuation for arbitrary wavelengths and frequencies.

The wave-like mechanical disturbances characterizing sound propagation represent a transport of energy, and ideally one might expect this energy to propagate through the medium. This ideal is never realized, of course, and energy is always absorbed in one way or another by the constituent particles of the system. Indeed, this is a beautifully simple manifestation of the molecular structure of matter. The propagation will certainly be affected by the basic particle-particle interactions, and so the attenuation of sound waves provides an important tool with which to study the basic properties of the medium itself. In addition, at very high frequencies (ultrasound) the mechanisms leading to sonic attenuation have a nonlocal character, thereby allowing for significant tests of theories of sound and hydrodynamics.

In the sequel we shall first review briefly the classical theory of sound, as well as the modern conventional approaches to a microscopic description. A linear theory of ultrasonic attenuation will then be developed, employing the formalism of irreversible processes from the preceding chapters. Although we shall include some simple applications in order to illustrate the formalism, emphasis is on the fundamental structure of the theory of sound and not on the detailed mechanisms of attenuation in specific systems.

A. The Classical Theory of Sound

We focus primarily on fluids—gas or liquid—and recall that two of the five hydrodynamic modes are sound modes. The classical theory then imposes certain restrictions on the number density of the system. The density can not be too high, on the one hand, for then three-body and higher interactions become important and microscopic calculations are intractable. On the other hand, the fluid can not be too dilute or the basis of linear hydrodynamics breaks down. That is, the macroscopic wavelengths of interest must be much longer than a mean-free-path, for hydrodynamics is a collision-dominated regime. When the system is extremely dilute special treatment is required, a point to which we shall return.

The classical theory of sound refers to small-amplitude disturbances, so that the basic equations are linear: Newton's equation of motion and the conservation laws, resulting in Euler's equation. Although it is most physical to describe the disturbance in the medium in terms of local pressure variations, it is mathematically convenient to proceed in an alternate manner. One always presumes the fluid to be irrotational—$\nabla \times \mathbf{v} = 0$—so that it is natural to define a *velocity potential* $\phi(\mathbf{x}, t)$ by writing the velocity field as $\mathbf{v} = \nabla \phi$. Then the equations of motion and continuity lead to the wave equation

$$\frac{\partial^2 \phi}{\partial t^2} = c_0^2 \nabla^2 \phi, \tag{9-1}$$

where we identify the speed of sound from

$$c_0^2 \equiv -\frac{V}{mn} \left(\frac{\partial P}{\partial V} \right)_S = \frac{1}{mn\kappa_S}, \tag{9-2}$$

and κ_S is the adiabatic compressibility. Moreover, each component of $\mathbf{v}$ also satisfies Eq.(9-1).

It is interesting to note that in the first theory of sound in gases, given by Newton in his *Principia* (1687), the speed of sound was thought to be related to the isothermal compressibility. Newton's theory was completely mechanical, so this is not too surprising, and the error was eventually corrected by Laplace (1816). The two are related by

$$\kappa_T = \frac{C_P}{C_V} \kappa_S, \tag{9-3}$$

where C_x is the heat capacity at constant x, and this ratio provides a measure of the error. Because fluids do not support shear, we are concerned principally with longitudinal waves consisting of alternate compressions and rarefactions. But in a solid it is possible to induce transverse waves as well, and in that case one must distinguish between longitudinal and transverse sound speeds—c_ℓ and c_t, respectively.

Now, a particular solution to Eq.(9-1) is an infinite plane wave, propagating unimpeded. But, as Laplace also knew, no sound wave propagates through a medium without losing energy. In 1845 Stokes found that sound waves are damped exponentially with an attenuation constant given by

$$\alpha = \frac{2\eta\omega^2}{3mnc_0^3}, \tag{9-4}$$

where η is the shear viscosity (Stokes, 1845). Later Kirchhoff (1868) showed that the contribution from heat conduction is the same order of magnitude and also proportional to ω^2. Physically, the loss of energy arises because the system is not in complete thermal equilibrium under passage of the sound wave. Where ΔP is positive the compression has not been isothermal, because the process is too fast for the temperature to equalize. It is closer to the truth to consider the process adiabatic, although some thermal energy must flow from higher to lower temperature. Hence, some energy from the wave goes into heat. Moreover, the heating must increase with frequency owing to shorter wavelengths, even though the periods are shortened. We see, then, that dissipation of the sound wave arises fundamentally because of a necessary coupling of density and energy fluctuations induced by the disturbance.

Formally, one can modify the classical theory to include dissipative processes by using the linearized Navier-Stokes equation, rather than Euler's equation, as the fluid equation of motion. If we define a friction constant

$$f \equiv \tfrac{4}{3}\eta + \varsigma,\qquad (9\text{--}5)$$

where ς is the bulk viscosity, then a re-derivation yields the modified wave equation for the velocity potential:

$$\frac{\partial^2 \phi}{\partial t^2} = c_0^2 \nabla^2 + \frac{f}{mn}\frac{\partial}{\partial t}(\nabla^2 \phi).\qquad (9\text{--}6)$$

In the spirit of linear hydrodynamics one presumes the effects of viscosity to be small, so that Eq.(9–6) still exhibits plane-wave solutions. Now, however, the wave motion is damped.

If the disturbance creating the sound waves is maintained by a continuous source at fixed frequency, then ω must be real, and can be either positive or negative. But for the disturbance to die out sufficiently far from the source it is then necessary for the propagation vector **k** to have a complex magnitude. In one dimension, with no loss of generality, we write

$$\bar{k} = \frac{2\pi}{\lambda} + i\alpha,\qquad (9\text{--}7)$$

and define the phase velocity as

$$c \equiv \frac{\lambda}{2\pi}\omega = \left[\mathrm{Re}\left(\frac{\bar{k}}{\omega} \right) \right]^{-1}.\qquad (9\text{--}8)$$

Note that α must be even and c odd in ω. These relations describe travelling waves, and the elementary solution to Eq.(9–6) in one dimension is seen to be, for positive z,

$$\phi(z,t) = e^{-\alpha z}\, e^{i(kz - \omega t)},\qquad (9\text{--}9)$$

where $k \equiv \mathrm{Re}(\bar{k})$. Let

$$R \equiv \frac{f}{mnc_0^2}.\qquad (9\text{--}10)$$

Then substitution into the wave equation allows us to identify the attenuation constant from

$$\alpha^2 = \frac{\omega^2}{2c_0^2(1 + \omega^2 R^2)}[(1 + \omega^2 R^2)^{1/2} - 1] \xrightarrow[R \to 0]{} 0,\qquad (9\text{--}11)$$

and the phase velocity through

$$c^2 = \frac{2(1 + \omega^2 R^2)}{1 + (1 + \omega^2 R^2)^{1/2}} \, c_0^2 \xrightarrow[R \to 0]{} c_0^2 . \tag{9-12}$$

If we set $\varsigma = 0$, which is exactly its value in an ideal monatomic gas, and again take $\omega R \ll 1$, then Stokes' result (9–4) is recovered. (Note that this limit is equivalent to the requirement $k^2 \gg \alpha^2$.)

The frequency dependence of α and c manifests the phenomenon of dispersion. Fourier transformation of the wave equation (9–6) yields the form

$$[\omega^2 - c_0^2 k^2 + F(\mathbf{k}, \omega)] \, \phi(\mathbf{k}, \omega) = 0 , \tag{9-13}$$

where now $F(\mathbf{k}, \omega)$ represents the completely general effects of dissipation. Because ϕ is not generally zero we deduce the dispersion relation

$$c_0^2 k^2 = \omega^2 + F(\mathbf{k}, \omega) , \tag{9-14}$$

allowing us to solve for k in terms of real ω, thereby determining c and α. We return to further discussion of these matters below.

As an aside, if the disturbance is not maintained, but merely evolves from some initial value, then $k = 2\pi/\lambda$ is real and λ is the wavelength of the disturbance. For the latter to decay after a sufficiently long time it is necessary that ω then be complex, and we write

$$\omega = \omega' - i\tau^{-1} , \tag{9-15a}$$

where τ is called the relaxation time. The phase velocity is

$$c = \frac{\lambda}{2\pi}\omega' = \mathrm{Re}\left(\frac{\omega}{k}\right) , \tag{9-15b}$$

and the solutions now represent standing waves. One obtains temporal damping, and only in the small-dissipation limit are the two attenuation constants simply related by $\tau \simeq \alpha c_0$. This general scenario of complex frequencies is that employed in the description of linear hydrodynamics earlier. For the most part we shall be interested in travelling waves generated by a driving source at real frequency ω, and which propagate with complex wave number k.

The classical theory is not yet complete, because the effects of heat conduction have not been included. This was first done by Kirchhoff (1868), and he found that

$$\alpha = \frac{\omega^2}{2mnc_0^3}\left[\tfrac{4}{3}\eta + \varsigma + \Lambda(C_V^{-1} - C_P^{-1})\right] , \tag{9-16}$$

where Λ is the thermal conductivity. This expression is valid in the long-wavelength, or hydrodynamic limit, and plainly represents a simple extension to the friction term in Eq.(9–6).

It is clear that the classical theory of sound involves a number of approximations, so that at this point a critique is in order. We hasten to emphasize, however, that there can be little quarrel with the linear approximation. In most common problems of acoustics this a quite realistic approximation, and the study of macrosonics as related to large-amplitude disturbances must be relegated to a separate theory of nonlinear acoustics. One is entitled to some expression of dissatisfaction, though, with a theory which is fundamentally limited to very long wavelengths and low frequencies, as is the case with that based on linear hydrodynamics. In addition, the classical theory presumes small dissipation, a limitation which now seems less realistic in view of broad current interest in ultrasonics and hypersonics (sound disturbances in the range of frequencies at the limit supportable by the medium).

As already noted, when the wavelength of sound is made significantly shorter, so that the effects of viscosity and heat conduction are no longer small, the validity of hydrodynamics itself becomes questionable and it is not at all clear that a wave equation can be derived in an unambiguous way. Hence, the presumption of the classical exponential damping becomes tenuous, an observation already made in connection with time correlations in fluids. Finally, at shorter wavelengths one must be skeptical about being able to calculate the dissipative parameters appearing in Eq.(9–16).

The attenuation described by Eqs.(9–10) and (9–16) has been well verified in monatomic gases at very low frequencies. As the frequency is raised, however, both the phase velocity and the absorption appear to increase monotonically, and the ω^2-dependence of the latter induces an enormous increase (e.g., Greenspan, 1950). A standard approach to extending the theory to higher frequencies and densities has been to obtain higher-order solutions to the Boltzmann equation by means of the Chapman-Enskog development, and then to derive a linearized hydrodynamics from which one can obtain a more reliable dispersion relation. One has in this way a means for incorporating particle-particle interactions directly into the expressions for the dissipative parameters. Some of these procedures have been reviewed in Chapter 1, and by Uhlenbeck and Ford (1963). Subsequently much more detail was provided by Foch and Ford (1970), whose development we sketch here.

Let us define the *Eucken number* as

$$g \equiv \frac{\Lambda}{C_V \eta} = \frac{2}{3} \frac{m\Lambda}{\kappa\eta},\tag{9–17}$$

and a quantity

$$\gamma \equiv \frac{\eta}{mnc_0},\tag{9–18}$$

where κ is Boltzmann's constant. In terms of the dimensionless parameters

$$\xi \equiv \omega\frac{\gamma}{c_0}, \qquad x \equiv \gamma k,\tag{9–19}$$

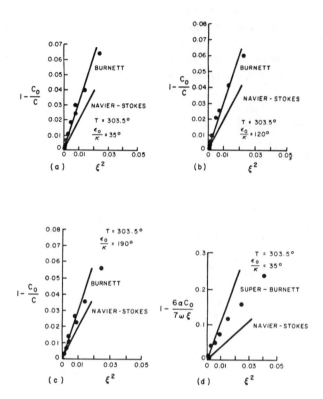

Fig. 9–1. Comparison of Eq.(9–20a) with the data of Greenspan (1956) in (a) neon, (b) argon, and (c) krypton; comparison of Eq.(9–20b) with the data for neon (d). The parameter ϵ_0/κ refers to the Lennard-Jones potential model. [*Reproduced with permission from Foch and Ford (1970).*]

the linearized hydrodynamics of the Boltzmann equation yields

$$\frac{c_0}{c} = \mathrm{Re}\left(\frac{x}{\xi}\right) = 1 - \left(\tfrac{2}{3} + \tfrac{2}{3}g - \tfrac{3}{50}g^2\right)\xi^2 + \cdots, \tag{9-20a}$$

$$\alpha\frac{c_0}{\omega} = \mathrm{Im}\left(\frac{x}{\xi}\right) = \left(\tfrac{2}{3} + \tfrac{1}{5}g\right)\xi - \left(\tfrac{20}{27} + \tfrac{14}{9}g - \tfrac{7}{75}g^2 - \tfrac{7}{250}g^3\right)\xi^3 + \cdots, \tag{9-20b}$$

where the leading term in α is just the Stokes-Kirchhoff contribution.

In higher order one obtains the so-called Burnett, and super-Burnett equations, and thus further correction terms depending on specific potential models. Foch

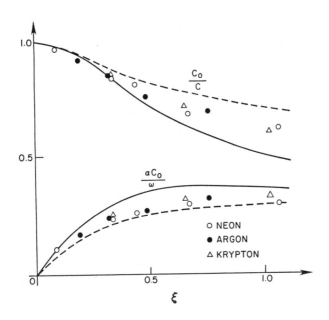

Fig. 9–2. The dispersion and absorption of sound as functions of the dimensionless parameter ξ, Eq.(9–19), compared with the extended data of Greenspan (1956). [*Reproduced with permission from Foch and Ford (1970).*]

and Ford also carry out similar calculations proceeding directly from the linearized Boltzmann equation, without going through the hydrodynamic equations, and obtain comparable results with different numerical factors. Figure 9–1 compares the results of these calculations with the low-density data of Greenspan (1956), and the deviations are obvious at all but the lowest frequencies. In Figure 9–2 the extended data at intermediate frequencies are exhibited along with the hydrodynamic results using the Navier-Stokes equations. We see that both the phase velocity and the attenuation constant appear to level off as the frequency is increased. Although the theory tends to follow the experimental results qualitatively, it is clear that quantitative agreement is lacking. Later efforts have extended the theory to higher frequencies for the case of Maxwell molecules (interaction potential proportional to r^{-5}), and the comparison with the data is taken up to $\xi = 1$ (Ford and Fuentes, 1972).

From the preceding discussion the following simple picture of sonic attenuation, at least in gases, emerges. The sound disturbance creates a pressure change in each cycle of the wave resulting in a reduction in volume which is not quite in phase with the change in density, so that particle interactions with the surrounding gas cause the compressed portion to lose energy through diffusion of heat and momentum. These irreversible effects of heat conduction and viscosity convert part of the organized motion of the sound mode into disorganized translational motion. This is referred to as a translational relaxation of the sonic pressure. The attenuation induced by these effects has been seen to agree quite well with experiments in dilute monatomic gases at low frequencies, and appears to be exponential. In liquids, however, the details of the molecular motion are not as well understood and the situation is not as clear, even at low frequencies.

One finds in many gases and liquids that α is *not* proportional to ω^2, as in Eq.(9–16), over a significant range of frequencies. For example, in diatomic and polyatomic gases viscosity and heat conduction are only a part of the attenuation mechanism, for there is also an irreversible transfer of energy from the sound wave to the internal degrees of freedom of the molecules. There are then additional relaxation processes in operation, and a detailed theory of sound propagation and attenuation can lead to a deep understanding of these molecular processes (e.g., Lambert, 1962).

Microscopic Connection

The conventional modern theory of acoustical attenuation focuses on the energy transfer due to complicated internal processes in the many-body system. It is useful to outline briefly this approach for purposes of later comparison. Let us take $E(z)$ to be the local energy density for a one-dimensional problem and calculate the rate of change of $E(z)$ with respect to z. This will be some functional of $E(z)$ itself, so that an expansion and linear approximation yields

$$\frac{dE(z)}{dz} \simeq -\alpha_I E(z) ,$$ (9–21a)

with solution

$$E(z) = E(0)e^{-\alpha_I z} .$$ (9–21b)

The linear damping constant α_I is related to the familiar constant by $\alpha_I = 2\alpha$, owing to the difference between intensity and amplitude. Thus, the presumption of exponential damping is inherent to the approach. If the net energy loss per unit volume across a slab in time δt is taken as

$$\delta E(z) \simeq E(z)\, 2\alpha\, \delta z ,$$

then the power dissipated per unit volume is

$$Q(z) \equiv \frac{\delta E(z)}{\delta t} = 2\alpha c_0 E(z) .$$ (9–22)

This expression is valid at $z = 0$ in particular, so that the attenuation per unit length is just

$$\alpha = \frac{Q}{2c_0 E} .$$ (9–23)

That is, one calculates Q as if no dissipation were taking place and then normalizes by the incident flux.

Calculation of α is thus reduced to calculation of Q from the microscopic properties of the system, which is usually carried out in principle by means of the theory of dynamical response. The system Hamiltonian is taken to be

$$\hat{H} = \hat{H}_0 + \hat{H}_1(t)\,, \tag{9-24a}$$

where $\hat{H}_0$ refers to the equilibrium system and

$$\hat{H}_1(t) \equiv \int d^3x'\, \hat{n}(\mathbf{x}',t)\phi(\mathbf{x}',t)\,. \tag{9-24b}$$

Let $\delta\hat{n} \equiv \hat{n} - \langle\hat{n}\rangle_0$ be the deviation of the density from equilibrium and interpret $\phi(\mathbf{x},t)$ as the velocity potential. Then, from the general discussion of linear response in Chapter 3, we have

$$\langle\delta\hat{n}(\mathbf{x},t)\rangle = \frac{1}{\hbar}\int_{-\infty}^{t} dt' \int d^3x'\, \phi(\mathbf{x}',t')\langle[\hat{n}(\mathbf{x},t),\hat{n}(\mathbf{x}',t')]\rangle_0\,. \tag{9-25}$$

The energy dissipation owing to the driving term $\hat{H}_1(t)$ is defined as

$$Q(t) \equiv \left\langle \frac{\partial\hat{H}}{\partial t}\right\rangle = \frac{i}{\hbar}\int_{-\infty}^{t} dt' \left\langle\left[\hat{H}_1(t), \frac{\partial\hat{H}_1(t)}{\partial t}\right]\right\rangle_0\,. \tag{9-26}$$

Because Q in Eq.(9–23) refers to a single cycle, we average over a wave cycle to define

$$Q \equiv \frac{1}{T}\int_0^T Q(t)\,dt$$
$$= \frac{1}{T}\int_0^T dt \int d^3x\, \langle\delta\hat{n}(\mathbf{x},t)\rangle \tfrac{\partial}{\partial t}\phi(\mathbf{x},t)\,, \tag{9-27}$$

after a short calculation. Here, T is the period of the disturbance. If we suppose the disturbance to be a plane wave, then ϕ must also have that form and a spatial Fourier transform yields

$$Q = \mathrm{Re}\left\{-i\omega\left[\frac{1}{T}\int_0^T dt\, e^{-i\omega t}\,\langle\delta\hat{n}(\mathbf{k},t)\rangle\right]\phi(\mathbf{k},\omega)\right\}\,, \tag{9-28}$$

where

$$\phi(\mathbf{x},t) \equiv \phi(\mathbf{k},\omega)e^{i(\mathbf{k}\cdot\mathbf{x}-\omega t)}\,. \tag{9-29}$$

For a harmonic response the cycle average is just $1/2$, and we have

$$Q = \tfrac{1}{2}\mathrm{Re}\left\{-i\omega\langle\delta\hat{n}(\mathbf{k},\omega)\rangle\phi(\mathbf{k},\omega)\right\}\,, \tag{9-30}$$

which is just the Joule heating owing to the interactions. Because $\phi(\mathbf{k}, \omega)$ is real the attenuation is then essentially the imaginary part of the linear response $\langle \delta \hat{n}(\mathbf{k}, \omega) \rangle$. This is now just an ordinary, but tedious exercise in the spirit of Chapter 4.

Considerable interest has centered on acoustic attenuation in metals, where the effects of the sound wave allow one to express the attenuation in terms of the electrical conductivity and to calculate the energy transferred to the electron system, say. In the case of metals there will be both transverse and longitudinal effects, of course, so that the above discussion of Q can only be taken to refer to the longitudinal attenuation, and $\hat{n}(\mathbf{x}, t)$ is interpreted as the charge density. Note how the theory is intrinsically related to the physical concept of density fluctuations.

In order to conclude this review of the conventional theory of sound, let us consider an interesting and rather useful application to ultrasonic attenuation in metals. Pippard (1955) has formulated the problem of sound attenuation for a free-electron model of normal metals, and Tsuneto (1961) reformulated it in terms of the linear response theory as outlined above. [A similar formulation has also been provided by Nagaoka (1961).] A sound wave propagating through a crystal distorts the lattice and, in particular, if impurities are present these will also move under influence of the disturbance. Because the free electrons scatter from the impurities they will be dragged along with the motion of those impurities. This 'collision drag' effect results in an additional electronic current which turns out to be rather important in the attenuation of sound at very high frequencies (Holstein, 1959). If τ is the electronic relaxation time, then this effect can be neglected for frequencies such that $\omega\tau \ll 1$, which is generally valid up to several hundred MHz. Let v_F be the Fermi velocity and $\ell = v_F\tau$ the electron mean free path. Then Pippard finds for longitudinal waves, and for $\omega\tau \ll 1$,

$$\alpha = \frac{Nmv_F}{c_0 n\ell}\left[\frac{1}{3}\frac{k^2\ell^2\tan^{-1}(k\ell)}{k\ell - \tan^{-1}(k\ell)} - 1\right], \qquad (9\text{--}31)$$

where c_0 is the longitudinal speed of sound and N is the total number of electrons of mass m and density n.

In the low-frequency region characterized by $k\ell \ll 1$, a straightforward expansion yields

$$\alpha \xrightarrow[k\ell \ll 1]{} \frac{4}{15}\frac{Nmv_F}{nc_0^3}\ell\omega^2, \qquad (9\text{--}32)$$

exhibiting the usual quadratic frequency dependence. At higher frequencies, but still maintaining $\omega\tau \ll 1$, one finds that

$$\alpha \xrightarrow[k\ell \gg 1]{} \frac{\pi}{6}\frac{Nmv_F}{nc_0^2}\omega, \qquad (9\text{--}33)$$

not only linear in the frequency, but independent of mean free path. When band structure is properly accounted for the essential behavior of Eq.(9–33) appears to agree quite well with the data on copper, say (Macfarlane and Rayne, 1967).

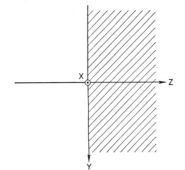

Fig. 9–3. Right half-space filled by the medium in which the sound disturbance propagates.

Ultrasonic attenuation in metals can yield a great deal of information on their structure, such as the nature of the Fermi surface. By introducing a magnetic field one probes further details by observing the magnetic-field dependence of the attenuation constant. At low temperatures, for example, α is an oscillating function of the field strength (Bömmel, 1954), a phenomenon known as the *magneto-acoustic effect*. In superconductors at very low temperatures the dominant attenuation mechanism is the destruction of Cooper pairs, and depairing provides an important means for determining temperature dependence and anisotropy of the energy gap.

B. Linear Theory of Sound Propagation and Attenuation

The preceding discussion demonstrates that ultrasonic attenuation provides an important tool for probing the structure of matter. But an understanding of acoustics over the entire frequency range is of great interest in its own right, so that it is very important that any fundamental description be unflawed both conceptually and mathematically. As Uhlenbeck and Ford (1963) have pointed out, sound propagation should properly be studied by introducing a disturbance at the boundary of a medium and simply describing how fast that disturbance is propagated and attenuated from first principles, free of preconceptions. We now turn to construction of such a theory, in a form first suggested by Snow (1967).

Consider a single-component medium filling completely the right half-space (rhs) from the xy-plane along the positive z-axis, as in Figure 9–3. In the unperturbed state the system is in thermal equilibrium, is uniform, and is described by a time-independent Hamiltonian $\hat{H}_0$. Hence, in that state the appropriate statistical operator is

$$\hat{\rho}_0 = \frac{1}{Z_0} e^{-\beta \hat{K}_0}, \qquad Z_0 = \text{Tr}\, e^{-\beta \hat{K}_0}, \tag{9-34}$$

and expectation values are written $\langle \hat{A} \rangle_0 = \text{Tr}(\hat{\rho}_0 \hat{A})$. As usual, $\hat{K}_0 = \hat{H}_0 - \mu \hat{N}$, where μ is the chemical potential and $\hat{N}$ the total number operator.

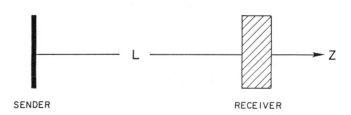

Fig. 9–4. Schematic illustration of the experimental arrangement often employed in measurements of acoustic attenuation in a fluid.

It is instructive to observe the usual experimental method for making precise measurements of acoustical phenomena (e.g., Greenspan, 1956), as illustrated in Figure 9–4. A piezoelectrically-driven quartz plate generates the sound waves along a plane, and the (attenuated) acoustic disturbance is picked up with an identical receiver. The distance L between source and receiver usually is considerably larger than a mean free path in the medium, by orders of magnitude—but not always (see below). Hence, it is quite realistic to envision the external disturbance so introduced by specifying the particle current on the boundary plane at $z = 0$. The system excited by the sound wave is then described by the statistical operator

$$\hat{\rho} = \frac{1}{Z} \exp\left\{ -\beta \,\hat{K}_0 + \int dx' \int dy' \int dt' \, \boldsymbol{\lambda}(x',y',t') \cdot \hat{\mathbf{J}}(x',y',0;t') \right\}, \quad (9\text{–}35a)$$

$$Z = \operatorname{Tr} \exp\left\{ -\beta \,\hat{K}_0 + \int dx' \int dy' \int dt' \, \boldsymbol{\lambda}(x',y',t') \cdot \hat{\mathbf{J}}(x',y',0;t') \right\}, \quad (9\text{–}35b)$$

with expectation values given by $\langle \hat{A} \rangle = \operatorname{Tr}(\hat{\rho}\hat{A})$. This is a boundary-value problem in which the particle-current density is specified on the xy-plane. (We shall find it notationally convenient in this chapter to denote the particle-current operator by $\hat{\mathbf{J}}$.)

In order to make the mathematics more tractable it is convenient to impose the infinite-volume limit and presume the z-component of current to be specified over the entire xy-plane, and for all time. Although there are no currents in the equilibrium system, current components at any time in the perturbed system in the right half-space are given by

$$\langle \hat{J}_\alpha(x,y,z;t) \rangle = \operatorname{Tr}[\hat{\rho}\,\hat{J}_\alpha(x,y,z;t)]. \quad (9\text{–}36)$$

Restriction to small-amplitude disturbances—corresponding to small departures from equilibrium—implies the linear approximation to be adequate:

$$\langle \hat{J}_\alpha(x,y,z;t) \rangle = \int_{-\infty}^{\infty} dx' \int_{-\infty}^{\infty} dy' \int_{-\infty}^{\infty} dt' \, \lambda(x',y',t') K_{J_\alpha J_z}(x',y',0,t';x,y,z,t).$$

$$(9\text{–}37)$$

Consistency requires this expression to reproduce the boundary condition at $z = 0$:

$$\langle \hat{J}_z(x,y,0;t) \rangle = \int_{-\infty}^{\infty} dx' \int_{-\infty}^{\infty} dy' \int_{-\infty}^{\infty} dt' \, \lambda(x',y',t') K_{J_z J_z}(x-x',y-y',0;t-t'),$$
(9–38)

where space-time uniformity in the equilibrium system has been made explicit.

Owing to linearity it is sufficient to consider the disturbance at the boundary to be a monochromatic plane wave. Thus, we write

$$\langle \hat{J}_z(x,y,0;t) \rangle = J e^{-i\omega t},$$
(9–39)

where J is a constant amplitude. Substitution of this boundary value into Eq.(9–38) allows one to solve the integral equation for $\lambda(x',y',t')$ immediately by Fourier transformation, and the Lagrange-multiplier function is determined directly by means of the driving term. We find that

$$\lambda(x,y,t) = \lambda_\omega \, e^{-i\omega t},$$
(9–40a)

with

$$\lambda_\omega^{-1} \equiv J^{-1} \int_{-\infty}^{\infty} dx \int_{-\infty}^{\infty} dy \, K_{J_z J_z}(x,y,0;\omega),$$
(9–40b)

and λ is actually independent of spatial variables. Owing to the form of the covariance function in Eq.(9–38), the current in the right half-space will also be independent of x and y:

$$\langle \hat{J}_\alpha(x,y,z;t) \rangle = \lambda_\omega \, e^{-i\omega t} \int_{-\infty}^{\infty} dx' \int_{-\infty}^{\infty} dy' \int_{-\infty}^{\infty} dt' \, e^{i\omega t'} K_{J_\alpha J_z}(x',y',z;t').$$
(9–41)

In view of these last observations it is useful to define a function

$$K_{J_\alpha J_z}(z,\omega) \equiv K_{J_\alpha J_z}(0,0,z;\omega)$$
$$= \int_{-\infty}^{\infty} \frac{dk_z}{2\pi} e^{ik_z z} K_{J_\alpha J_z}(k_z,\omega),$$
(9–42)

where $K_{J_\alpha J_z}(k_z,\omega) \equiv K_{J_\alpha J_z}(0,0,k_z;\omega)$. Then Eq.(9–37) for the current in the perturbed system can be rewritten as

$$J_\alpha(z,t) \equiv \langle \hat{J}_\alpha(0,0,z;t) \rangle = J_\alpha(z) e^{-i\omega t},$$
(9–43)

and

$$J_\alpha(z) \equiv \lambda_\omega K_{J_\alpha J_z}(z,\omega).$$
(9–44)

In this notation we can rewrite Eq.(9–40b) as

$$\lambda_\omega^{-1} = J^{-1} K_{J_z J_z}(0,\omega).$$
(9–45)

Thus, the amplitude of the sound wave relative to that of the initial disturbance is just

$$\frac{J_\alpha(z)}{J} = \frac{K_{J_\alpha J_z}(z,\omega)}{K_{J_z J_z}(0,\omega)}\,. \tag{9-46}$$

These last five equations comprise a very general theory of sound propagation and attenuation in the customary small-amplitude approximation. The only non-general specification has been the type and duration of initial disturbance. This boundary condition is easily generalized, but at the expense of making transparent the ensuing description. Application of a monochromatic plane wave at the boundary results in a disturbance which propagates through the system harmonically, but with an apparent attenuation along the positive z-axis given by $J_\alpha(z)$. An analysis of the spatial decay, therefore, depends on the detailed structure of the current-current covariance function, and *only* on that. A similar statement also applies to the problem of extracting the general wave form.

Attribution of some special forms to the covariance function provides insight into the predictions made by Eq.(9-43). For example, suppose that

$$K_{J_z J_z}(k_z,\omega) = 2\pi g(\omega)\,\delta(k_z - k_0)\,. \tag{9-47a}$$

Then, with the help of Eq.(9-42), the z-component of current is just

$$J_z(z) = J\,e^{ik_0 z}\,, \tag{9-47b}$$

and the initial plane wave propagates with no attenuation.

As another example, consider a covariance function of Lorentzian form, such as

$$K_{J_z J_z}(k_z,\omega) = \frac{\alpha f(\omega)}{\alpha^2 + (k_z - k_0)^2}\,. \tag{9-48a}$$

A similar calculation yields

$$J_z(z) = J\,e^{ik_0 z}\,e^{-\alpha|z|}\,, \tag{9-48b}$$

which exhibits the classical exponential attenuation. Although the Lorentzian form of the covariance function in Eq.(9-48a) provides at least a sufficient condition for exponential decay of the sound wave, there clearly is no obvious requirement for the attenuation to be exponential in general.

It is worth noting that the number density itself could have been predicted in the above discussion, merely by replacing $\hat{J}_\alpha$ with $\hat{n}$. In a similar manner we find that

$$n(z,t) \equiv \langle \hat{n}(0,0,z;t)\rangle - \langle \hat{n}\rangle_0 = n(z)\,e^{-i\omega t}\,, \tag{9-49a}$$

where

$$n(z) \equiv \lambda_\omega K_{n J_z}(z,\omega)\,. \tag{9-49b}$$

But from Table 6–2 one notes that the covariance function K_{nJ_z} is always directly proportional to only the density-density covariance function K_{nn}. Consequently, we can always write

$$K_{nJ_z}(z,\omega) = \frac{\omega}{2\pi} \int_{-\infty}^{\infty} \frac{dk_z}{k_z} e^{ik_z z} K_{nn}(k_z,\omega). \tag{9-50}$$

The variation in density, $n(z,t)$, is directly related to the correlated propagation of density fluctuations, and it is precisely this correlation of fluctuations that makes intelligible speech possible—if not intelligent.

Finally, one could have prescribed the spatial variation of current at $z = 0$, $t = 0$, rather than the presumed driving condition. In that case standing waves are obtained, with temporal attenuation in the form of a current $J_\alpha(t)$. The ensuing mathematical development is quite similar to the above and will not be pursued further here.

SOME GENERAL CONSIDERATIONS

Although the preceding theory is rather general, it is not entirely complete. There are some additional features which must be appended to it in order to make it so, and a major one of these concerns dispersion and the relation between attenuation and phase velocity. In order to develop these points it is useful to reformulate the problem in a way that emphasizes more directly the physical processes taking place in the medium, while at the same time continuing to view those processes as a linear response to an external disturbance.

Longitudinal sound waves consist of alternate regions of compression and rarefactions in the medium, and in order to describe these it is useful to define a *condensation function* $s(t)$ as the fractional change in density owing to pressure changes. If we view the fluid as having compressional elasticity only—that is, it effectively obeys Hooke's law— then $s = \kappa_S \Delta P$, where κ_S is the adiabatic compressibility. We can then generalize this relationship so as to describe the linear response of the system to a disturbance in the form of a pressure wave:

$$s(t) = \int K(t-t') P(t') \, dt', \tag{9-51}$$

and we have deliberately omitted the limits of integration for the moment. Here, $K(t-t')$ is the linear response function defining the generalized adiabatic compressibility.

Although not defined explicitly as usual in terms of the expectation value of a commutator, $K(t)$ must still have the same causal properties as any linear response function if Eq.(9–51) is to describe the physical situation corresponding to a sonic disturbance. In particular, $K(t)$ must be real, and the choice of integration limits must be such as to enforce causality. But this immediately puts the entire analysis of Chapter 3 at our disposal, and it is then useful to conduct the discussion in terms of the Fourier transform of the generalized compressibility:

$$K(\omega) = K'(\omega) + iK''(\omega), \tag{9-52}$$

where K' and K'' are both real functions of ω, K' is even, and K'' is odd. We know, however, that these functions also constitute a pair of Hilbert transforms, as in Eqs.(3–32). In Problem 9.1 the reader is asked to show that these latter equations can also be written, in the present notation, as

$$K'(\omega) = \frac{2}{\pi} \mathrm{P} \int_0^\infty \frac{uK''(u)}{u^2 - \omega^2}\, du \,, \tag{9-53a}$$

$$K''(\omega) = -\frac{2}{\pi} \mathrm{P} \int_0^\infty \frac{\omega K'(u)}{u^2 - \omega^2}\, du \,. \tag{9-53b}$$

These are the Kramers-Kronig relations for longitudinal acoustic disturbances, and are a result only of linearity and causality—they are simply general laws of physics.

In a real system one expects the response, and hence the compressibility function, to fall off rapidly at high frequencies, so that the integrals will converge. That this is the case mathematically, as well, has been demonstrated by Weaver and Pao (1981). The necessary asymptotic behavior follows from the representation of the wave number $k(\omega)$ as a Herglotz function.

Equations (9–53) can be utilized for relating attenuation and phase velocity if we insist on acoustic wave propagation, which then requires a relation of the form noted in Eq.(9–2). Thus, in the frequency plane we require the compressibility to satisfy the relation

$$k^2(\omega) = \omega^2 \rho_0 K(\omega) \,, \tag{9-54}$$

where $\rho_0 \equiv mn_0$ is the mass density of the medium. For real ω this equation requires k to be complex, so that as in Eq.(9–7) we write

$$k(\omega) = \frac{\omega}{c(\omega)} + i\alpha(\omega) \,, \tag{9-55}$$

introducing the phase velocity through its Fourier transform $c(\omega)$, and the attenuation coefficient α in the same way. If the disturbance propagates through the medium as a plane wave, for example, then α will emerge as an exponential attenuation constant. But clearly this is not a general requirement, so that $\alpha(\omega)$ is simply a parameter describing dissipation of the sound wave.

Now substitute Eq.(9–55) into Eq.(9–54) and separate real and imaginary parts. The result is a set of equations determining $c(\omega)$ and $\alpha(\omega)$, and relating them to the compressibility:

$$\frac{\omega^2}{c^2} - \alpha^2 = \omega^2 \rho_0 K'(\omega) \,, \tag{9-56a}$$

$$2\frac{\alpha}{c} = \omega\rho_0 K''(\omega) \,. \tag{9-56b}$$

These are broad generalizations of Eqs.(9–11) and (9–12) but, as with those equations, the standard presumption that the real part of k dominates the magnitude allows us to uncouple them. Thus, for $\alpha \ll \omega/c$, one finds that

$$c(\omega) = [\rho_0 K'(\omega)]^{-1/2} \,, \tag{9-57a}$$
$$\alpha(\omega) = \tfrac{1}{2}\rho_0 c(\omega)\omega K''(\omega) \,. \tag{9-57b}$$

These expressions can then be coupled to the Kramers-Kronig equations (9–53) so as to provide either c or α at all frequencies if the other is known at all frequencies.

Although this last assertion is true in principle, it is of limited practical value owing to the obvious nonlocality of Eqs.(9–53). Unless the dominant contributions to the integrals occur over a small frequency band, the evaluations are very difficult. For this reason it is important to develop nearly-local relationships between c and α, which might then be employed to predict ultrasonic properties of substances in which precise measurements are difficult to perform. One such set of relationships has been derived by O'Donnell, $et\ al$ (1978, 1981) which appear to be quite accurate in the absence of resonances. It is then possible to use results of this kind to distinguish those features of ultrasonic propagation determined by general laws of physics from those depending on the particular mechanisms responsible for that propagation.

C. Ideal Gases, I: Boltzmann Statistics

Sound propagation possesses at least one feature which distinguishes it from other transport processes, in that it can be studied at least formally in systems of noninteracting particles. That is, because of its nonzero compressibility the ideal gas can support sound waves. For example, in the ideal Boltzmann system $PV = N\kappa T$, and PV^{γ} is a constant, with $\gamma \equiv C_P/C_V = \kappa_T/\kappa_S$. Hence, from Eq.(9–2), the speed of sound is

$$c_0 = \left(\frac{\gamma \kappa T}{m} \right)^{1/2} . \qquad (9\text{--}58)$$

In an ideal monatomic gas, say, $\gamma = 5/3$, so that c_0 is decidedly nonzero. Thus, it may be useful to employ such simple models to gain an understanding of how the formalism functions in specific, albeit unrealistic, models.

According to Eq.(9–46) the propagation of sound is to be understood by studying the current-current covariance function, and from Table 6–2 this quantity is given in a homogeneous isotropic system by

$$K_{J_i J_j}(\mathbf{k}, \omega) = \frac{\omega^2}{k^2} k_i k_j\, K_{nn}(\mathbf{k}, \omega) + \kappa_{ij} X_3(k, \omega) , \qquad (9\text{--}59)$$

where k_i is a unit vector and κ_{ij} the unit transverse tensor. Both K_{nn} and X_3 depend only on the magnitude of $\mathbf{k}$ and are even functions of ω. Substitution of Eq.(9–59) into Eq.(9–42) reveals by direct calculation that X_3 does not contribute to the integral. Moreover, it is found that, with these boundary conditions, longitudinal driving generates no shear components in any system, and for a uniform condensate there is no contribution to K_{JJ} from the ground state of the ideal Bose system. One can therefore omit subscripts in these models and characterize the longitudinal sound propagation from Eqs.(9–42)-(9–46) as follows:

$$J(z,t) = \frac{K(z)}{K(0)} e^{-i\omega t} \equiv J(z) e^{-i\omega t} , \qquad (9\text{--}60)$$

with

$$K(z) \equiv K_{J_z J_z}(z, \omega) = \int_{-\infty}^{\infty} \frac{dk}{2\pi} e^{ikz} \frac{\omega^2}{k^2} K_{nn}(k, \omega) . \qquad (9\text{--}61)$$

For the ideal Boltzmann gas the limiting form of Eq.(6–148) is appropriate:

$$K_{nn}^0(\mathbf{k}, \omega)_{\mathrm{B}} = n_0 \frac{(2\pi\beta m)^{1/2}}{|k|} e^{-\frac{1}{2}\beta m \omega^2/k^2} . \qquad (9\text{--}62)$$

It is important here to note that the factor of k^{-1} must be considered an absolute value, because it was originally extracted from a δ-function in arriving at Eq.(9–62). This would not be important, except that here k is *not* the magnitude of $\mathbf{k}$, but k_z. Substitution into Eq.(9–61) then yields

$$\frac{K(z)}{K(0)} = 2F[\omega z (\beta m/2)^{1/2}] , \qquad (9\text{--}63)$$

where we define a characteristic function

$$F(s) \equiv \int_0^{\infty} \cos(sy) e^{-1/y^2} \frac{dy}{y^3} \quad \xrightarrow[s \to 0]{} \quad \frac{1}{2}$$

$$\xrightarrow[s \to \infty]{} \quad C \exp\{-\tfrac{3}{2}(s/2)^{2/3}\} . \qquad (9\text{--}64)$$

This function is a special case of a class known as *Abramowitz integrals* (Abramowitz, 1953), and some of their properties are presented by Abramowitz and Stegun (1964, pp.1001-1003).

As an aside, note that $K_{JJ}(\mathbf{k}, \omega)$ also describes the temporally attenuated current $J_z(t)$ if spatial variation is prescribed on the boundary plane. The Fourier transform would then be in the frequency variable, and a sharp peak in the integrand would project out a value of ω describing the standing-wave response of the medium. In fact, the frequency response falls off very rapidly as a Gaussian and there is an almost immediate time decay of the current. As expected, no hydrodynamic modes are excited.

Evaluation of the integral in Eq.(9–63) does not appear feasible in closed form, but it has been evaluated numerically. The result is plotted in Figure 9–5, and we see that the amplitude decays very rapidly. Thus, the ideal Boltzmann gas initially responds to a sonic disturbance in a way predicted by the classical theory of sound, but propagation is destroyed in extremely short order. In argon at room temperature, for example, $s \simeq (2.8 \times 10^{-5})\omega z$, so that at low frequencies and close in to the source *some* propagation might be expected, but with almost monotonic attenuation. At 20 kHz one finds that at $z = 10$ cm the amplitude has decayed to about $1/10$ its original value.

Any possible propagation would be determined by a sharp peak in the covariance function (9–59), which we have plotted in Figure 9–6 as a function of wavenumber. The parameter p was defined in Chapter 6 by writing $p^2 = \frac{1}{4}\beta(\hbar^2 k^2/2m)$, and here $K \equiv \frac{1}{4}\beta\hbar\omega$. There is a reasonably sharp peak at very small s with a maximum

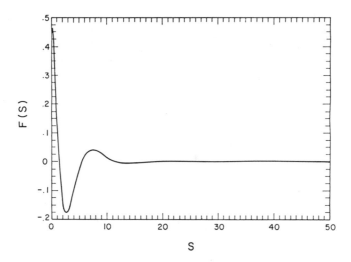

Fig. 9–5. Computer evaluation of the function $F(s)$, Eq.(9–64), depicting the rapid damping of sound in an ideal Boltzmann gas.

at $k_0 z = \sqrt{2/3}\, s$. If one adopts this point as a selected wavelength, then this yields a sound speed

$$c_0 = \left(\frac{3\kappa T}{m}\right)^{1/2},\qquad (9\text{–}65)$$

which does not differ greatly from the value for a monatomic gas given by the classical theory, Eq.(9–58). (Selection of a single point, however, also selects a particular z, so that this result is more characteristic of a standing wave.)

One can not expect to find significant sound propagation in an ideal gas, of course, for such organized behavior must correspond to strong collision dominance. Moreover, the classical dissipative processes leading to attenuation are absent in this system. On the one hand, it might be expected that this scenario would be a good model for dilute gases. On the other hand, though, a sonic disturbance of any appreciable amplitude and frequency would also be expected to push such a system far from equilibrium very quickly, and successive wavefronts would not be encountering the same physical conditions. The linear approximation of the present theory would hardly apply. This difficulty of describing sound propagation in dilute systems in terms of small departures from equilibrium has also been noted explicitly by Kahn and Mintzer (1965), and is just a special case of the general problem of validity of linear response theories. One is impelled to inquire further, therefore, as to the possible meaning of this model.

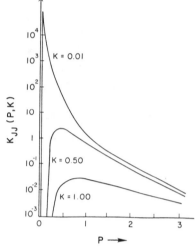

Fig. 9–6. Current-current covariance function for the ideal Boltzmann gas.

Physical Interpretation

Ultrasound in a very dilute gas actually possesses some history, both experimentally and theoretically, and the system is often called a *Knudsen gas* (Knudsen, 1934). The so-called Knudsen regime can be defined in terms of the ratio of the nominal wavelength of the perturbing sonic disturbance, $\lambda_0 \equiv 2\pi c_0/\omega$, to the mean free path, Λ. Over thirty years ago a series of experiments of this type were carried out in a framework much like that of Figure 9–4, in which the distance z from source to receiver was a good deal less than a mean free path. These two parameters are related through the experiments themselves, and so define a two-dimensional region as indicated in Figure 9–7. Greenspan (1950, 1956) made measurements in He at 11 MHz, whereas Meyer and Sessler (1957) worked in argon at $\omega z \simeq 3 \times 10^5$ cm/s. Both sets of experiments were carried out at densities $n \simeq 3 \times 10^{14} \text{cm}^{-3}$ and pressures of about 10^{-5} atm. This experimental work has been reviewed by Greenspan (1965), and Sessler (1965), and some of the data will be presented below. One curious feature that arises in this region is that both the wavenumber and attenuation parameter appear to be functions of distance from the source plane.

In Figure 9–7 we employ a characteristic parameter $r \equiv P/\omega\mu$, where $P = n_0\kappa T$ and μ is a viscosity coefficient. This parameter is also proportional to the ratio of collision frequency to sound frequency, so that in the Knudsen region $\omega_c \ll \omega$. In turn, $\omega \gtrsim \omega_c$ is precisely where the usual solutions to the Boltzmann equation break down. Consequently, the above experiments led to a concomitant flurry of theoretical activity pointed toward finding new solutions to that equation valid in

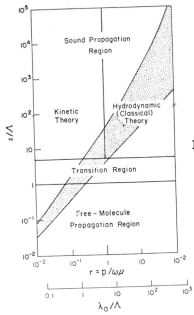

Fig. 9–7. Regions of free-molecule and sound propagation. The shaded region is that in which measurements have been made. [*Reproduced with permission from Sessler (1965).*]

the Knudsen region. Most of this work has been reviewed by Cercignani (1975) and compared carefully with experiments. Figures 9–8 and 9–9 exhibit what appear to be the best fits to the data, from which readers can draw their own conclusions.

It is not at all clear that this problem of 'free-molecule propagation' is really a bonafide matter of sound propagation and attenuation. The boundary condition in both theory and experiment can be thought of as enforced by an oscillating plate, so that particles striking the plate on an outward stroke lose net momentum, whereas they gain on an inward stroke. These are the only collisions in the system, and they result in an enforced diffusion process. That is, the current is actually a diffusion current, and would be expected to die out rather rapidly in just a short distance from the source. Thus, the attenuation mechanism for any kind of organized motion is just a result of phase shifts in the diffused currents. There is no sound propagation or attenuation in the ordinary sense, and it is only collisions with the container walls that maintain any semblance of equilibrium. One suspects, in fact, that the experimental data are only a result of phase-mixing at the receiver.

Despite these observations, though, there is good reason to understand the Knudsen regime, as well as the transition region and on into that of classical wave propagation. First of all, knowledge of the entire spectrum of response to sonic

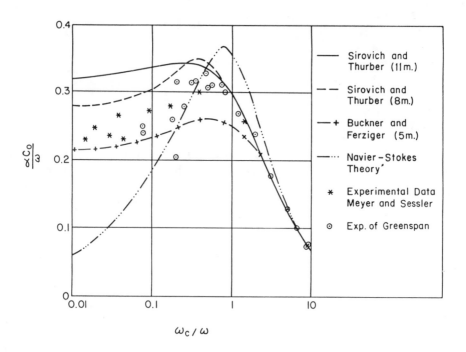

Fig. 9–8. Comparison of some standard theoretical predictions for the attenuation parameter with experiment, indicating the number of moments used in the approximations. [*Reproduced with permission from Cercignani (1975).*]

disturbances in a medium is required to ascertain the validity of a general theory. Secondly, we have seen that the Knudsen system is rather germane to an understanding of the validity of linear response theory itself, so that deeper insight into the former may lead to more definite criteria for the latter. Almost all theoretical work to date has been based on the Boltzmann equation, with its demonstrated difficulties. Thus, it is useful to examine this problem from the viewpoint of the present theory—whose structure in fact corresponds to the experiments themselves. In addition, it is of some interest to investigate whether the inclusion of interactions can clarify the transition from the free-molecule to classical propagation domains, and so we turn here to a semi-quantitative sketch of such a construction.

AN INTERACTION MODEL

A logical first step toward studying actual sound propagation in an ordinary fluid is to introduce particle interactions by means of a perturbation expansion of

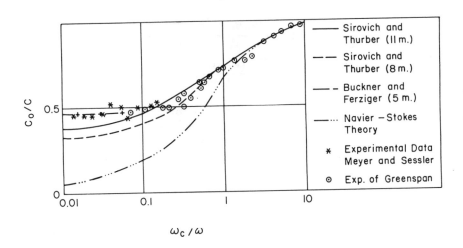

Fig. 9–9. Comparison of some theoretical predictions for phase speed with experiment, indicating the number of moments used in the approximations. [*Reproduced with permission from Cercignani (1975).*]

the covariance function. To leading order, and in the Boltzmann limit, we find from Appendix C that

$$K_{nn}^1(\mathbf{k},\omega)_{\mathrm{B}} = K_{nn}^0(\mathbf{k},\omega)_{\mathrm{B}}[1 - \beta n v(k)], \qquad (9\text{–}66)$$

where $K_{nn}^0(\mathbf{k},\omega)_{\mathrm{B}}$ is given in Eq.(9–62), and

$$v(k) \equiv \int e^{i\mathbf{k}\cdot\mathbf{r}}\, V_2(r)\, d^3r\,. \qquad (9\text{–}67)$$

Unfortunately, when Eq.(9–66) is employed in Eq.(9–61) one finds effects which differ very little from the free-particle system. This is not too surprising, because perturbation theory is intrinsically related to low-density behavior and one can not expect to see collision-dominated hydrodynamic modes excited. Equation (9–66) actually describes propagation only from one particle to another, rather than collective sound modes. It appears, therefore, that one must find a higher-density model of the covariance function that is more closely related to the normal hydrodynamic situation in order to exhibit ordinary sound propagation and attenuation.

A first step in this direction is to re-order the perturbation series in a way that emphasizes more strongly the collective effects in the medium—such as in the ring sum of Appendix C. Among other things, the sum over ring diagrams continues the covariance function to higher densities for short-range interactions. Thus, from

Eq.(C–34) we consider the leading-order ring sum in the Boltzmann limit,

$$K^r_{nn}(\mathbf{k},\omega)_B = n\frac{(2\pi\beta m)^{1/2}}{|k|}\frac{e^{-\frac{1}{2}\beta m\omega^2/k^2}}{1+f(k)}$$

$$= \frac{K^0_{nn}(\mathbf{k},\omega)_B}{1+f(k)}, \qquad (9\text{–}68)$$

and the first term in the expansion of the denominator leads back to Eq.(9–66). The function $f(k)$ was defined in Eq.(C–26), appeared in its essential aspects in Eq.(C–4) and (C–5), and throughout Appendix C was evaluated in the limit $\beta\hbar\omega \ll 1$.

In order to pursue the goals of the present discussion we shall *not* employ this inequality in evaluating $f(k)$ here, for reasons which shall become clear presently. We rewrite this function as

$$f(k) = \frac{2}{\sqrt{\pi}}\beta n v(k)\int_0^\infty y^2 e^{-y^2} I(y)\, dy, \qquad (9\text{–}69)$$

where, with $\varsigma \equiv k\lambda_T y/\sqrt{\pi}$,

$$I(y) \equiv \varsigma^{-1}\int_{-\varsigma}^{\varsigma}\frac{1-e^{-z}e^{-\lambda_T^2 k^2/2\pi}}{z-\beta\hbar\omega+\beta\hbar^2 k^2/2m}\, dz,$$

$$\xrightarrow[\text{Boltzmann}]{} \varsigma^{-1}\int_{-\varsigma}^{\varsigma}\frac{1-e^{-z}}{z-\beta\hbar\omega}. \qquad (9\text{–}70)$$

Note that the contributing values of ς can not be very large, owing to the cut-off in Eq.(9–69), so that z is also restricted to small values. But the second line in Eq.(9–70) is not quite independent of $\hbar$, and if we expand the exponential we see that every term will yield a convergent integral in Eq.(9–69). Thus, only the linear term in z will be independent of $\hbar$. Hence, the true Boltzmann limit is given by

$$I(y) \simeq \varsigma^{-1}\int_{-\varsigma}^{\varsigma}\frac{z\, dz}{z-\beta\hbar\omega}, \qquad (9\text{–}71)$$

which exists as a Cauchy principal value:

$$I(y) = 2 + q\ln\left|\frac{1-q/y}{1+q/y}\right| + O(\hbar), \qquad (9\text{–}72)$$

and q is defined by

$$q^2 \equiv \frac{1}{2}\beta m\frac{\omega^2}{k^2} = \frac{1}{4}\frac{(\beta\hbar\omega)^2}{\beta\epsilon(k)}, \qquad (9\text{–}73)$$

with $\epsilon(k) \equiv \hbar^2 k^2/2m$ a free-particle energy.

When the limit $\beta\hbar\omega \ll 1$ is enforced only the first term on the right-hand side of Eq.(9–72) contributes, and in that case one recovers the expressions (C–34), or

(9–66). Consideration of ultrasonic frequencies in this model, however, requires us to retain the full expression for $I(y)$. Substitution into Eq.(9–69) and a change of variables to $x \equiv y/q$ then yields

$$f(k) = \beta n v(k)[1 + g(q)], \qquad (9\text{–}74)$$

where

$$g(q) \equiv \frac{2}{\sqrt{\pi}} q^3 \int_0^\infty x \ln \left| \frac{1-x}{1+x} \right| e^{-q^2 x^2} \, dx, \qquad (9\text{–}75)$$

and q is defined in Eq.(9–73).

Let us observe carefully two points. First, q is just the characteristic argument of the exponential functions appearing in all covariance functions for a simple fluid in the Boltzmann limit, as in Eq.(9–68). But it would certainly be out of order to expand those exponentials for small q—the essential behavior near $k = 0$ would be obliterated. Thus, it may be of value to retain other functions of q in their full forms, at least for awhile. Second, note that the integral in Eq.(9–75) is well behaved, for the logarithm is an integrable singularity. Nevertheless, one must carry out the evaluation with some care, because the reality of $K_{nn}(\mathbf{k}, \omega)$ requires $g(q)$ to be real as well.

In Problem 9.2 the reader is asked to verify that $g(q)$ is effectively Dawson's function (e.g., Spanier and Oldham, 1987; Chapter 42),

$$g(q) = -2qe^{-q^2} \int_0^q e^{t^2} \, dt, \qquad (9\text{–}76)$$

and to prove the following results:

$$g(q) = \sum_{n=1}^\infty \frac{(-2q^2)^n}{(2n-1)!!}$$

$$\xrightarrow[q \to \infty]{} -1 - \sum_{n=1}^\infty \frac{(2n-1)!!}{(2q^2)^n}. \qquad (9\text{–}77)$$

Figure 9–10 is a plot of $g(q)$ in which the abscissa is taken as q^2 so as to better exhibit the full features of the curve.

We now substitute our model covariance function (9–68) into the sound integral of Eq.(9–61) to obtain

$$K(z) = 2\frac{n_0 \omega^2}{\sqrt{\pi}} \left(\frac{\beta m}{2} \right)^{1/2} \int_{-\infty}^\infty e^{ikz} \frac{e^{-k_1^2/k^2}}{1 + f(k)} \frac{dk}{k^3}, \qquad (9\text{–}78)$$

where $k_1^2 \equiv \beta m \omega^2/2$. Note that, owing to the denominator, the integrand is no longer necessarily an even function of k. The relevant parameter of Eq.(9–64) is thus $s = k_1 z$. According to Eq.(9–60) we must then study the integral

$$I(z) \equiv \int_{-\infty}^\infty e^{ikz} \frac{e^{-k_1^2/k^2}}{1 + f(k)} \frac{dk}{k^3}, \qquad (9\text{–}79)$$

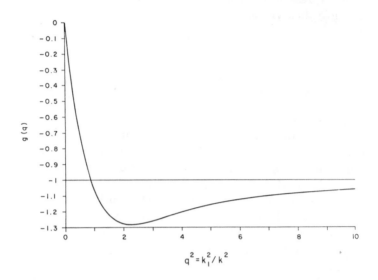

Fig. 9-10. Computer evaluation of the function $g(q)$, Eq.(9-76).

in terms of which the spatial current is $J(z) = I(z)/I(0)$.

If the model is to exhibit any propagation features at all, then the integrand in Eq.(9-79) will have to possess a rather narrow and large peak. In turn, this requires the denominator to have zeros somewhere on the real axis, or at least have a resonance. Thus, we examine the structure of

$$1 + f(k) = 1 + D[1 + g(q)]. \qquad (9\text{-}80)$$

A typical two-body potential for this model is a soft-core, square-well interaction, for which we can choose parameter values that provide a good fit to argon at 0 °C (e.g., Hirschfelder, *et al*, 1954). Upon Fourier transformation one finds no detectable difference between $v(k)$ and $v(0)$ before the integrand has decayed effectively to zero. This is also the case for a screened-Coulomb potential, say. Consequently, with almost no error we can set the density parameter $D = \beta n v(0)$ in Eq.(9-80).

The function $g(q)$ has an absolute minimum at $q_0 \simeq 1.5$, where $g(q_0) \simeq 1.285$, as indicated in Figure 9-10. (At room temperature for argon, this corresponds to $k \simeq 18\,\mathrm{cm}^{-1}$.) Accordingly, the function $1 + f(k)$ will not vanish for real variables unless D exceeds some critical value, D_c. We define this value as

$$D_c \equiv [1 - |g(q_0)|]^{-1} \simeq 3.51\,, \qquad (9\text{-}81)$$

for which there appears a double root at $q = q_0$. For $D > D_c$ this splits into two real roots, whereas for $D < D_c$ we have two complex-conjugate roots. It is useful to characterize these a bit more explicitly.

Expand $g(q)$ about $q = q_0$, noting that $g'(q_0) = 0$. This provides an expansion for $[1 + f(k)]$ as well, and so for q near q_0 the zeros of the denominator are found from

$$(q - q_0)^2 \simeq -2 \frac{1 + D[1 - |g(q_0)|]}{Dg''(q_0)}. \qquad (9\text{-}82)$$

Note that $g''(q_0) > 0$. Hence, in the two regions of interest,

$$q \simeq \begin{cases} q_0 \pm i\gamma, & D < D_c \\ q_0 \pm \gamma, & D > D_c \end{cases}, \qquad (9\text{-}83\text{a})$$

where

$$\gamma \equiv \left| 2 \frac{1 + D[1 - |g(q_0)|]}{Dg''(q_0)} \right|^{1/2}. \qquad (9\text{-}83\text{b})$$

One is now in a position to investigate the behavior of the integral $I(z)$, Eq.(9-79), as D varies over positive values. (For D to be negative requires $v(0)$ to be so, and this generally yields an unstable interaction in the many-body system.)

Qualitatively, for $0 \leq D < D_c$ the integrand has two complex-conjugate poles located well off the real axis in the first and fourth quadrants of the q-plane—$\gamma \to \infty$ as $D \to 0$. When D approaches D_c from below these poles move onto the real axis at $q = q_0$, and as D increases further the poles split into two real poles (q^-, q^+) that move left and right, respectively. The next step is to examine their physical nature.

Although it is very difficult to evaluate $I(z)$ analytically, we can analyze it for large z by means of a saddle-point integration, and thereby gain at least qualitative information about the possibility of propagation. Toward this end we change variables from k to q, and then to $t \equiv q/\nu$, where $\nu \equiv (s/2)^{1/3} = (k_1 z/2)^{1/3}$. Then,

$$I(z) = \nu \int_0^\infty F(\nu t) e^{-\nu^2 (t^2 - i2/t)} \, dt, \qquad (9\text{-}84)$$

where

$$\nu F(\nu t) \equiv \frac{\nu^2}{k_1^2} \frac{t}{1 + D[1 + g(\nu t)]}, \qquad (9\text{-}85)$$

and we are interested in the behavior for large ν. In the various regions of interest we see that

$$\nu F(\nu t) \xrightarrow[D \ll D_c]{} \frac{\nu^2}{k_1^2} t,$$

$$\xrightarrow[D \lesssim D_c]{} \frac{2}{Dk_1^2 g''(q_0)} \frac{t}{(t - t_0)^2 + \gamma^2/\nu^2},$$

$$\xrightarrow[D \gtrsim D_c]{} \frac{2}{Dk_1^2 g''(q_0)} \frac{t}{(t - t_0)^2 - \gamma^2/\nu^2}, \qquad (9\text{-}86)$$

with $t_0 \equiv q_0/\nu$, and γ given by Eq.(9-83b).

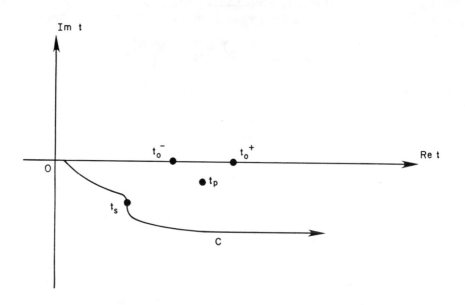

Fig. 9–11. Contour in the t-plane for asymptotic evaluation of the integral $I(z)$ by the method of steepest descents. The point t_s is the saddle point, and the positions of the relevant sound poles are labeled t_p, t_0^-, t_0^+.

The integrand possesses saddle points where the derivative of $t^2 - i2/t$ vanishes, and the only pertinent one here is

$$t_s = \frac{\sqrt{3}}{2} - \frac{i}{2}. \qquad (9\text{–}87)$$

Owing to the singularity at the origin, we deform the contour off the real axis into that labeled C in Figure 9–11 as follows: begin at the origin $t = 0$ and proceed a distance $\epsilon > 0$ before departing the real axis; pass through the saddle point t_s along a path $\mathrm{Re}\, t = constant$; finally, proceed to infinity in a direction $\mathrm{Im}\, t = constant$. This choice is dictated by the method of steepest descents itself (e.g., Bleistein and Handelsman, 1975).

There is no point in burdening the reader with details of the saddle-point integration, for the model is qualitative at best—those details are readily supplied at any rate. When $D \ll D_c$ there are no poles anywhere near the contour C, the denominator is effectively unity, and $I(z)/I(0)$ is just the function $F(s)$ plotted in Figure 9–5. As $\nu \to \infty$ we find that

$$I(z) \simeq \frac{\sqrt{\pi}}{2k_1^2} \left(\frac{s}{2}\right)^{1/3} e^{-\frac{3}{2}(s/2)^{2/3}}, \qquad (9\text{–}88)$$

and we have suppressed oscillating factors. Clearly, this contribution comes completely from the saddle point.

As $D \to D_c^-$ the pole $t_p = (q_0 - i\gamma)/\nu$ in the lower half-plane approaches the real axis in the neighborhood of the saddle point. Thus, if the pole is in close enough, as the contour is deformed away from the real axis it sweeps this pole and we pick up a residue at $t = t_p$ as part of the contribution to the integral. For large ν, in addition to the term of Eq.(9–88), we acquire the additional contribution

$$I(z)_p \simeq -\frac{4\pi}{Dk_1^2 g''(q_0)} \frac{(q_0 - i\gamma)}{2\gamma} e^{-(q_0^2 - \gamma^2) + 2i\gamma q_0}$$

$$\times e^{ik_0 z - \alpha_0 z}, \qquad (9\text{–}89)$$

where

$$k_0 \equiv \frac{q_0 k_1}{q_0^2 + \gamma^2}, \qquad \alpha_0 \equiv \frac{k_1 \gamma}{q_0^2 + \gamma^2}. \qquad (9\text{–}90)$$

We emphasize that we are still in the region $D \lesssim D_c$. Note here that $\alpha_0 \propto \omega$, commensurate with the essentially non-hydrodynamic character of the model.

When $D \gtrsim D_c$ the integral must be considered a principal-value integral, so that we obtain a contribution from the saddle point plus one-half the residues at the two real poles. In this case,

$$I(z) \simeq \frac{2\pi i}{Dk_1^2 g''(q_0)} e^{-(q_0^2 + \gamma^2)} e^{ik_0 z}, \qquad (9\text{–}91)$$

and the propagation is completely undamped. At this point the model has been pushed to its limits, for there is no way in which the classical attenuation mechanisms can arise. The point $D = D_c$ in argon at room temperature, for example, corresponds to a density of about 10^{17} cm^{-3}. The original experiments in the Knudsen regime were carried out at about 10^{14} cm^{-3}, so that the model extends well into the transition region. Note, however, that the asymptotic analysis we have given here corresponds to *very large* $k_1 z$, and so can not be related to the data—neither k_0 nor α_0 depend on z.

The model can be investigated in further detail by means of computer calculations, though, and in Figures 9–12 through 9–15 we present evaluations of $J(z)$ for various values of D. The onset of propagation as $D \to D_c$ is evident. One *can* compare these results with the data, but to maintain the same scenario it is necessary to presume exponential damping. The logarithm of the amplitude as a function of the distance yields the slope of a least-squares fit to the peak-to-peak straight line. Phase speed is determined by obtaining an effective wavelength from several zero crossings of the curves.

The numerical results obtained in this way actually tend to agree with the data partway into the transition region (Havermann, 1983)—but this is surely fortuitous, so we shall not present these comparisons. Our reluctance here stems from the completely unrealistic nature of the model. That is, the entire pole structure

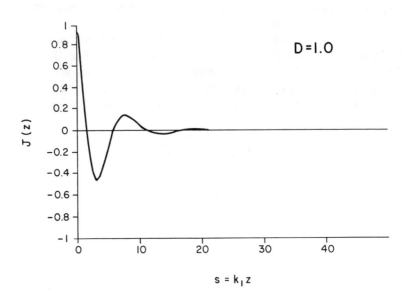

Fig. 9–12. Computer evaluation of the acoustic current $J(z)$ as determined by the model of Eq.(9–78), for $D = 1.0$.

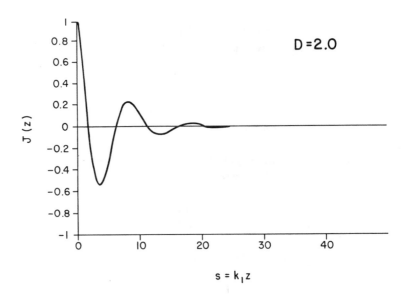

Fig. 9–13. Computer evaluation of the acoustic current $J(z)$ as determined by the model of Eq.(9–78), for $D = 2.0$.

obtained from the ring-sum approximation to K_{nn} arises from *not* enforcing the limit $\beta\hbar\omega \ll 1$, and so depends crucially on $\beta\hbar\omega$. This simply can not be a reasonable procedure in the Boltzmann limit, as a short numerical estimate confirms: at room temperature $\beta\hbar\omega \simeq 1$ corresponds to a frequency of about 10^7 MHz, which is out of the realm of possibility!

Although a more realistic model is clearly called for, it is equally clear that the present theory has the capability of describing propagation and attenuation of sound over the entire frequency range. Once again, the detailed calculation of system covariance functions is at the heart of the matter. As noted following Eq.(9–4), for example, the classical attenuation is related to a coupling of fluctuations in number density and energy density, so that one expects some kind of *mode coupling* to be important, as in Chapter 7. It is further expected that a construction going beyond the elementary structure of the ring sum will complete in an entirely satisfactorily way the description of ultrasonic attenuation begun here. Unfortunately, this may have to await a new generation of computers.

D. Ideal Gases, II: Quantum Statistics

Because no other transport processes can take place, one concludes that it is not possible to study sound propagation satisfactorily in a collisionless gas. None of the classical attenuation mechanisms is operative, and the absence of interactions impedes any significant propagation. Nevertheless, one might expect the quantum fluids to be richer in content, for quantum statistics introduces exchange forces into the system.

In order to examine these possibilities, we recall that the covariance function appropriate for the ideal quantum fluids is given by Eq.(6–145), where we note that the ground-state contribution for the Bose system is to be omitted in the present context. Substitution into Eq.(9–61) then yields

$$K(z) = \varepsilon \frac{m(2S+1)}{4\pi^2\hbar^3\beta^2} qp \int_0^\infty \frac{dp}{p^3} \cos(rp) \log \left| \frac{1 - \varepsilon e^{\beta\mu - (q+p)^2}}{1 - \varepsilon e^{\beta\mu - (q-p)^2}} \right| , \qquad (9\text{–}92)$$

where $r \equiv (8m/\beta\hbar^2)^{1/2} z$, and we recall that the remaining notation was established in Eq.(6–140):

$$q^2 = \tfrac{1}{2}\beta m \frac{\omega^2}{k^2}, \quad p^2 = \beta \frac{\hbar^2 k^2}{8m}, \quad qp \equiv K = \tfrac{1}{4}\beta\hbar\omega . \qquad (9\text{–}93)$$

The frequency behavior in the quantum case is essentially the same as in the Boltzmann system, with no particular frequency being singled out. Hence, no hydrodynamic modes can be sustained in the quantum gases either. Moreover, the asymptotic behavior of $K(z)$ for large r is proportional to $F(s)$, Eq.(9–64), and at low temperatures r is actually extraordinarily large, even for small values of z. Thus, even the ideal quantum gases do not appear particularly promising for a study of sound propagation. Indeed, numerical evaluation of $K(z)$ in the Bose case reveals

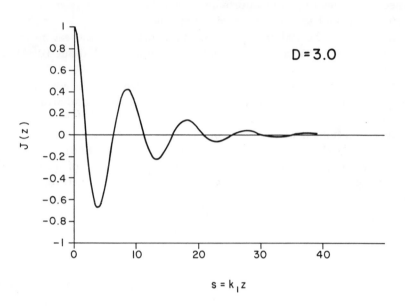

Fig. 9–14. Computer evaluation of the acoustic current $J(z)$ as determined by the model of Eq.(9–78), for $D = 3.0$.

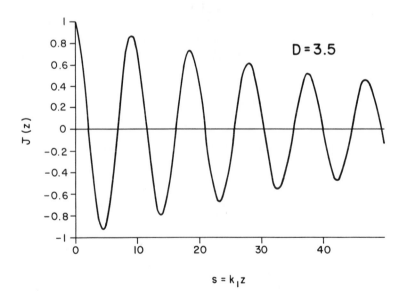

Fig. 9–15. Computer evaluation of the acoustic current $J(z)$ as determined by the model of Eq.(9–78), for $D = 3.5$.

no significant propagation for any degree of degeneracy. For large degeneracy there are some persistent small-amplitude oscillations superimposed on the Boltzmann curve at low frequencies, and at higher frequencies the damping is very rapid and effectively independent of the degree of degeneracy.

The very-low-temperature Fermi system behaves quite differently, however, as suggested by Figure 9–16, wherein K_{jj} is shown as a function of p. One notes that a sharp peak is possible at low frequencies and very large values of βE_F, where E_F is the Fermi energy. That is, from the integrand of Eq.(9–92), with $\varepsilon = -1$, we are led to consider the function

$$f(p, K) \equiv \frac{1}{p^3} \ln \frac{1 + e^{\beta E_F - (p - K/p)^2}}{1 + e^{\beta E_F - (p + K/p)^2}} . \tag{9-94}$$

When $\beta E_F \gg K$ this function has a sharp peak at

$$p_0 \equiv \frac{K}{(\beta E_F)^{1/2}} \ll 1, \tag{9-95}$$

which can be seen as follows. For large p we have $f \propto 4Kp^{-3}$, and f falls off rapidly; for small p the behavior is $f \propto p^{-3} \exp(-K^2/p^2)$, and f again falls off rapidly as p approaches zero. Only when $p \simeq p_0$ does this analysis fail, so that $f(p, K)$ rises rapidly to a peak at p_0 and then falls off as p^{-3}.

The integral in Eq.(9–92) for the degenerate Fermi system can now be written

$$\int_0^\infty \cos(rp) f(p, K) \, dp, \tag{9-96}$$

which is evaluated in the separate intervals $(0, p_0)$ and (p_0, ∞). When $\beta E_F \gg K$ we can approximate the requisite integrals to obtain

$$\frac{K(z)}{K(0)} \simeq \cos(rp_0) - rp_0 \sin(rp_0)$$

$$\xrightarrow[\beta E_F \gg K]{} \cos(rp_0). \tag{9-97}$$

In turn, this implies the dispersion law $kz \equiv rp_0$, or

$$\omega = kv_F, \qquad v_F \equiv (2E_F/m)^{1/2} - \hbar k_F/m, \tag{9-98}$$

and v_F is called the Fermi velocity. That is, for $\beta E_F \gg K$ the disturbance propagates at the Fermi velocity which, for example, is on the order of 2 m/sec in metals at room temperature. The extent of this propagation and its physical origin will be discussed presently.

When the degeneracy is moderate the condition $K \gg 1$ again leads to the peak at $\hbar\omega = \hbar^2 k^2/2m$, which does not yield propagation or a dispersion law. For small K we find, as in the Boltzmann case, $c_0 = (9\kappa T/8m)^{1/2}$. This is similar to the classical result, but again the disturbance is strongly damped.

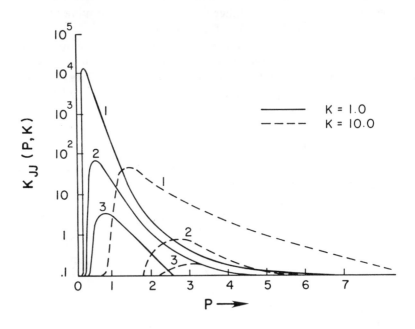

Fig. 9–16. The current-current covariance function for an ideal Fermi fluid at two different frequencies. The curves correspond to various values of βE_F, as follows: (1) 100, (2) 10, (3) 1.

We conclude, therefore, that significant propagation can take place at certain frequencies in the completely degenerate Fermi system. Numerical evaluation of $K(z)$ demonstrates the effects of degeneracy over a broad frequency range in the ideal Fermi gas. At low ultrasonic frequencies ($\omega \simeq 10^8$ Hz, $K = 10^{-3}$) and large degeneracy ($\beta E_F \simeq 10^5$), the disturbance propagates with significant amplitude for extraordinarily large values of r (large z), as in Figure 9–17, although it eventually damps out. What is most interesting is that this propagation and ultimate damping *do* have a physical interpretation, as first elaborated by Snow (1967).

LANDAU THEORY OF THE FERMI LIQUID

Perspective regarding the propagating sound mode in the completely degenerate Fermi system is gained from a brief study of Landau's theory of the Fermi liquid (Landau, 1957). A useful review of this theory has been provided by Pines and Noziéres (1966), and upon working out the details one finds a number of collective modes in the completely degenerate neutral Fermi system. These modes involve coherent motion of all particles in the system, are characterized by nonvanishing

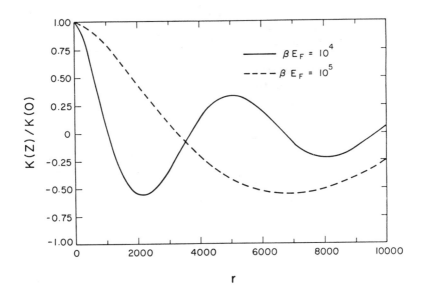

Fig. 9–17. Propagation of sound in the highly-degenerate ideal Fermi system, where $r = (8m/\beta\hbar^2)^{1/2} z$.

density fluctuations, and can be described in terms of quasi-particles.

Let ω be the frequency characterizing a particular mode, and suppose it is possible to describe the system in terms of a quasi-particle collision frequency ν. In the region $\omega \ll \nu$ there are many quasi-particle collisions in a time ω^{-1}, so that we have a collision-dominated situation known as the hydrodynamic regime, as discussed earlier. Collisions occur often enough to provide the restoring force necessary to support sound waves and, for reasons to become clear presently, we refer to this phenomenon as *first sound*.

The opposite limit, $\omega \gg \nu$, is called the collisionless regime. In this region, as long as the quasi-particle interactions are repulsive, there occurs a well-defined collective propagation mode called *zero sound*, which is a high-frequency counterpart of ordinary, or first sound. This mode arises from organized fluctuations of the particle density.

In the intermediate region—$\omega \simeq \nu$—there are sufficient collisions to destroy the zero sound mode, but not enough to support undamped first sound. One expects, then, that a maximum in the attenuation of sound waves is attained in the transition from first to zero sound. Figure 9–18 illustrates the qualitative damping of sound waves as a function of frequency at fixed temperature. At very low frequencies there is little damping, and what there is corresponds to the viscous Stokes damping of Eq.(9–4). Of course, because the viscosity generally goes as T^{-2} at very low

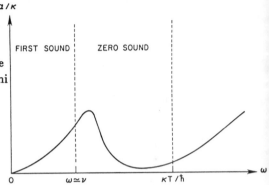

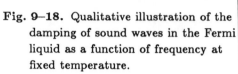

Fig. 9–18. Qualitative illustration of the damping of sound waves in the Fermi liquid as a function of frequency at fixed temperature.

temperatures, the absorption of sound should be very large in any event at these temperatures.

Some qualitative clarification is gained by recalling that in the very-low temperature Fermi system the particles fill the Fermi sphere up to E_F as $T \to 0$. If a sound wave is to undergo absorption by excitation of quasi-particles of momentum $\hbar \mathbf{k}$, then the disturbance must provide an excitation energy of at least

$$E_{\text{ex}} = \frac{\hbar^2}{2m}(\mathbf{k}_F + \mathbf{k})^2 - \frac{\hbar^2 k_F^2}{2m} = \frac{\hbar^2 k^2}{2m} + \frac{\hbar}{m}(\mathbf{k}_F \cdot \mathbf{k}). \tag{9-99}$$

Explicitly, $\mathbf{k}_F \cdot \mathbf{k} = k k_F \cos\theta$, and we note that energy conservation requires that $E_{\text{ex}} = \hbar c_0 k$. Then, as $T \to 0$, we have $v_F \cos\theta = c_0 - (\hbar k / 2m) \simeq c_0$, or

$$\cos\theta \simeq \frac{c_0}{v_F}. \tag{9-100}$$

For $c_0 < v_F$ the absorption condition can always be realized and there is little propagation. But if $c_0 = v_F$ the phase space available for absorption shrinks to zero, and zero sound becomes the only viable mode of propagation.

Additional understanding of the transition from first to zero sound can be obtained by defining parameters λ_i, such that for first sound

$$\lambda_1 \equiv \frac{\omega/k}{v_F} = \frac{c_0}{v_F}. \tag{9-101}$$

Define also a number F_0 as the scattering-length parameter for s-wave collisions, which is dominant in both the extremes $\omega \gg \nu$ and $\omega \ll \nu$. The quasi-particle interactions are repulsive when $F_0 > 0$, whereas the range $-1 < F_0 < 0$ characterizes weak attractions, and $F_0 < -1$ describes strong attractions. When $F_0 \simeq 0$ there are effectively no interactions. From the Landau theory one finds for first sound

$$\lambda_1 = \left[\frac{1 + F_0}{3}\right]^{1/2}, \tag{9-102}$$

so that in the collisionless limit $c_0 \simeq v_F/\sqrt{3}$.

For zero sound one finds the corresponding, but qualitatively different relation

$$\frac{\lambda_0}{2} \ln \frac{\lambda_0 + 1}{\lambda_0 - 1} - 1 = \frac{1}{F_0}. \tag{9-103}$$

Hence, when $F_0 > 0$ one obtains a solution $\lambda_0 > 1$, characterizing an undamped zero-sound wave. If $-1 < F_0 < 0$ the solution is complex, and the zero-sound wave is damped. When $F_0 < -1$ the wave becomes unstable. In general,

$$\lambda \longrightarrow \begin{cases} (F_0/3)^{1/2}, & F_0 \gg 1 \\ 1, & |F_0| \ll 1 \end{cases}. \tag{9-104}$$

There is a strong suggestion, then, that the propagating sound mode found in Eq.(9–98) is just that of zero sound, because a noninteracting system is certainly collisionless. In the completely degenerate system of noninteracting fermions at very low temperatures the particles have effectively settled into the Fermi sea. The system develops a stiffness making it a much better conductor of sound than the 'mushy' Bose or Boltzmann fluids, and this is just what one might expect on the basis of the Pauli principle.

Problems

9.1 Derive the form of the Kramers-Kronig relations given in Eqs.(9–53) from Eqs.(3–32).

9.2 Demonstrate the equivalence of the two representations for $g(q)$ that are given by Eqs.(9–75) and (9–76) by means of the following steps.

(a) Show that the form (9–75) is equivalent to the principal value

$$g(q) = -\frac{2q^2}{\sqrt{\pi}} \, \mathrm{P} \int_0^\infty \frac{e^{-t^2}}{q^2 - t^2} \, dt,$$

for real q. This is evident from discussions of error functions.

(b) Define an analytic function of z,

$$K(z) \equiv \int_0^\infty \frac{e^{-t^2}}{z^2 - t^2} \, dt, \quad \mathrm{Im}\, z > 0,$$

which is *not* a principal value, but is an integral representation of the complex error function $w(z)$:

$$K(z) = \frac{\pi}{2iz} \, w(z), \quad \mathrm{Im}\, z > 0$$

$$= \frac{\pi}{2iz} \, e^{-z^2} \, \mathrm{erfc}(-iz).$$

Show that the limit of $K(z)$ on the real axis, $K(q)$, is just the expression (9-76)—after some algebra.

(c) Finally, establish the first line of Eq.(9-77) by successive integrations by parts, and the second by reference to the well-known asymptotic expansion of $\text{erfc}(z)$.

REFERENCES

Abramowitz, M.: 1953, 'Evaluation of the Integral $\int_0^\infty e^{-u^2-x/u}\,du$', J. Math. and Phys. **32**, 188.

Abramowitz, M., and I.A. Stegun (eds.): 1964, Handbook of Mathematical Functions, AMS 55, Natl. Bur. Stds., Washington.

Bleistein, N., and R.A. Handelsman: 1975, Asymptotic Expansions of Integrals, Holt, Rinehart, and Winston, New York.

Bömmel, H.E.: 1954, 'Ultrasonic Attenuation in Superconducting Lead', Phys. Rev. **96**, 220.

Buckner, J.K., and J.H. Ferziger: 1966, 'Linearized Boundary Value Problem for a Gas and Sound Propagation', Phys. Fluids **9**, 2315.

Cercignani, C.: 1975, Theory and Application of the Boltzmann Equation, Scottish Academic Press, Edinburgh.

Foch, J.D., Jr., and G.W. Ford: 1970, 'The Dispersion of Sound in Monoatomic Gases', in J. De Boer and G.E. Uhlenbeck (eds.), Studies in Statistical Mechanics, Vol. V, North-Holland, Amsterdam.

Foch, J.D., Jr., and M.L. Fuentes: 1972, 'Improved Kinetic Theory of Sound Propagation', Phys. Rev. Letters **28**, 1315.

Greenspan, M.: 1950, 'Propagation of Sound in Rarefied Helium', J. Acoust. Soc. Am. **22**, 568.

Greenspan, M.: 1956, 'Propagation of Sound in Five Monatomic Gases', J. Acoust. Soc. Am. **28**, 644.

Havermann, R.A.: 1983, unpublished.

Hirschfelder, J.O., C.F. Curtiss, and R.B. Bird: 1954, Molecular Theory of Gases and Liquids, Wiley, New York.

Holstein, T.D.: 1959, 'Theory of Ultrasonic Absorption in Metals: the Collision-Drag Effect', Phys. Rev. **113**, 479.

Kirchhoff, G.: 1868, 'Über den Einfluss der Wärmeleitung in einem Gase auf die Schallbewegung', Pogg. Ann. **134**, 177.

Knudsen, M.: 1934, The Kinetic Theory of Gases, Methuen, London.

Lambert, J.D.: 1962, 'Relaxation in Gases', in D.R. Bates (ed.), Atomic and Molecular Processes, Academic Press, New York, p.783.

Landau, L.D.: 1957, 'Oscillations in a Fermi Liquid', Sov. Phys. JETP **5**, 101.

Laplace, P.S.: 1816, 'Sur la vitesse du son dans l'air et dans l'eau', Ann. Phys. Chim. **3**, 238.

Macfarlane, R.E., and J.A. Rayne: 1967, 'Ultrasonic Attenuation in the Noble Metals', Phys. Rev. **162**, 532.

Meyer, E., and G. Sessler: 1957, 'Schallausbreitung in Gases bei hohen Frequenzen und sehr niedrigen Drucken', *Z. Physik* **149**, 15.

Nagaoka, Y.: 1961, 'Theory of Ultrasonic Attenuation in Metals', *Prog. Theor. Phys.* **26**, 589.

Newton, I.: 1687, *Principia Mathematica Naturalis Philosophiae*, London.

O'Donnell, M., E.T. Jaynes, and J.G. Miller: 1978, 'General Relationships Between Ultrasonic Attenuation and Dispersion', *J. Acoust. Soc. Am.* **63**, 1935.

O'Donnell, M., E.T. Jaynes, and J.G. Miller: 1981, 'Kramers-Kronig Relationship Between Ultrasonic Attenuation and Phase Velocity', *J. Acoust. Soc. Am.* **69**, 696.

Pines, D., and P. Noziéres: 1966, *The Theory of Quantum Liquids*, Benjamin, New York.

Pippard, A.B.: 1955, 'Ultrasonic Attenuation in Metals', *Phil. Mag.* **46**, 1104.

Sessler, G.M.: 1965, 'Free-Molecule Propagation in Rarefied Gases', *J. Acoust. Soc. Am.* **38**, 974.

Sirovich, L., and J.K. Thurber: 1969, 'Wave Propagation and Other Spectral Problems in Kinetic Theory', *J. Math. Phys.* **10**, 239.

Snow, J.A.: 1967, 'Sound Absorption in Model Quantum Systems', *Ph.D. thesis*, Washington Univ.. St. Louis, MO (unpublished).

Spanier, J., and K.B. Oldham: 1987, *An Atlas of Functions*, Hemisphere Publ., Washington Chap.42.

Stokes, G.: 1845, 'On the Theories of the Internal Friction of Fluids, and of the Equilibrium and Motion of Elastic Solids', *Trans. Camb. Phil. Soc.* **8**, 287.

Tsuneto, T.: 1961, 'Ultrasonic Attenuation in Superconductors', *Phys. Rev.* **121**, 402.

Uhlenbeck, G.E., and G.W. Ford: 1963, *Lectures in Statistical Mechanics*, Am. Math. Soc., Providence, RI (*Lectures in Applied Mathematics*, Vol.I).

Epilogue

In the preceding chapters we have attempted to lay a coherent, systematic foundation for nonequilibrium statistical mechanics, as well as provide a number of specific applications in order to illustrate the theory. Most of this work rests in the central part of a certain spectrum of physical problems, in the sense that many more applications can and should be made, and one knows essentially how to proceed, though the going is nontrivial. At one end of this spectrum the issues have been clear and well understood for almost a century, and one has great confidence that the treatment of equilibrium problems is well in hand. This special case of the theory has been presented in some detail in Volume I. At the other, effectively open, end of the spectrum, however, one is looking into virtually uncharted territory where the possibilities for application are limitless. But it is just such a view that also emphasizes the need for a great deal of further development.

Although there are a number of problems requiring attention in order to expand from these foundations, three stand out as having some immediacy: an understanding of the analytical structure of covariance functions, extension of the theory to highly nonlinear scenarios, and investigation of the role of irregular motion and 'chaos' in the general many-body system. Of the three, perhaps the first is of most immediate interest, for it strongly affects that part of nonequilibrium phenomena which is best understood and not quite at the *far* end of the spectrum. We have discussed these calculational difficulties at some length in the text, as the occasion arose, and little more need be added here. It is quite clear at this point that development of systematic—probably nonperturbative—calculational techniques for correlation functions is imperative for further progress. Rapid advances in modern *computational* techniques may, in the long run, lead to resolution of this problem.

NONLINEARITIES

By 'nonlinear' we have meant, and continue to mean, nonlinear departures from equilibrium. This is most often manifested by nonlinearities in terms of the Lagrange-multiplier functions, as in the Navier-Stokes equations. We shall return presently to the macroscopic consequences of this latter example, but for the moment let us focus on the difficulties of evaluating expectation values in other than linear approximation. These difficulties, of course, are intimately related to the extreme nonlinearity of exponential operators and, despite the clever development of a number of useful identities, we have not been able to master the calculus of these operators. Increased activity in the study and application of path integrals, however, shows promise of eventually providing relief in this area.

It is rather disappointing that the functional theory presented here in principle describes systems arbitrarily far from equilibrium, yet at the present time is mathematically tractable only in its expanded form—and even then one is faced with the problem of calculating covariance functions, and higher correlations. This situation is somewhat reminiscent of that faced by the general theory of relativity until approximately thirty years ago. For problems other than those which were fairly simple and possessed obvious symmetries, the nonlinearity of Einstein's equations mandated linear and 'post-Newtonian' approximations in order to make progress. But new ideas on how to analyze the full nonlinear equations, and what different questions should possibly be asked, led to a revival of interest in the theory at a time when new observational capabilities were requiring just that. In the same vein, there is a clear need to view the nonequilibrium expectation value from a different perspective, which perhaps will generate new questions to be asked about it.

Unfortunately, one of the severe underlying difficulties in moving away from the linear approximation is a certain ambiguity in the definition of the nonequilibrium state itself. From our experience with phenomenological theories—linear hydrodynamics, say—the variables specified by the five microscopic conservation laws appear sufficient to specify the state. As one departs farther from equilibrium it is not at all clear what constitutes a 'complete' set of state variables. One would guess, of course, that these same variables are adequate, but they are folded nonlinearly into new sets of variables appropriate for describing various sets of collective modes in the system. Moreover, these sets will vary from problem to problem. Uncovering how this is done is just the task of finding different ways to view the full nonlinear problem.

IRREGULAR MOTION

It has been known since the work of Poincaré that the number of integrable dynamical systems is severely limited, and that nonlinear equations of motion possess solutions exhibiting highly irregular behavior for given parameter values. Not until relatively recently, however, with the enormous advances in computers, has it been possible to study this behavior in any detail. Examples of deterministic dynamical systems in which irregular or 'chaotic' motions can occur are now commonplace. How do these developments bear, if at all, on the statistical mechanics of nonequilibrium processes?

Let us focus first on the microscopic particle equations of motion and enquire about possible observable effects should these equations be nonlinear and exhibit irregular solutions. But by 'observable' we mean 'macroscopic', and the quintessential aspect of the statistical view is that we are neither able nor wish to follow the microscopic motion. Given macroscopic information, we ask only macroscopic questions about the system, a strategy dictated by our ignorance of microscopic initial conditions. This ignorance forces the use of a probability distribution over the possible microstates of the system consistent with that macroscopic information, and the only requirement the particle dynamics places upon us is that of being able to specify those possible alternatives, or states.

Even for relatively simple Hamiltonians this last task is not usually trivial, and requires thoughtful approximation. But the theory demands that it be possible only in principle—actual calculations, of course, require more than this. Presumably nothing changes here if the particle equations of motion are highly nonlinear. If one is unable to enumerate the possible states of the system *in principle*, then we are no longer discussing deterministic physics. As long as this spectrum can be presented in principle, however difficult in practice, then it matters little how irregular that microscopic motion may be. That is actually the point of statistical mechanics!

One can reach entirely different conclusions, however, with respect to macroscopic motions of a system if the governing equations are nonlinear, for we can and do follow the trajectories in this case. An example of major interest in the present context is that of the full Navier-Stokes equations in the conventional convected frame, which are nonlinear in the fluid velocity:

$$ mn_0\Big(\partial_t v_i + \big[(\mathbf{v}\cdot\nabla)\mathbf{v}\big]_i\Big) = -\partial_j P\,\delta_{ij} + \partial_k p_{ik} + mn_0 F_i\,, \qquad \text{(E–1)} $$

$\nabla\cdot\mathbf{v} = 0$, and all the quantities involved have been defined earlier. These are rather difficult either to solve or to analyze in general, of course, but one can approximate them without destroying the essential nonlinearity in particular applications. The now-classic example of this procedure begins with the attempt by Saltzman (1962) to model the Rayleigh-Bénard instability in two dimensions by Fourier expansion and truncation into a set of ordinary differential equations. Shortly thereafter these equations were adopted by Lorenz (1963) as a model for the unpredictable behavior of the weather, and studied extensively by him—with remarkable results. These equations for convection of the fluid are

$$ \frac{dx}{dt} = \sigma(y - x)\,, $$
$$ \frac{dy}{dt} = rx - y - xz\,, $$
$$ \frac{dz}{dt} = xy - bz\,, \qquad \text{(E–2)} $$

where $x(t)$ is proportional to the amplitude of convective motion, $y(t)$ and $z(t)$ are proportional to two temperature modes, σ is called the *Prandtl number* (the ratio of kinematic viscosity to the thermal diffusivity), r is the *Rayleigh number* in units of its critical value (the convective analog of the Reynolds number, see below), and b is a constant related to the wavenumber of the fundamental mode.

The set of equations (E–2) is completely deterministic, so that we can study (via computer) the trajectories generated from various initial conditions by fixing σ and b at the values adopted by Lorenz, say, and varying r. For $r < 1$ all trajectories are attracted to a stable solution at the origin of the variables in Eq.(E–2): $x = y = z = 0$. If r exceeds unity by much the model is no longer physically realistic, but nevertheless still worth studying. One finds two stable solutions for $1 < r < 13.9$, to which all trajectories are attracted, and in the region $13.9 < r < 24.1$ a complicated transition begins to take place. For $r > 24.1$ all trajectories are attracted toward a subspace in which they wander 'chaotically' forever. That is, the motion is highly irregular and essentially unpredictable. This subspace is called a *strange attractor*.

We hesitate to use the word 'chaos' here, because it tends to convey an impression of motion which is not deterministic. In reality, the motion here is no more chaotic than that of the particles in an equilibrium gas—they all obey well-defined equations of motion. But 'chaos' now assumes a more technical meaning—namely, the result of an extraordinary sensitivity to initial conditions. In the chaotic regime it is virtually impossible to specify initial conditions precisely enough to be sure of the ensuing trajectory, and it is in this sense we employ the above phrase 'essentially unpredictable'.

The importance of these results lies with the possibility of finally being able to describe turbulence in some detail as a solution of the Navier-Stokes equations, say. Let $R \equiv v\ell/\nu$ denote the *Reynolds number*, where ν is the kinematic viscosity, and $\mathbf{v}$ and ℓ are a characteristic velocity and length, respectively, for the system. It has been known for over 100 years that smooth laminar flow will become unstable and cascade into turbulence when R exceeds some critical value. This is a *macroscopic* phenomenon, and so would seem to be outside the purview of statistical mechanics. That is, the role of the latter should cease with the complete derivation of the macroscopic equations of motion and provision for calculation of the relevant parameters.

But completely developed turbulence is more than just chaotic motion, and the phenomena uncovered by study of the Lorenz equations only provide us with a beginning. The onset of chaos may well signal the approach to a turbulent state, which is intrinsically nonequilibrium and collective in nature. As the parameters of the macroscopic equations continue to change, and full turbulence develops, one realizes that the number of *macroscopic* degrees of freedom has increased enormously. There are now a great many possible trajectories available to the system, but it is very difficult to know which is taken owing to the extreme sensitivity to initial conditions. Although the system state may well be described by only a few macroscopic variables—or 'supermacroscopic' variables—just what that state may be is difficult to determine exactly. It is as if one did not really know the precise initial conditions.

At this point everything begins to sound familiar, as in Chapter 1 of Volume I, and we begin to see the unfolding of a new kind of statistical mechanics, on a higher level. There is, as yet, no definite means for determining what the new variables should be in most cases, nor has there emerged much insight into exactly what is meant by a 'reproducible experiment' in these various scenarios. One has yet to catalog the type of macroscopic information that is needed in order to construct a probability distribution over trajectories, an exercise surely required owing to the uncertainty associated with initial conditions.

In this writer's view the beautiful construct called statistical mechanics is about to move up onto a new level and enter an entirely new era in connection with these macroscopic systems. Despite the fact that serious calculational problems remain at the lower level—a situation familiar from the relation between classical and quantum electrodynamics—the prospects for this next phase are exciting, and may even shed some light on the old problems. The story is far from over!

REFERENCES

Saltzman, B.: 1962, 'Finite Amplitude Free Convection as an Initial Value Problem—I', *J. Atmos. Sci.* **19**, 329.

Lorenz, E.N.: 1963, 'Deterministic Nonperiodic Flows', *J. Atmos. Sci.* **20**, 130.

Appendices

Appendix A

Operator Calculus and Identities

$\mathbf{W}$e gather together in this appendix a number of useful identities based on the calculus of linear operators, and which have been employed in the text. The following notation is employed for the commutator of two operators on an appropriate Hilbert space:

$$[\hat{A}, \hat{B}] \equiv \hat{A}\hat{B} - \hat{B}\hat{A} = -[\hat{B}, \hat{A}], \qquad (A–1a)$$

whereas the anticommutator is written

$$\{\hat{A}, \hat{B}\} \equiv \hat{A}\hat{B} + \hat{B}\hat{A}. \qquad (A–1b)$$

It should be emphasized that in these and subsequent expressions a state vector is tacitly presumed to stand to the right of all operators. The commutator is associative and satisfies the following algebraic relations:

$$[\hat{A}, \hat{B}\hat{C}] = [\hat{A}, \hat{B}]\hat{C} + \hat{B}[\hat{A}, \hat{C}], \qquad (A–2a)$$
$$[\hat{A}\hat{B}, \hat{C}] = [\hat{A}, \hat{C}]\hat{B} + \hat{A}[\hat{B}, \hat{C}]. \qquad (A–2b)$$

From elementary quantum mechanics one obtains for non-commuting $\hat{A}$ and $\hat{B}$ the familiar 'zipper' principle,

$$e^{\hat{A}} \hat{B} e^{-\hat{A}} = \hat{B} + [\hat{A}, \hat{B}] + \frac{1}{2!}[\hat{A}, [\hat{A}, \hat{B}]]$$
$$+ \frac{1}{3!}\left[\hat{A}, [\hat{A}[\hat{A}, \hat{B}]]\right] + \cdots. \qquad (A–3)$$

OPERATOR IDENTITIES

In a seminal review of exponential-operator calculus, Wilcox has shown how a great many important operator theorems can be generated from the following identity:

$$\frac{\partial}{\partial\lambda}e^{-\beta\hat{H}} = -\int_0^\beta e^{-(\beta-u)\hat{H}} \frac{\partial\hat{H}}{\partial\lambda} e^{-u\hat{H}}\, du, \qquad (A–4)$$

where β and λ are parameters and $\hat{H} = \hat{H}(\lambda)$ is an arbitrary operator. One readily proves Eq.(A–4) by showing that both sides of the equation satisfy the first-order differential equation

$$\frac{\partial}{\partial\beta}\hat{F}(\beta) + \hat{H}\,\hat{F}(\beta) = -e^{-\beta\hat{H}}\frac{\partial\hat{H}}{\partial\lambda}, \qquad (A–5)$$

with the initial condition $\hat{F}(0) = 0$.

As a first example of the kinds of identities derivable from Eq.(A–4), we have the following due originally to Feynman (1951), and Karplus and Schwinger (1948):

$$e^{\hat{A}+\hat{B}} = \exp\left[\int_0^1 e^{x\hat{A}} \hat{B} e^{-x\hat{A}} \, dx\right] e^{\hat{A}}, \tag{A–6}$$

$$e^{-(\hat{A}+\hat{B})} = e^{-\hat{A}} \exp\left[-\int_0^1 e^{x\hat{A}} \hat{B} e^{-x\hat{A}} \, dx\right], \tag{A–7}$$

$$e^{\hat{A}+\hat{B}} = e^{\hat{A}} \left[\hat{1} + \int_0^1 e^{-x\hat{A}} \hat{B} e^{x(\hat{A}+\hat{B})} \, dx\right], \tag{A–8}$$

for arbitrary $\hat{A}$ and $\hat{B}$.

Another important identity is obtained from a similarity transformation of Eq.(A–4) involving parameters λ and β. Let

$$\hat{H}(\lambda) = e^{\lambda\hat{A}} \hat{B} e^{-\lambda\hat{A}},$$
$$e^{-\beta\hat{H}} = e^{\lambda\hat{A}} e^{-\beta\hat{B}} e^{-\lambda\hat{A}}. \tag{A–9}$$

Then,

$$[\hat{A}, e^{-\beta\hat{B}}] = -\int_0^\beta e^{-(\beta-u)\hat{B}} [\hat{A}, \hat{B}] e^{-u\hat{B}} \, du. \tag{A–10}$$

This last expression can be put into a simpler form by means of the Kubo transform of an operator $\hat{B}$ with respect to the operator $\hat{A}$ employed in Chapters 3 and 5:

$$\overline{\hat{B}} \equiv \int_0^1 e^{-x\hat{A}} \hat{B} e^{x\hat{A}} \, dx. \tag{A–11}$$

Therefore, for arbitrary $\hat{A}$ and $\hat{B}$, the content of of Eq.(A–10) is

$$[e^{\hat{A}}, \hat{B}] = e^{\hat{A}} [\hat{A}, \overline{\hat{B}}]. \tag{A–12}$$

With the aid of these techniques Eq.(A–3) can be put into an occasionally more useful form:

$$e^{-\beta\hat{A}} \hat{B} e^{\beta\hat{A}} = \hat{B} + \int_0^\beta e^{-\lambda\hat{A}} [\hat{B}, \hat{A}] e^{\lambda\hat{A}} \, d\lambda. \tag{A–13}$$

Equations (A–4) and (A-11) can be employed to obtain a useful derivative formula when $\hat{H}$ has the form $\hat{A} + \lambda\hat{B}$. We find that

$$\frac{\partial}{\partial\lambda} e^{\hat{A}+\lambda\hat{B}} = e^{\hat{A}+\lambda\hat{B}} \int_0^1 e^{-x(\hat{A}+\lambda\hat{B})} \hat{B} e^{x(\hat{A}+\lambda\hat{B})} \, dx$$

$$= e^{\hat{A}+\lambda\hat{B}} \, \overline{\hat{B}}, \tag{A–14}$$

where $\overline{B}$ is here defined with respect to the *full* operator $\hat{A} + \lambda\hat{B}$. As an example of this result, in equilibrium statistical mechanics one is often interested in the expectation value of an operator $\hat{Q}$,

$$\langle\hat{Q}\rangle \equiv \frac{\mathrm{Tr}\left[e^{-\beta\hat{H}}\,\hat{Q}\right]}{\mathrm{Tr}\left[e^{-\beta\hat{H}}\right]}\,, \tag{A–15}$$

where $\hat{H} = \hat{H}_0 + \lambda\hat{V}$. With the notation $\delta\hat{V} \equiv \hat{V} - \langle\hat{V}\rangle\hat{1}$ for the deviation, we readily find from the above that

$$\frac{\partial\langle\hat{V}\rangle}{\partial\lambda} = -\int_0^\beta \langle e^{u\hat{H}}\,\delta\hat{V}\,e^{-u\hat{H}}\,\delta\hat{V}\rangle\,du$$

$$= -\beta\langle\overline{\delta\hat{V}\,\delta\hat{V}}\rangle\,. \tag{A–16}$$

A final example of the power of Eq.(A–4) is proof of the Baker-Campbell-Hausdorff formula for arbitrary operators $\hat{A}$ and $\hat{B}$:

$$e^{\hat{A}}\,e^{\hat{B}} = \exp\left\{\hat{A} + \hat{B} + \frac{1}{2}[\hat{A},\hat{B}] + Z_3 + \cdots + Z_n + \cdots\right\}, \tag{A–17}$$

where

$$Z_3 = \frac{1}{12}\left[\hat{A},[\hat{A},\hat{B}]\right] + \frac{1}{12}\left[[\hat{A},\hat{B}],\hat{B}\right], \tag{A–18}$$

and Wilcox (1967) has provided recursion formulas for higher Z_n. When $\hat{A}$ and $\hat{B}$ both commute with their commutator we obtain the familiar form

$$e^{\hat{A}}\,e^{\hat{B}} = e^{\hat{A}+\hat{B}+[\hat{A},\hat{B}]/2}\,. \tag{A–19}$$

TIME-EVOLUTION OPERATORS

A dynamical system of N particles may be described by a Hamiltonian. In that event the time development of the system is governed by the Schrödinger equation:

$$i\hbar\partial_t|\psi(t)\rangle = \hat{H}\,|\psi(t)\rangle\,, \tag{A–20}$$

where we shall here suppress explicit spatial dependence of state vectors. Although $\hat{H}$ and other operators may possess an explicit time dependence, the view in the Schrödinger picture is that operators do not develop in time with the dynamics of the system. Because the relation (A–20) is linear we expect it can be solved by the further linear relation

$$|\psi(t)\rangle = \hat{U}(t,t_0)|\psi(t_0)\rangle\,, \tag{A–21}$$

where t_0 is fixed and $\hat{U}(t,t_0)$ is called the time-evolution operator—it also defines a unitary transformation to the Heisenberg picture, where operators *do* develop in time.

As is readily demonstrated, $\hat{U}$ possesses the group property,

$$\hat{U}(t_2, t_1)\hat{U}(t_1, t_0) = \hat{U}(t_2, t_0), \quad t_2 > t_1 > t_0. \tag{A-22}$$

If $\hat{H}$ is Hermitian $\hat{U}$ is unitary:

$$\hat{U}^{\dagger}(t, t_0) = \hat{U}^{-1}(t, t_0) = \hat{U}(t_0, t), \tag{A-23}$$

so that

$$\hat{U}^{\dagger}(t, t_0)\hat{U}(t, t_0) = \hat{1}, \tag{A-24}$$

where $\hat{1}$ is the identity operator. That is, the equation of motion induces a unitary one-to-one mapping of the Hilbert space onto itself.

Substitution of Eq.(A–21) into (A–20) yields an equation of motion for $\hat{U}(t, t_0)$:

$$i\hbar \frac{d\hat{U}}{dt} = \hat{H}\,\hat{U}, \tag{A-25}$$

which also follows from Stone's theorem in Hilbert space. With the initial condition $\hat{U}(t_0, t_0) = \hat{1}$ the differential equation is readily converted into an integral equation:

$$\hat{U}(t, t_0) = \hat{1} - \left(\tfrac{i}{\hbar}\right) \int_{t_0}^{t} \hat{H}(t')\hat{U}(t', t_0)\, dt'. \tag{A-26}$$

If $\hat{H} \neq \hat{H}(t)$, one verifies by direct substitution into Eq.(A–25) that the solution is

$$\hat{U}_0(t, t_0) = \exp\left[-\tfrac{i}{\hbar}(t - t_0)\,\hat{H}\right], \tag{A-27}$$

where the subscript 0 will always denote this special (but common) case of a time-independent Hamiltonian.

When $\hat{H}$ depends explicitly on the time it is extremely difficult to obtain exact solutions of either Eq.(A–25) or Eq.(A–26). One can, however, obtain a series solution by iterating the integral equation (A–26):

$$\begin{aligned}
\hat{U}(t, t_0) = \hat{1} &+ \left(-\tfrac{i}{\hbar}\right) \int_{t_0}^{t} \hat{H}(t')\, dt' \\
&+ \left(-\tfrac{i}{\hbar}\right)^2 \int_{t_0}^{t} dt_1 \int_{t_0}^{t_1} dt_2\; \hat{H}(t_1)\,\hat{H}(t_2) \\
&+ \cdots.
\end{aligned} \tag{A-28}$$

Such a series is not usually of great practical value, because the successive terms contain powers of the *total* Hamiltonian, which can not generally be considered small in any sense. But we shall nevertheless find Eq.(A–28) to have considerable theoretical utility. Note that the order of the operators in this expansion is quite important, because $\hat{H}$ may not commute with itself at different times. A more compact notation, as well as an explicit indication of this ordering requirement, is secured by introduction of a *time-ordering operator* T:

$$T[\hat{F}(t')\hat{F}(t'')] \equiv \begin{cases} \hat{F}(t')\hat{F}(t''), & t' \geq t'' \\ \hat{F}(t'')\hat{F}(t'), & t'' \geq t' \end{cases} . \qquad (A\text{-}29)$$

Then,

$$\hat{U}(t,t_0) = \hat{1} + \sum_{n=1}^{\infty} \frac{1}{n!} \left(-\frac{i}{\hbar}\right)^n \int_{t_0}^{t} dt_1 \cdots \int_{t_0}^{t} dt_n \, T[\hat{H}(t_1)\cdots\hat{H}(t_n)]$$

$$= T \exp\left[-\frac{i}{\hbar}\int_{t_0}^{t}\hat{H}(t')\,dt'\right], \qquad (A\text{-}30)$$

providing an exact *formal* solution to the integral equation (A–26).

If the time-dependent Hamiltonian can be separated such that $\hat{H} = \hat{H}_0 + \hat{H}'(t)$, with $\hat{H}_0 \neq \hat{H}_0(t)$, then an alternative solution may be found useful if $\hat{H}'(t)$ is in some sense small. One easily verifies that

$$\hat{U}(t,t_0) = \hat{U}_0(t,t_0)\left[\hat{1} - \frac{i}{\hbar}\int_{t_0}^{t}\hat{H}_1(t')\hat{U}_0^{\dagger}(t',t_0)\hat{U}(t',t_0)\,dt'\right] \qquad (A\text{-}31)$$

satisfies Eq.(A–25) with the standard initial condition, where we have defined

$$\hat{H}_1(t) \equiv \hat{U}_0^{\dagger}(t,t_0)\,\hat{H}'(t)\hat{U}_0(t,t_0). \qquad (A\text{-}32)$$

Iteration of this integral equation then yields, to leading order,

$$\hat{U}(t,t_0) \simeq \hat{U}_0(t,t_0)\left\{\hat{1} - \frac{i}{\hbar}\int_{t_0}^{t} dt_1 \, \hat{U}_0^{\dagger}(t_1,t_0)\,\hat{H}'(t_1)\hat{U}_0(t_1,t_0)\right\}, \qquad (A\text{-}33a)$$

$$\hat{U}^{\dagger}(t,t_0) \simeq \left\{\hat{1} + \frac{i}{\hbar}\int_{t_0}^{t} dt_1 \, \hat{U}_0^{\dagger}(t_1,t_0)\,\hat{H}'(t_1)\hat{U}_0(t_1,t_0)\right\}\hat{U}_0^{\dagger}(t,t_0). \qquad (A\text{-}33b)$$

A further useful result can be obtained at this point if $\hat{H}$ contains some arbitrary parameter λ. Direct differentiation in the second line of Eq.(A–30) yields

$$\frac{\partial}{\partial\lambda}\hat{U}(t,t_0) = -\frac{i}{\hbar}\hat{U}(t,t_0)\int_{t_0}^{t}\hat{U}^{\dagger}(t',t_0)\frac{\partial\hat{H}(t')}{\partial\lambda}\hat{U}(t',t_0)\,dt'. \qquad (A\text{-}34)$$

Although operators do not generally evolve in this picture, the one exception is the statistical operator, because it is defined in terms of projection operators. Thus,

$$\hat{\rho}(t) = \hat{U}(t,t_0)\hat{\rho}(t_0)\hat{U}^{\dagger}(t,t_0). \qquad (A\text{-}35)$$

A more explicit equation of motion is obtained by direct differentiation and use of Eq.(A-25):

$$i\hbar\frac{d\hat{\rho}}{dt} = [\hat{H}, \hat{\rho}],\tag{A-36}$$

with appropriate initial conditions. If $\hat{H} = \hat{H}_0 + \hat{H}'(t)$, and $[\hat{H}_0, \hat{\rho}(0)] = 0$, these equations are equivalent to the following integral equation:

$$\hat{\rho}(t) = \hat{\rho}(0) - \frac{i}{\hbar}\int_{t_0}^{t} dt_1\, e^{-i(t-t_1)\hat{H}_0/\hbar}\,[\hat{H}'(t_1), \hat{\rho}(t_1)]e^{i(t-t_1)\hat{H}_0/\hbar},\tag{A-37}$$

as is readily verified by direct differentiation of $\hat{\rho}(t)$. Clearly, if $\hat{\rho}$ commutes with $\hat{H}$, then $\hat{\rho}$ is a constant of the motion.

If $\hat{H} \neq \hat{H}(t)$, the solution to Eq.(A-36) is just

$$\hat{\rho}(t) = e^{-it\hat{H}/\hbar}\,\hat{\rho}(0)e^{it\hat{H}/\hbar},\tag{A-38}$$

where for convenience we have set $t_0 = 0$. In this case it is occasionally useful to define the so-called *Liouville operator* L in terms of its effect on other operators:

$$\hat{L}\hat{A} \equiv \frac{1}{\hbar}[\hat{H}, \hat{A}].\tag{A-39}$$

Equation (A-36) then yields as an equivalent equation of motion

$$i\dot{\hat{\rho}} = \hat{L}\hat{\rho},\tag{A-40}$$

with solution

$$\hat{\rho}(t) = e^{-it\hat{L}}\,\hat{\rho}(0).\tag{A-41}$$

REFERENCES

Feynman, R.P.: 1951, 'An Operator Calculus Having Applications in Quantum Electrody-namics', *Phys. Rev.* **84**, 108.

Karplus, R., and J. Schwinger: 1948, 'A Note on Saturation in Microwave Spectroscopy', *Phys. Rev.* **73**, 102.

Wilcox, R.M.: 1967, 'Exponential Operators and Parameter Differentiation in Quantum Physics', *J. Math. Phys.* **8**, 962.

Appendix B

Diagrammatic Analysis
of
Covariance Functions

A very general perturbation theory for covariance functions involving field operators pertaining to a simple fluid was developed in Chapter 6. Those are the operators of Table 6–1, and the formal perturbation series was expressed in Eq.(6–121). In order to analyze the expansion in any detail it is necessary to develop further a systematic scheme for evaluating terms in the series to arbitrary order, and for arbitrary spin-independent spherically-symmetric two-body interactions. We shall outline such a scheme in this appendix.

As throughout, we shall presume that the unperturbed system is homogeneous and that it is a valid procedure to take the infinite-volume limit, so that the plane-wave representation is appropriate. All the microscopic field operators of interest then have the following general form [e.g., Eqs.(6–124)-(6–133)]:

$$\hat{A}(\mathbf{x}, t) = \frac{1}{V} \sum_{\mathbf{k}_1 \mathbf{k}_2} e^{i(\mathbf{k}_2 - \mathbf{k}_1) \cdot \mathbf{x}} f(\mathbf{k}_1, \mathbf{k}_2) \hat{a}^\dagger_{\mathbf{k}_1 \alpha}(t) \hat{a}_{\mathbf{k}_2 \alpha}(t)$$

$$+ \frac{1}{V^2} \sum_{\substack{\mathbf{k}_1 \mathbf{k}_2 \\ \mathbf{k}_3 \mathbf{k}_4}} e^{i(\mathbf{k}_4 + \mathbf{k}_3 - \mathbf{k}_1 - \mathbf{k}_2) \cdot \mathbf{x}} g(\mathbf{k}_1, \mathbf{k}_2, \mathbf{k}_3, \mathbf{k}_4)$$

$$\times \hat{a}^\dagger_{\mathbf{k}_1 \alpha}(t) \hat{a}^\dagger_{\mathbf{k}_2 \beta}(t) \hat{a}_{\mathbf{k}_3 \alpha}(t) \hat{a}_{\mathbf{k}_4 \beta}(t)$$

$$\equiv \hat{A}_f(\mathbf{x}, t) + \hat{A}_g(\mathbf{x}, t), \tag{B–1}$$

where Greek indices represent spin variables and are implicitly summed over their $(2S+1)$ values. The function $f(\mathbf{k}_1, \mathbf{k}_2)$ emerges as merely a factor in the covariance functions and, for all operators of interest here, possesses the following symmetry property:

$$f(\mathbf{k}_1, \mathbf{k}_2) = f(\mathbf{k}_2, \mathbf{k}_1). \tag{B–2}$$

The quantity g is a linear functional of the interaction potential, and both quantities are tabulated for the field operators of interest in Table B–1. We also recall that the interaction functions in this table are defined as the scalar

$$v(\mathbf{k}) \equiv \int e^{i \mathbf{k} \cdot \mathbf{r}} V_2(r) \, d^3 r = v(-\mathbf{k}), \tag{B–3}$$

Operator	$f(\mathbf{k}_1, \mathbf{k}_2)$	$g(\mathbf{k}_1, \mathbf{k}_2, \mathbf{k}_3, \mathbf{k}_4)$
$\hat{n}(\mathbf{x}, t)$	1	0
$\hat{\jmath}(\mathbf{x}, t)$	$\frac{\hbar}{2m}(\mathbf{k}_1 + \mathbf{k}_2)$	0
$\hat{h}(\mathbf{x}, t)$	$\frac{\hbar^2}{2m}(\mathbf{k}_1 \cdot \mathbf{k}_2)$	$\frac{1}{2}v(\mathbf{k}_3 - \mathbf{k}_2)$
$\hat{q}_i(\mathbf{x}, t)$	$\frac{\hbar^3}{4m^2}(\mathbf{k}_1 \cdot \mathbf{k}_2)(\mathbf{k}_1 + \mathbf{k}_2)_i$	$\frac{\hbar}{4m}(\mathbf{k}_2 + \mathbf{k}_3)_i\, v(\mathbf{k}_4 - \mathbf{k}_1)$
		$-\frac{\hbar}{8m}(\mathbf{k}_2 + \mathbf{k}_3)_j\, V_{ij}(\mathbf{k}_1, \mathbf{k}_2, \mathbf{k}_3, \mathbf{k}_4)$
$\hat{T}_{ij}(\mathbf{x}, t)$	$\frac{\hbar^2}{4m}[k_{2i}k_{2j} + k_{2i}k_{1j}$	$-\frac{1}{4}V_{ij}(\mathbf{k}_1, \mathbf{k}_2, \mathbf{k}_3, \mathbf{k}_4)$
	$+ k_{2j}k_{1i} + k_{1i}k_{1j}]$	

Table B–1. Fock-space structure of the microscopic field operators for a simple fluid.

and the tensor

$$V_{ij}(\mathbf{k}_1, \mathbf{k}_2, \mathbf{k}_3, \mathbf{k}_4) \equiv \int d^3r \left[\frac{r_i r_j}{r}\frac{dV_2(r)}{dr} \right] e^{(i/2)(\mathbf{k}_2 + \mathbf{k}_4 - \mathbf{k}_1 - \mathbf{k}_3)\cdot\mathbf{r}}$$
$$\times \int_{-1}^{+1} d\lambda\, e^{(i\lambda/2)(\mathbf{k}_3 + \mathbf{k}_4 - \mathbf{k}_1 - \mathbf{k}_2)\cdot\mathbf{r}}$$
$$= V_{ji}. \tag{B-4}$$

The covariance function of any two of the operators described in Table B–1 can be expressed as a sum over Feynman-type diagrams. It is convenient to carry out the analysis in terms of spatial Fourier transforms, so that the analytic expressions for these diagrams contain factors of the *free-particle analytic propagator*:

$$G^0_{\alpha\beta}(\mathbf{k}, \tau - \tau') \equiv e^{(\tau - \tau')\epsilon^0(\mathbf{k})}[\theta_C(\tau' - \tau) + \epsilon\nu(\mathbf{k})]\delta_{\alpha\beta}, \tag{B-5}$$

defined for arbitrary complex τ and τ' on the previously-considered contour C of Figure 6–1. It is not possible to carry out the Fourier time transform at present because of the difficulty in developing a Fourier inversion theorem along the contour C. We shall see, however, that a principal bonus of the diagrammatic analysis is that one can eventually effect the complete transform term by term.

With reference to Eq.(6–118), we see that the contractions arising from application of Wick's theorem in the perturbation expansion are evaluated as follows:

$$\left[\hat{a}_{\mathbf{k}'\alpha}(\tau')\hat{a}^\dagger_{\mathbf{k}\beta}(\tau)\right] = G^0_{\alpha\beta}(\mathbf{k}, \tau - \tau')\delta_{\mathbf{kk}'}, \tag{B-6a}$$

$$\left[\hat{a}^\dagger_{\mathbf{k}'\alpha}(\tau')\hat{a}_{\mathbf{k}\beta}(\tau)\right] = \epsilon G^0_{\alpha\beta}(\mathbf{k}, \tau' - \tau)\delta_{\mathbf{kk}'}. \tag{B-6b}$$

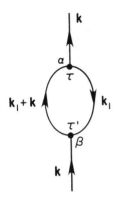

Fig. B–1. The simplest C-diagram, which also represents the basic structure of C-diagrams generally.

For equal 'times',
$$G^0_{\alpha\beta}(\mathbf{k},0) = \varepsilon\nu(\mathbf{k})\delta_{\alpha\beta}. \tag{B-7}$$

Perhaps the quickest way to grasp the essential aspects of the diagrammatic analysis is to recall the expression (6–139) for the free-particle density-density covariance function, and rewrite it as

$$K^0_{nn}(\mathbf{k},t-t') = \sum_{\alpha,\beta} \int_0^\beta \frac{ds}{\beta} \int \frac{d^3k_1}{(2\pi)^3} \varepsilon G^0_{\alpha\beta}(\mathbf{k}_1+\mathbf{k},\tau'-\tau) G^0_{\beta\alpha}(\mathbf{k}_1,\tau-\tau'), \tag{B-8}$$

where only the spatial variable has been Fourier transformed. This expression can be represented by a diagram, as illustrated in Figure B–1, in which each internal line (connected at both ends) represents a free-particle propagator, Eq.(B–5). This diagram characterizes the basic structure of what we shall call covariance diagrams, or *C-diagrams*. After defining these diagrams below we can write the covariance function for two field operators $\hat{A}(\mathbf{x}',t')$ and $\hat{B}(\mathbf{x},t)$ as

$$K_{AB}(\mathbf{k},t-t') = \sum_{n=0}^\infty \left[\begin{array}{c} \text{all connected} \\ n\text{th-order C-diagrams} \end{array} \right]_{AB} \tag{B-9}$$

The eventual proof of this expression follows the standard treatment of diagrammatic expansions (e.g., Fetter and Walecka, 1971), and will be omitted here.

The rules for constructing C-diagrams and their corresponding mathematical expressions follow.

(1) An nth-order C-diagram has one incoming and one outgoing directed external solid line, n wiggly directed interaction lines, $2n+2$ directed solid internal lines, $2n$ simple vertices, and 2 generalized vertices. Each nth-order diagram is topologically distinct, including line directions, from all other diagrams. Connected diagrams are entirely interconnected by lines.

(2) Three, and only three, lines attach to each vertex: three solid, or two solid and a wiggly.

(3) The external solid lines are both labeled with the momentum variable $\mathbf{k}$, whereas each solid internal line and wiggly interaction line is labeled with a directed momentum variable $\mathbf{k}_i$. Momentum is conserved at each vertex, thereby reducing the number of independent momentum variables.

(4) Each simple vertex is drawn as a circular dot and labeled with a 'time' variable τ_i and a spin variable α. The generalized vertices are drawn as solid squares and labeled τ and τ', where the outgoing and incoming solid external lines attach, respectively. The two vertices connected by any wiggly interaction line are labeled by identical variables τ_i.

In order to construct the mathematical expression corresponding to a particular C-diagram, we must return momentarily to a discussion of the operator structure described by Eq.(B–1). One sees that the above rules actually apply to the diagrammatic expansion of covariance functions $K_{A_f B_g}$, say. Thus, from the point of view of the perturbation expansion there are many more covariance functions to consider than just the original fifteen of Table 6–2. To a given order, a complete expression of an arbitrary covariance function will take the form

$$K_{AB} = K_{A_f B_f} + K_{A_f B_g} + K_{A_g B_f} + K_{A_g B_g}. \tag{B-10}$$

Thus, with each diagram on the right-hand side of Eq.(B–9) is associated an expression formed with the following rules.

(5) Associate a factor

$$G^0_{\alpha\beta}(\mathbf{k}_i, \tau_m - \tau_n) = \delta_{\alpha\beta} G^0(\mathbf{k}_i, \tau_m - \tau_n)$$

with each internal directed solid line, such that the line runs from τ_m to τ_n, and with each wiggly interaction line associate a factor $v(\mathbf{k}_i)$, Eq.(B–3).

(6) A factor of either $f(\mathbf{k}_i, \mathbf{k}_i + \mathbf{k})$ or $g(\mathbf{k}_i, \mathbf{k}_j, \mathbf{k}_m, \mathbf{k}_n)$, called *vertex functions* and defined in Table B–1, is associated with each of the vertices labeled τ and τ' such that the function associated with the operator $\hat{A}$ in K_{AB} is associated with τ'. When a vertex carries a function f it becomes a simple vertex, whereas the complicated situation involving vertex functions g will be discussed further below.

(7) Assign to each nth order diagram a factor $(-1)^n$ and a factor ε^{N_c}, where N^c is the number of closed loops in the diagram and $\varepsilon = +1$ for bosons, and -1 for fermions. A closed loop is a path which closes on itself by following the arrows on solid internal lines.

(8) With each diagram is associated an overall numerical factor S^{-1}, where the *symmetry number* S is the total number of permutations of the internal momentum variables associated with the solid lines that leave the diagram topologically unchanged, including the positions of these variables relative to one another.

(9) Sum over all spin indices, and integrate over all internal variables as follows:

$$\int \frac{d^3 k_i}{(2\pi)^3} \int_0^\beta d\tau_j \, ,$$

along the contour C. Finally, integrate over the real part of τ':

$$\int_0^\beta \frac{ds}{\beta} \, .$$

The need for generalized vertices and vertex functions g arises from the complicated form of the general operator expression (B–1). Because g itself is a linear functional of the interaction, its appearance as a factor raises the *effective* order of the diagram. In order to evaluate a diagram containing a function g, one must first also expand the generalized vertex diagrammatically before applying the above rules. That is, the final evaluation of a diagram concerns only simple vertices. We shall postpone a detailed diagrammatic study of generalized vertices.

Aside from the complicated rule (6), the major steps in the proof of Eq.(B–9) are verification of rules (7) and (8). In the former the factor of $(-1)^n$ arises from Eq.(6–114), whereas a factor of ε arises from a closed loop because this corresponds to contractions in opposite directions and hence a commutation of operators. In Eq.(6–114) a factor of $(n!)^{-1}$ occurs owing to the $n!$ possible time orderings, and there is a factor of $1/2$ for each interaction operator. Thus, an nth-order diagram possesses an intrinsic numerical factor of magnitude $(2^n n!)^{-1}$. But each interaction line can be drawn in two possible directions in a diagram and, owing to Eq.(B–3), two diagrams are equivalent if they differ only in the directions of wiggly lines. Therefore, we draw the wiggly lines in C-diagrams in arbitrary directions so long as momentum is conserved at each vertex, thereby cancelling the factor of 2^{-n}. One can permute the vertex labels in $n!$ ways, also, but some of these permutations will lead to identical contributions under the 'time' and momentum integral signs, so that the factor of $(n!)^{-1}$ is not completely cancelled. Rather, there are only $n!/S$ different contributions, and we have rule (8). One might wish to calculate S by permuting the 'time' labels, but our convention of fixing the positions of τ and τ' makes this awkward, so that it is better to define S by means of momentum variables.

The only zero-order C-diagram is illustrated in Figure B–2a, and if the generalized vertices are replaced by simple vertices the diagram of Figure B–1 is regained, as well as the analytical free-particle expression of Eq.(B–8) for the density-density covariance function. All other diagrams are just modifications, in one way or another, of this basic structure. For example, a typical first-order diagram is depicted in Figure B–2b. In Figure B–3 we exhibit all six first-order C-diagrams with simple vertices, including their symmetry numbers. The only zero-order C-diagram with simple vertices is that of Figure B–1.

It is useful to examine a specific diagram in order to understand a bit better the diagrammatic expansion and subsequent evaluation procedures. The diagram of

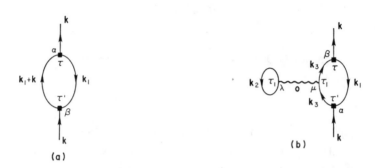

Fig. B–2. (a) The only zero-order C-diagram, and (b) a typical first-order C-diagram.

Figure B–3a is one of the first-order contributions to the density-density covariance function when $f = 1$ at both vertices, τ and τ'. By employing the above rules, and recalling the definition of the free-particle analytic propagator, one finds the corresponding expression for the contribution from that particular diagram:

$$
K_{nn}(\mathbf{k}, t - t')_a = -(2S + 1)^2 \, v(0) \int \frac{d^3 k_2}{(2\pi)^3} \, \nu(\mathbf{k}_2) \int_0^\beta \frac{ds}{\beta} \int_0^\beta d\tau_1 \int d^3 k_1
$$
$$
\times \, [1 + \varepsilon \nu) \mathbf{k}_1)] e^{(\tau' - \tau)[\epsilon^0 (\mathbf{k}_1 + \mathbf{k})]} \, \nu(\mathbf{k}_1 + \mathbf{k})
$$
$$
\times \, [\theta_C(\tau - \tau_1) + \theta_C(\tau_1 - \tau') + \varepsilon \nu(\mathbf{k}_1 + \mathbf{k})] . \tag{B–11}
$$

If one performs the τ_1-integration and then attempts to carry out the Fourier time transform, an immediate difficulty is encountered: the step-functions produce linear factors $(\tau' - \tau)$, which in turn lead to derivatives of δ-functions. Higher-order terms will produce polynomials in $(\tau' - \tau)$ and the ensuing calculation becomes more than a little awkward.

The source of this problem is readily uncovered by noting that the variable τ_1 appears *only* in the step-functions in the expression (B–11). That is, the terms containing τ_1 in the exponentials of the propagators cancel, and this also occurs in diagrams (b), (c), and (d) of Figure B–3. Note, also, that in diagrams (a) and (b) zero momentum is transferred in the interaction. These kinds of diagram will play a unique role over and over again throughout the theory and physically depict forward-scattering processes. This situation arises in most diagrammatic expansions and is representative of a class of so-called self-energy problems. Thus, one is led to a summation of an infinite set of selected diagrams such that structures of the form appearing in (a)-(d) of Figure B–3 no longer appear explicitly. Although these procedures are common in the theory of Green functions, say, the process is essential in the present case in order to make *any* meaningful calculations. It also turns out to be essential in other contexts in the calculation of

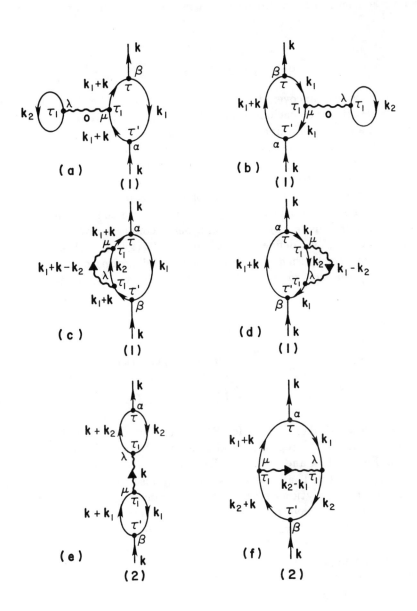

Fig. B–3. The six first-order C-diagrams with simple vertices.

ground-state quantities. The problem here, of course, arises from the need to delay taking Fourier time transforms because we are working on the contour C.

SELF-ENERGY ANALYSIS

The difficulty just uncovered clearly has to do with the iteration of certain diagrammatic structures on each internal solid line of a C-diagram. Thus, it is useful to consider just a single solid line and the various structures which can arise on it, which amounts to studying the modifications of the analytic propagator $G_{\alpha\beta}^0$ owing to particle interactions. In Figures B–4 and B–5 we exhibit all the first- and second-order diagrammatic structures of this type which can arise in C-diagrams, grouping them so that they fall into certain classes possessing common characteristics.

These diagrams are called self-energy insertions, and they possess the obvious structure of sub-diagrams to any C-diagram, being connected by means of one line in and one line out. If we define the *self-energy* $P_{\mu\lambda}^*(\mathbf{k}_1, \tau_1 - \tau_2)$ as the sum over all self-energy insertions, then the effects of these diagrams can be summarized by replacing $G_{\alpha\beta}^0$ on each solid line by the following new line factor, or analytic propagator:

$$G_{\alpha\beta}(\mathbf{k}_1, \tau' - \tau) \equiv G_{\alpha\beta}^0(\mathbf{k}_1, \tau' - \tau)$$

$$+ \sum_{\mu,\lambda} \int_0^\beta d\tau_1 \int_0^\beta d\tau_2 \, G_{\alpha\mu}^0(\mathbf{k}_1, \tau' - \tau_1)$$

$$\times P_{\mu\lambda}^*(\mathbf{k}_1, \tau_1 - \tau_2) G_{\lambda\beta}^0(\mathbf{k}_1, \tau_2 - \tau) \,. \qquad \text{(B–12)}$$

By inserting the analytical expression for $P_{\mu\lambda}^*$, according to the rules for C-diagrams, one verifies that Eq.(B–12) reproduces all the contributions from diagrams of the type shown in Figures B–4 and B–5. The diagrammatic representation of Eq.(B–12) is shown in Figure B–6a. If all possible self-energy insertions are included in $P_{\mu\lambda}^*$, then $G_{\alpha\beta}$ is the true analytic propagator for the many-body system.

It is convenient to convert Eq.(B–12) into an integral equation for $G_{\alpha\beta}$. In order to do this we first observe the structure of the diagrams in Figure B–4, c–f. A *proper* self-energy insertion is defined as one which can not be separated into two pieces by cutting a single solid internal line. The *proper self-energy* $P_{\mu\lambda}(\mathbf{k}_1, \tau_1 - \tau_2)$ is the sum over all proper self-energy insertions. With this definition Eq.(B–12) can be replaced by the integral equation

$$G_{\alpha\beta}(\mathbf{k}_1, \tau' - \tau) = G_{\alpha\beta}^0(\mathbf{k}_1, \tau' - \tau)$$

$$+ \sum_{\mu,\lambda} \int_0^\beta d\tau_1 \int_0^\beta d\tau_2 \, G_{\alpha\mu}^0(\mathbf{k}_1, \tau' - \tau_1)$$

$$\times P_{\mu\lambda}(\mathbf{k}_1, \tau_1 - \tau_2) G_{\alpha\beta}(\mathbf{k}_1, \tau_2 - \tau) \,, \qquad \text{(B–13)}$$

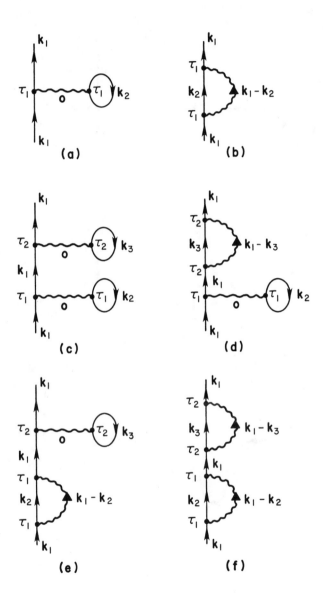

Fig. B–4. The first-order and simplest second-order diagrammatic structures which can be iterated onto a solid line of a C-diagram.

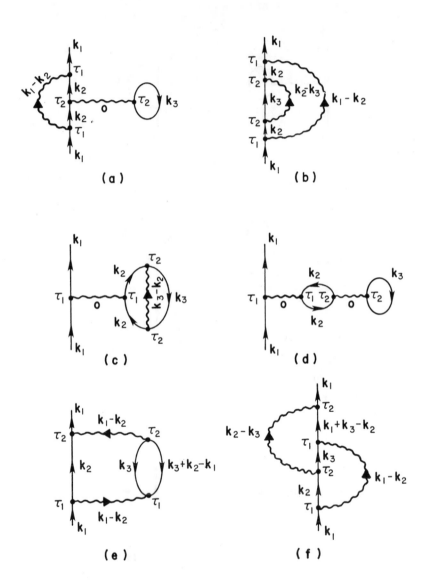

Fig. B–5. The remaining second-order diagrammatic structures which can be iterated onto a solid line in a C-diagram.

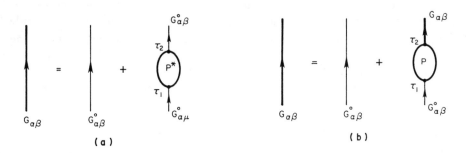

Fig. B–6. (a) Diagrammatic representation of Eq.(B–12). (b) Diagrammatic representation of the integral equation (B–13).

which is known as Dyson's equation, and is to be considered along the contour C. The equation is represented diagrammatically in Figure B–6b.

Examination of the self-energy diagrams reveals that the above difficulty with Fourier time transforms arises in those cases where the proper self-energy takes the form

$$P_{\mu\lambda}(\mathbf{k}_1, \tau_1 - \tau_2) = P_1(\mathbf{k}_1)\delta_{\mu\lambda}\,\delta_C(\tau_1 - \tau_2)\,, \tag{B-14}$$

and the δ-function is defined as the derivative of a step-function discontinuity along the contour C. In this case Dyson's equation (B–13) simplifies to

$$G_{\alpha\beta}^1(\mathbf{k}_1, \tau' - \tau) = G_{\alpha\beta}^0(\mathbf{k}_1, \tau' - \tau)$$

$$+ P_1(\mathbf{k}_1) \sum_\mu \int_0^\beta d\tau_1\, G_{\alpha\mu}^0(\mathbf{k}_1, \tau' - \tau_1)$$

$$\times G_{\mu\beta}^1(\mathbf{k}_1, \tau_1 - \tau)\,, \tag{B-15}$$

which has a remarkably simple solution. Define a *renormalized single-particle energy*

$$\epsilon^1(\mathbf{k}) \equiv \epsilon^0(\mathbf{k}) - P_1(\mathbf{k}) = \frac{\hbar^2 k^2}{2m} - P_1(\mathbf{k}) - \mu\,, \tag{B-16}$$

and a new free-particle-type, but *renormalized momentum distribution*

$$\nu^1(\mathbf{k}) \equiv \frac{ze^{-\beta\epsilon^1(\mathbf{k})}}{1 - \varepsilon ze^{-\beta\epsilon^1(\mathbf{k})}}\,, \tag{B-17}$$

where $z \equiv \exp(\beta\mu)$ is still the fugacity. Of course, this is only an approximation to the correct equilibrium momentum distribution. The solution to the integral equation (B–15) is then the *renormalized analytic propagator*

$$G_{\alpha\beta}^1(\mathbf{k}, \tau' - \tau) \equiv e^{(\tau' - \tau)\epsilon^1(\mathbf{k})}[\theta_C(\tau - \tau') + \varepsilon\nu^1(\mathbf{k})]\delta_{\alpha\beta}\,. \tag{B-18}$$

This solution is verified by direct substitution into Eq.(B–15) for the separate cases $\tau' > \tau, \tau > \tau'$.

The preceding summation procedure has the effect of casting the entire diagrammatic theory into a renormalized form, simply by replacing $G^0_{\alpha\beta}$ in rule (5) by $G^1_{\alpha\beta}$ and omitting all proper self-energy diagrams connected to solid lines at initial and final vertices carrying the same 'time' label τ_i. Note that $P_1(\mathbf{k})$ excludes diagrams such as those of Figure B–5, e and f. In fact, by noting that $P_1(\mathbf{k})$ inserts line factors of G^1 into all the solid lines of its proper diagrams, one recognizes G^1 as just the Hartree-Fock approximation to the analytic propagator G, and G^1 is determined self-consistently (e.g., Fetter and Walecka, 1971). It must be emphasized, however, that the procedure we have carried out here is no approximation, but just a necessary re-arrangement of the theory. The Hartree-Fock approximation neglects an infinite number of diagrams, such as those of Figure B–5, e and f, and corresponds physically to a single-particle model in which each particle is considered to move in a background potential effectively produced by all other particles. Although we exclude diagrams such as those of Figure B–5, e and f, from $P_1(\mathbf{k})$, we do not omit them from the theory. Explicitly, the Hartree-Fock self-energy is

$$P_1(\mathbf{k}) = -\varepsilon \int \frac{d^3k_1}{(2\pi)^3} \, G^1(\mathbf{k}_1, 0)[(2S+1)v(0) + \varepsilon v(\mathbf{k} - \mathbf{k}_1)]. \qquad (\text{B–19})$$

INTERACTION MODELS

It is clear that the perturbation scheme depends fundamentally on the scalar interaction function $v(\mathbf{k})$, Eq.(B–3), which is just the Fourier transform of the spherically-symmetric, spin-independent two-body potential. For well-behaved potentials it is a simple matter to evaluate this function, but for many realistic interactions $v(\mathbf{k})$ will not exist. As an example, under some circumstances the Lennard-Jones (6,12)-type interaction,

$$V_2(r) = \frac{\lambda}{r^6} + \frac{\mu}{r^{12}}, \qquad (\text{B–20})$$

is very useful in fitting various data. But it is obvious that the Fourier transform of this function does not exist. This is only one example from a list of interaction models whose Fourier transforms are divergent, so that eventually it is necessary to re-shape the theory into a form which resolves these difficulties.

Experimental data overwhelmingly enforce the conclusion that a realistic potential must contain a hard core. Such a model is described by

$$V_2(r) = \begin{cases} +\infty, & r < a \\ u(r), & r > a \end{cases}. \qquad (\text{B–21})$$

and occasionally one sets $u(r) = 0$, which yields a purely repulsive, infinite hard core. In this case $v(\mathbf{k})$ is not merely divergent, it is simply undefined. Physically,

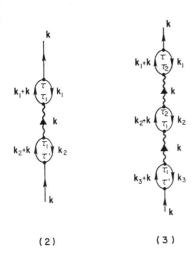

Fig. B–7. First- and second-order ring diagrams exhibiting the fundamental ring structure.

(2)   (3)

the hard core is not really an interaction, but more a boundary condition on the wavefunction, so that it is impossible to utilize $v(\mathbf{k})$ directly in these models.

A different problem arises with long-range potentials, such as the Coulomb interaction:

$$V_2(r) = \frac{e^2}{r} \quad \longrightarrow \quad v(\mathbf{k}) = \frac{4\pi e^2}{k^2}. \tag{B–22}$$

Although the singularity at $k = 0$ may be overcome in a volume integral, there exist diagrams in which these factors of k^{-2} pile up and lead to highly divergent integrals. These are processes in which the interacting particles exchange the same momentum in successive interactions. Clearly one must develop methods for overcoming this difficulty.

What we shall show in the following is that these problems can be overcome in the context of perturbation theory, at least, and so are probably resolvable in general. Thus, we now turn to some of the more pragmatic problems of applications, for which we construct several approximate but useful diagrammatic techniques. These serve, for example, to render the theory meaningful when the two-body interaction contains a hard core, or an iterated divergence for some values of $\mathbf{k}$.

RING DIAGRAMS

In Figure B–7 we exhibit two diagrams with simple vertices having a similar structure. This structure is somewhat ubiquitous in the diagrammatic theory and, for obvious reasons, these are called *ring diagrams*. Their most prominent physical

Fig. B–8. Diagrammatic representation of the sum over all ring diagrams, Eq.(B–23).

characteristic is that they transfer the same momentum in each successive inter-action. If the interaction model is that of the Coulomb potential of Eq.(B–22), say, then the individual factors of k^{-2} are iterated into a series of highly-divergent expressions when the momentum-space integrals are evaluated.

The apparent difficulty uncovered here can be overcome by studying the sum over all ring diagrams, as illustrated in Figure B–8. We see that the physical effect of this summation procedure is to replace the 'bare' interaction with an effective interaction, which one hopes is well behaved. But even if $v(\mathbf{k})$ induces no divergence problems, the summation procedure may also be useful in its own right.

From the rules for C-diagrams the sum of Figure B–8 is found to have the following expression:

$$
\begin{aligned}
U_{\overline{\tau}}(\mathbf{k}; \tau', \tau) &\equiv v(\mathbf{k})\delta_{\mathrm{C}}(\tau' - \tau) \\
&\quad - \varepsilon v^2(\mathbf{k}) \sum_{\mu, \lambda} \int \frac{d^3 k_1}{(2\pi)^3} G^0_{\lambda\mu}(\mathbf{k}_1, \tau - \tau') G^0_{\lambda\mu}(\mathbf{k}_1 + \mathbf{k}, \tau' - \tau) \\
&\quad + \varepsilon^2 v^3(\mathbf{k}) \sum_{\substack{\mu, \lambda \\ \gamma, \xi}} \int \frac{d^3 k_1}{(2\pi)^3} \int \frac{d^3 k_2}{(2\pi)^3} \int_0^\beta d\tau_1 \, G^0_{\lambda\mu}(\mathbf{k}_1, \tau_1 - \tau') \\
&\quad \times G^0_{\mu\lambda}(\mathbf{k}_1 + \mathbf{k}, \tau' - \tau_1) G^0_{\xi\gamma}(\mathbf{k}_2, \tau - \tau_1) G^0_{\gamma\xi}(\mathbf{k}_2 + \mathbf{k}, \tau_1 - \tau) \\
&\quad - \cdots \\
&\equiv v(\mathbf{k}) L_{\overline{\tau}}(\mathbf{k}; \tau', \tau) \,,
\end{aligned}
\qquad (B\text{–}23)
$$

where $L_{\bar{\tau}}$ is defined by the integral equation

$$L_{\bar{\tau}}(\mathbf{k}; \tau', \tau) = \delta_C(\tau' - \tau) - \varepsilon(2S + 1)v(\mathbf{k}) \int_0^{\bar{\tau}} d\tau_1 \int \frac{d^3 k_1}{(2\pi)^3} G^0(\mathbf{k}_1, \tau - \tau_1)$$
$$\times G^0(\mathbf{k}_1 + \mathbf{k}, \tau_1 - \tau) L_{\bar{\tau}}(\mathbf{k}; \tau', \tau_1). \qquad \text{(B-24)}$$

The use of $G^0_{\alpha\beta}$ rather than $G^1_{\alpha\beta}$ is immaterial for the present discussion, and overall spin indices are omitted because they have no effect on the following development. For notational convenience we define a function

$$N(\mathbf{k}; \tau', \tau) \equiv (2S + 1) \int \frac{d^3 k_1}{(2\pi)^3} G^0(\mathbf{k}_1, \tau - \tau') G^0(\mathbf{k}_1 + \mathbf{k}, \tau' - \tau), \qquad \text{(B-25)}$$

in terms of which the above integral equation becomes

$$L_{\bar{\tau}}(\mathbf{k}; \tau', \tau) = \delta_C(\tau' - \tau) - \varepsilon v(\mathbf{k}) \int_0^{\bar{\tau}} d\tau_1 N(\mathbf{k}; \tau, \tau_1) L_{\bar{\tau}}(\mathbf{k}; \tau', \tau_1), \qquad \text{(B-26)}$$

along the contour C. Substitution of the solution for $L_{\bar{\tau}}$ into Eq.(B-23) then determines the effective interaction $U_{\bar{\tau}}(\mathbf{k}; \tau', \tau)$, which is the sum over all ring diagrams.

Observe that we have replaced β by $\bar{\tau}$ in the upper limit of the integral in Eq.(B-26), which calls for some explanation. This does *not* mean that Eq.(B-26) is to be taken as a Volterra integral equation, rather than Fredholm. Indeed, $\bar{\tau}$ is to be considered constant here and τ is the independent variable. The reason for this modification has to do with the problem of reproducing the correct symmetry numbers in the iterated ring diagrams. One notes from Figure B-7 and the corresponding mathematical expressions that these diagrams all possess a cyclic symmetry under permutation of the internal momentum variables which are integrated. The appropriate symmetry numbers are introduced by means of the following theorem, proof of which is left to the reader: if $f(\tau_1, \ldots, \tau_m)$ is a cyclic function of its m variables, then

$$\frac{1}{m} \int_0^{\beta} d\tau_m \int_0^{\beta} d\tau_{m-1} \cdots \int_0^{\beta} d\tau_1 f(\tau_1, \ldots, \tau_m) =$$
$$\int_0^{\beta} d\tau_m \int_0^{\tau_m} d\tau_{m-1} \cdots \int_0^{\tau_m} d\tau_1 f(\tau_1, \ldots, \tau_m), \qquad \text{(B-27)}$$

along C. Iteration of the integral equation (B-26) is then seen to reproduce exactly the correct symmetry numbers for each diagram in Figure B-8.

This treatment of ring diagrams is just a special case of a summation technique leading to the idea of a general polarization insertion. An effective interaction can be obtained by summing all possible diagrams along the interaction lines, but the general procedure is more than required for immediate applications here.

The integral equation (B-26) for $L_{\bar{\tau}}(\mathbf{k}; \tau', \tau)$ is rather difficult to solve in general. For the electron gas at very low temperatures the solution for a similar problem was first found by Gell-Mann and Brueckner (1957), and for the fully ionized gas the equation can be solved formally to all orders in the fugacity (e.g., Grandy and Mohling, 1965). It is useful to study the exact solution for a special case, in order to obtain a physical feeling for the type of solution the equation admits. Thus, we shall consider the Boltzmann limit and first investigate the form of the kernel $N(\mathbf{k}; \tau', \tau)$ in this approximation.

One first writes out the explicit forms of the analytic propagators in Eq.(B–25) and then takes the Boltzmann limit by retaining only leading-order terms in the fugacity. In this high-temperature, low-density limit it is known that $z \ll 1$, as well as $\beta \hbar^2 k^2 / 2m = (k^2 \lambda_T^2)/4\pi \ll 1$, where λ_T is the thermal wavelength. Thus, in first approximation one can neglect exponentials except for those necessary to carry out the momentum integration in Eq.(B–25). This results in the leading-order approximation $N(\mathbf{k}; \tau', \tau) \simeq \varepsilon z (2S + 1) \lambda_T^{-3}$, in the Boltzmann limit. The integral equation (B–26) then becomes

$$L_{\bar\tau}(\mathbf{k}; \tau', \tau) \simeq \delta_{\mathrm{C}}(\tau' - \tau) - z \frac{(2S + 1)}{\lambda_T^3} v(\mathbf{k}) \int_0^{\bar\tau} d\tau_1 \, L_{\bar\tau}(\mathbf{k}; \tau', \tau_1) \,. \tag{B–28}$$

This approximate equation can be solved trivially by iteration and direct summation, and we find in the Boltzmann limit

$$L_{\bar\tau}(\mathbf{k}; \tau', \tau) \simeq \delta_{\mathrm{C}}(\tau' - \tau) - \bar\theta_{\mathrm{C}}(\bar\tau - \tau) \frac{z(2S + 1)\lambda_T^{-3}\, v(\mathbf{k})}{1 + \bar\tau z(2S + 1)\lambda_T^{-3}\, v(\mathbf{k})} \,, \tag{B–29}$$

where $\bar\theta_{\mathrm{C}}(\tau) = 1$ if $\tau \geq 0$, and zero otherwise.

An important example using this last result is provided by the Coulomb interaction, Eq.(B–22). We first define the *Debye wavelength* as

$$\begin{aligned} \lambda_D &\equiv [4\pi\beta e^2 (2S + 1) z \lambda_T^{-3}]^{-1/2} \\ &\simeq [4\pi\beta e^2 n_0]^{-1/2} \,, \end{aligned} \tag{B–30}$$

where in the high-temperature, low-density limit we employ the approximation $z \simeq n_0 \lambda_T^3 / (2S + 1)$. From Eq.(B–23) the effective interaction is then

$$U_{\bar\tau}(\mathbf{k}; \tau', \tau) \simeq \frac{4\pi e^2}{k^2} \left[\delta_{\mathrm{C}}(\tau' - \tau) - \bar\theta_{\mathrm{C}}(\tau' - \tau) \frac{\kappa T}{\bar\tau \kappa T + k^2 \lambda_D^2} \right] \,. \tag{B–31}$$

Thus, we see that the summation of ring diagrams leads to a screened Coulomb interaction in this case, where λ_D is the screening length. In the medium the long-range Coulomb potential has effectively a short-range influence.

LADDER DIAGRAMS

The summation of ring diagrams resolves the difficulty with potential functions leading to iterated divergences, but there remains the problem of incorporating singular interactions into the theory. When the two-body potential has no Fourier

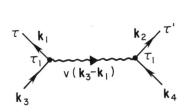

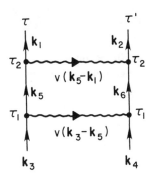

Fig. B–9. The basic structure of ladder diagrams illustrated by first- and second-order structures.

transform the perturbation theory apparently breaks down. This is a serious difficulty because the prime example is the hard-sphere interaction, which forms an important component of many realistic model systems. Yet, this example provides a clue to the resolution of the difficulty, for the hard-sphere interaction can be more realistically interpreted as a boundary condition on the two-body wavefunction. Hence, one might expect the product $V_2\psi$ to remain finite, for ψ vanishes precisely where V_2 becomes infinite. Because the scattering amplitude for hard spheres is always well defined, it may be both possible and useful to reformulate the theory in terms of scattering parameters.

Reference to the theory of Green functions reveals that this goal can be achieved by again summing a selected class of diagrams to obtain an effective interaction in place of the 'bare' interaction. These are the so-called *ladder diagrams*, whose structure is depicted in Figure B–9. We consider only those diagrams in which interactions occur repeatedly between two solid lines pointed in the same direction within a C-diagram. Thus, the exchange interaction occurs in the form of the second- and third-order ladder diagrams of Figure B–10, for example. Note that these structures all have symmetry numbers of unity.

It is convenient to define an interaction vertex function

$$\mathcal{V}_{\tau_1}^{\tau\tau'}(\mathbf{k}_3-\mathbf{k}_1) \equiv v(\mathbf{k}_3-\mathbf{k}_1)G^0(\mathbf{k}_1,\tau_1-\tau)G^0(\mathbf{k}_3,\tau_1-\tau'),\qquad (\text{B-32})$$

where we omit spin indices because they have no effect on the subsequent analysis

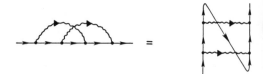

Fig. B–10. Second- and third-order exchange ladder diagrams.

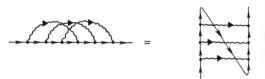

and introduce no factors of $(2S + 1)$. Although momentum conservation requires that $\mathbf{k}_2 = \mathbf{k}_4 + \mathbf{k}_3 - \mathbf{k}_1$, it is useful just to note this tacitly for the moment and re-impose the requirement later. The sum over all ladder diagrams is exhibited in Figure B–11, to which corresponds the explicit expression

$$
\begin{aligned}
S_{\tau_1}^{\tau\tau'}(\mathbf{k}_3{-}\mathbf{k}_1) = {} & \mathcal{V}_{\tau_1}^{\tau\tau'}(\mathbf{k}_3{-}\mathbf{k}_1) \\
& - \int \frac{d^3 k_5}{(2\pi)^3} \int_0^\beta d\tau_2 \, \mathcal{V}_{\tau_1}^{\tau_2\tau_2}(\mathbf{k}_3{-}\mathbf{k}_5) \mathcal{V}_{\tau_2}^{\tau\tau'}(\mathbf{k}_5{-}\mathbf{k}_1) \\
& + \int \frac{d^3 k_5}{(2\pi)^3} \int \frac{d^3 k_6}{(2\pi)^3} \int_0^\beta d\tau_2 \int_0^\beta d\tau_3 \, \mathcal{V}_{\tau_1}^{\tau_2\tau_2}(\mathbf{k}_3{-}\mathbf{k}_5) \\
& \qquad \times \mathcal{V}_{\tau_2}^{\tau_3\tau_3}(\mathbf{k}_5{-}\mathbf{k}_6) \mathcal{V}_{\tau_3}^{\tau\tau'}(\mathbf{k}_6{-}\mathbf{k}_1) \\
& - \cdots .
\end{aligned}
\tag{B-33}
$$

One readily verifies that this is just the iterated form of the following integral equation:

$$
S_{\tau_1}^{\tau\tau'}(\mathbf{k}_3{-}\mathbf{k}_1) = \mathcal{V}_{\tau_1}^{\tau\tau'}(\mathbf{k}_3{-}\mathbf{k}_1) - \int \frac{d^3 k_5}{(2\pi)^3} \int_0^\beta d\tau_2 \, \mathcal{V}_{\tau_1}^{\tau_2\tau_2}(\mathbf{k}_3{-}\mathbf{k}_5) S_{\tau_2}^{\tau\tau'}(\mathbf{k}_5{-}\mathbf{k}_1). \tag{B-34}
$$

Although this expression formally accomplishes the summation of ladder diagrams, it is rather difficult to solve owing to the fact that it is a *double* integral equation, in both momentum and 'time' variables.

A similar equation was encountered in some previous work by Mohling (1961), though, and we find that the integral equation can be effectively separated. To do this it is first necessary to temporarily suppress explicit momentum conservation at the vertices and rewrite the matrix element of the interaction as

$$
\langle \mathbf{k}_1 \mathbf{k}_2 | \hat{V}_2 | \mathbf{k}_3 \mathbf{k}_4 \rangle = \frac{1}{V} \delta_{\mathbf{k}_1+\mathbf{k}_2, \mathbf{k}_3+\mathbf{k}_4} \, v(\mathbf{k}_3{-}\mathbf{k}_1). \tag{B-35}
$$

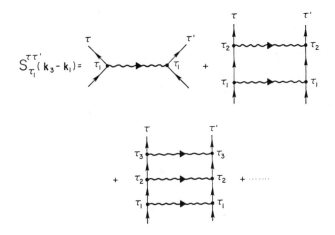

$$S_{\tau_1}^{\tau\tau'}(k_3 - k_1) =$$

Fig. B–11. Diagrammatic representation of the sum over all ladder diagrams.

The functions of Eqs.(B–32) and (B–33) are then replaced by the new vertex function

$$^{\tau\tau'}\begin{pmatrix} k_1 k_2 \\ k_3 k_4 \end{pmatrix}_{\tau_1} \equiv e^{-\tau_1(\epsilon_3 + \epsilon_4)} e^{\tau\epsilon_1} e^{\tau'\epsilon_2} \, \mathcal{V}_{\tau_1}^{\tau\tau'}(k_3 - k_1) \delta_{k_1 + k_2, + k_3 + k_4}, \qquad \text{(B–36)}$$

and the *pair function*

$$^{\tau\tau'}\begin{bmatrix} k_1 k_2 \\ k_3 k_4 \end{bmatrix}_{\tau_1} \equiv e^{-\tau_1(\epsilon_3 + \epsilon_4)} e^{\tau\epsilon_1} e^{\tau'\epsilon_2} \, S_{\tau_1}^{\tau\tau'}(k_3 - k_1) \delta_{k_1 + k_2, k_3 + k_4}, \qquad \text{(B–37)}$$

respectively. The Kronecker δ-functions will automatically introduce momentum conservation into the final results.

Combination of these quantities with Eq.(B–34) yields the following integral equation for the pair-function:

$$^{\tau_1 \tau_2}\begin{bmatrix} k_1 k_2 \\ k_3 k_4 \end{bmatrix}_{\tau_0} = \, ^{\tau_1 \tau_2}\begin{pmatrix} k_1 k_2 \\ k_3 k_4 \end{pmatrix}_{\tau_0}$$

$$- \sum_{k_5 k_6} \int_0^\beta ds \, ^{\tau_1 \tau_2}\begin{bmatrix} k_1 k_2 \\ k_5 k_6 \end{bmatrix}_s \, ^{ss}\begin{pmatrix} k_5 k_6 \\ k_3 k_4 \end{pmatrix}_{\tau_0}, \qquad \text{(B–38)}$$

where explicitly

$$^{\tau_1 \tau_2}\begin{pmatrix} k_1 k_2 \\ k_3 k_4 \end{pmatrix}_{\tau_0} = \langle k_1 k_2 | V | k_3 k_4 \rangle e^{\tau_0(\epsilon_1 + \epsilon_2 - \epsilon_3 - \epsilon_4)}$$

$$\times \, [\theta_C(\tau_1 - \tau_2) + \epsilon\nu(k_1)][\theta_C(\tau_2 - \tau_0) + \epsilon\nu(k_2)], \qquad \text{(B–39)}$$

and the Kronecker δ-functions provide us with the additional convenience of utilizing momentum sums rather than integrals.

One readily verifies the following solution to the integral equation (B–38):

$$
{}^{\tau_1 \tau_2}\begin{bmatrix} \mathbf{k}_1 \mathbf{k}_2 \\ \mathbf{k}_3 \mathbf{k}_4 \end{bmatrix}_{\tau_0} = e^{\tau_0(\epsilon_1 + \epsilon_2 - \epsilon_3 - \epsilon_4)} \Big\{ [\theta_C(\tau_1 - \tau_0) + \varepsilon \nu(\mathbf{k}_1)][\theta_C(\tau_2 - \tau_0) + \varepsilon \nu(\mathbf{k}_2)]
$$
$$
\times \langle \mathbf{k}_1 \mathbf{k}_2 | T | \mathbf{k}_3 \mathbf{k}_4 \rangle
$$
$$
- [\theta_C(\tau_1 - \tau_2) + \varepsilon \nu(\mathbf{k}_1)] \sum_{\mathbf{k}_5 \mathbf{k}_6}{}' \frac{F(\mathbf{k}_1 \mathbf{k}_2 | \mathbf{k}_5 \mathbf{k}_6 | \mathbf{k}_3 \mathbf{k}_4)}{\epsilon_1 + \epsilon_2 - \epsilon_5 - \epsilon_6}
$$
$$
\times e^{(\tau_2 - \tau_0)(\epsilon_1 + \epsilon_2)} G^0(\mathbf{k}_5, \tau_0 - \tau_2) G^0(\mathbf{k}_6, \tau_0 - \tau_2)
$$
$$
- [\theta_C(\tau_2 - \tau_1) + \varepsilon \nu(\mathbf{k}_2)] \sum_{\mathbf{k}_5 \mathbf{k}_6}{}' \frac{F(\mathbf{k}_1 \mathbf{k}_2 | \mathbf{k}_5 \mathbf{k}_6 | \mathbf{k}_3 \mathbf{k}_4)}{\epsilon_1 + \epsilon_2 - \epsilon_5 - \epsilon_6}
$$
$$
\times e^{(\tau_1 - \tau_2)(\epsilon_1 + \epsilon_2)} G^0(\mathbf{k}_5, \tau_0 - \tau_1) G^0(\mathbf{k}_6, \tau_0 - \tau_1) \Big\},
$$

$$(\text{B–40})$$

where the primes on the sums indicate that principal values are to be taken when they are converted to integrals. (The author is indebted to Professor R. Inguva for construction of this solution.) The important feature of this solution is that the momentum functions T and F are defined by the following integral equations, *independent* of 'time' variables :

$$
\langle \mathbf{k}_1 \mathbf{k}_2 | T | \mathbf{k}_3 \mathbf{k}_4 \rangle = \langle \mathbf{k}_1 \mathbf{k}_2 | V | \mathbf{k}_3 \mathbf{k}_4 \rangle
$$
$$
+ \sum_{\mathbf{k}_5 \mathbf{k}_6}{}' \langle \mathbf{k}_1 \mathbf{k}_2 | T | \mathbf{k}_5 \mathbf{k}_6 \rangle \langle \mathbf{k}_5 \mathbf{k}_6 | V | \mathbf{k}_3 \mathbf{k}_4 \rangle \frac{P(\mathbf{k}_5, \mathbf{k}_6)}{\epsilon_1 + \epsilon_2 - \epsilon_5 - \epsilon_6},
$$

$$(\text{B–41})$$

$$
F(\mathbf{k}_1 \mathbf{k}_2 | \mathbf{k}_5 \mathbf{k}_6 | \mathbf{k}_3 \mathbf{k}_4) = \langle \mathbf{k}_1 \mathbf{k}_2 | T | \mathbf{k}_5 \mathbf{k}_6 \rangle \langle \mathbf{k}_5 \mathbf{k}_6 | V | \mathbf{k}_3 \mathbf{k}_4 \rangle
$$
$$
- \sum_{\mathbf{k}_7 \mathbf{k}_8}{}' F(\mathbf{k}_1 \mathbf{k}_2 | \mathbf{k}_7 \mathbf{k}_8 | \mathbf{k}_5 \mathbf{k}_6) \langle \mathbf{k}_5 \mathbf{k}_6 | V | \mathbf{k}_3 \mathbf{k}_4 \rangle
$$
$$
\times P(\mathbf{k}_7, \mathbf{k}_8) \left[\frac{1}{\epsilon_1 + \epsilon_2 - \epsilon_7 - \epsilon_8} + \frac{1}{\epsilon_7 + \epsilon_8 - \epsilon_5 - \epsilon_6} \right]
$$
$$
+ \sum_{\mathbf{k}_7 \mathbf{k}_8}{}' F(\mathbf{k}_1 \mathbf{k}_2 | \mathbf{k}_5 \mathbf{k}_6 | \mathbf{k}_7 \mathbf{k}_8) \langle \mathbf{k}_7 \mathbf{k}_8 | V | \mathbf{k}_3 \mathbf{k}_4 \rangle
$$
$$
\times \frac{P(\mathbf{k}_7, \mathbf{k}_8)}{\epsilon_5 + \epsilon_6 - \epsilon_7 - \epsilon_8},
$$

$$(\text{B–42})$$

where the quantity

$$
P(\mathbf{k}_1, \mathbf{k}_2) \equiv 1 + \varepsilon \nu(\mathbf{k}_1) + \varepsilon \nu(\mathbf{k}_2) \tag{B–43}
$$

is the so-called Pauli factor. Its physical significance is that in the very-low-temperature Fermi system it manifests the exclusion principle. At $T = 0$ it becomes

$1 - \theta[\mu_0 - \omega(\mathbf{k}_1)] - \theta[\mu_0 - \omega(\mathbf{k}_2)]$, where μ_0 is the fermi energy for the noninteracting system and $\omega(\mathbf{k}) = \hbar^2 k^2/2m$. Thus, $P(\mathbf{k}_1, \mathbf{k}_2)$ at $T = 0$ is $+1$ if two particles with momenta $\mathbf{k}_1$ and $\mathbf{k}_2$ are above the fermi sea, -1 if they are both below, and zero otherwise. At high temperatures $P(\mathbf{k}_1, \mathbf{k}_2) = 1 + O(z)$.

If $P(\mathbf{k}_5, \mathbf{k}_6)$ is approximated by unity and a transformation is made to center-of-mass coordinates, Eq.(B–41) is just that determining the usual T-matrix of scattering theory (e.g., Laughlin and Scott, 1968)—i.e., the half-off-shell, standing-wave T-matrix. Note, also, that F does not contribute to the pair function until second order in the interaction, at least.

Some useful identities relating T and F are readily derived. Multiplication of Eq.(B–42) by $P(\mathbf{k}_5, \mathbf{k}_6)(\epsilon_1 + \epsilon_2 - \epsilon_5 - \epsilon_6)^{-1}$ and summation over $\mathbf{k}_5$ and $\mathbf{k}_6$ yields the sum rule

$$\sum_{\mathbf{k}_5 \mathbf{k}_6}{}' \frac{F(\mathbf{k}_1\mathbf{k}_2|\mathbf{k}_5\mathbf{k}_6|\mathbf{k}_3\mathbf{k}_4)}{\epsilon_1 + \epsilon_2 - \epsilon_5 - \epsilon_6} P(\mathbf{k}_5, \mathbf{k}_6) =$$

$$= \sum_{\mathbf{k}_5 \mathbf{k}_6}{}' \frac{\langle \mathbf{k}_1\mathbf{k}_2|T|\mathbf{k}_5\mathbf{k}_6\rangle \langle \mathbf{k}_5\mathbf{k}_6|V|\mathbf{k}_3\mathbf{k}_4\rangle}{\epsilon_1 + \epsilon_2 - \epsilon_5 - \epsilon_6} P(\mathbf{k}_5, \mathbf{k}_6). \quad \text{(B–44)}$$

Substitution into Eq.(B–41) then gives the relation

$$\langle \mathbf{k}_1\mathbf{k}_2|T|\mathbf{k}_3\mathbf{k}_4\rangle = \langle \mathbf{k}_1\mathbf{k}_2|V|\mathbf{k}_3\mathbf{k}_4\rangle$$

$$+ \sum_{\mathbf{k}_5 \mathbf{k}_6}{}' \frac{F(\mathbf{k}_1\mathbf{k}_2|\mathbf{k}_5\mathbf{k}_6|\mathbf{k}_3\mathbf{k}_4)}{\epsilon_1 + \epsilon_2 - \epsilon_5 - \epsilon_6} P(\mathbf{k}_5, \mathbf{k}_6), \quad \text{(B–45)}$$

which can be used to reduce the integral equation (B–42) to the simpler form

$$F(\mathbf{k}_1\mathbf{k}_2|\mathbf{k}_5\mathbf{k}_6|\mathbf{k}_3\mathbf{k}_4) = \langle \mathbf{k}_1\mathbf{k}_2|V|\mathbf{k}_5\mathbf{k}_6\rangle \langle \mathbf{k}_5\mathbf{k}_6|V|\mathbf{k}_3\mathbf{k}_4\rangle$$

$$+ \sum_{\mathbf{k}_7 \mathbf{k}_8}{}' \frac{P(\mathbf{k}_7, \mathbf{k}_8)}{\epsilon_5 + \epsilon_6 - \epsilon_7 - \epsilon_8} [F(\mathbf{k}_1\mathbf{k}_2|\mathbf{k}_5\mathbf{k}_6|\mathbf{k}_7\mathbf{k}_8)$$

$$\times \langle \mathbf{k}_7\mathbf{k}_8|V|\mathbf{k}_3\mathbf{k}_4\rangle$$

$$+ F(\mathbf{k}_1\mathbf{k}_2|\mathbf{k}_7\mathbf{k}_8|\mathbf{k}_5\mathbf{k}_6) \langle \mathbf{k}_5\mathbf{k}_6|V|\mathbf{k}_3\mathbf{k}_4\rangle]. \quad \text{(B–46)}$$

Similarly, one can write the integral equation for F entirely in terms of T, but it is much more complicated.

Because the solutions for T and F can be expressed in terms of scattering parameters, we have resolved in principle the difficulty of incorporating singular potentials into the perturbation theory. Upon solving the above integral equations to the desired order, one obtains from Eq.(B–40) the pair function to the same order. That result is then substituted into the following expression for the sum over all ladder diagrams:

$$S_{\tau_1}^{\tau\tau'}(\mathbf{k}_3 - \mathbf{k}_1) = e^{\tau_1(\epsilon_3 + \epsilon_4)} e^{-\tau\epsilon_1} e^{-\tau'\epsilon_2} \left. \begin{bmatrix} \mathbf{k}_1\mathbf{k}_2 \\ \mathbf{k}_3\mathbf{k}_4 \end{bmatrix}_{\tau_1}^{\tau\tau'} . \quad \text{(B–47)}$$

Fig. B–12. Diagrammatic effect of the summation over all ladder diagrams.

Finally, where one inserted the sum over all ladder diagrams into C-diagrams, the expression of Eq.(B–32) is replaced by that of Eq.(B–47), momentum conservation is explicitly regained by setting $\mathbf{k}_2 = \mathbf{k}_3 + \mathbf{k}_4 - \mathbf{k}_1$, and spin Kronecker δ-functions are inserted. The entire calculation has the effect illustrated in Figure B–12. We see, then, that the summation of ladder diagrams leads to a description in terms of the two-body, free-particle scattering operator T, which in turn determines the *effective* scattering operator F in the medium. Thus, the present procedure is analogous to the well-known treatment by means of Galitskii's integral equations (Galitskii, 1958; Beliaev, 1958), in which the two-body scattering *amplitude* is used to determine the effective scattering *amplitude* in the medium (e.g., Fetter and Walecka, 1971).

GENERALIZED VERTICES

As mentioned above, the need for generalized vertices arises from the complicated form of the general operator expression in Eq.(B–1). One way to account for these complications is to represent the generalized vertex diagrammatically, and Figure B–13 illustrates the scheme at a vertex labeled τ', say. The symbols are explained in the figure caption. We remark that once the generalized vertex is expanded diagrammatically in this way, then only simple vertices are involved and the vertices labeled τ and τ' now carry *either* a factor f, *or* an additional directed dashed line conserving momentum. In the latter case the dashed line carries a factor g, and it is noted that the three-line rule at vertices τ and τ' can now be violated by addition of a dashed line. This factor g carried by the dashed line now raises the effective order in the interaction of the resulting diagram.

Thus, in evaluating an nth-order diagram one must first expand the generalized vertices to the desired order, as indicated in Figure B–13. Factors are assigned to

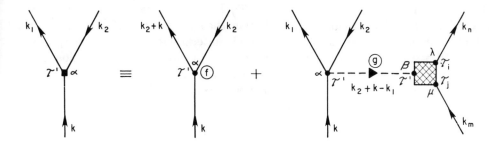

Fig. B–13. Diagrammatic representation of a generalized vertex. The cross-hatched square can represent any diagrammatic structure consisting of simple vertices.

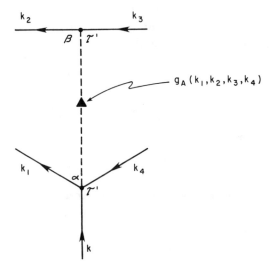

Fig. B–14. The convention for assigning dashed-line factors in expanded diagrams arising from generalized vertices.

the resulting diagram in the same way as indicated in the original rules, with the following additions:

(10) Include a factor of ε for each dashed line.

(11) If a simple vertex at τ' carries a dashed line, then assign a factor $g(\mathbf{k}_1, \mathbf{k}_2, \mathbf{k}_3, \mathbf{k}_4)$ for the appropriate operator in accordance with the convention illustrated in Figure B–14, and similarly for the vertex at τ.

As an example, Figure B–15 shows the four possible lowest-order modifications of the type shown in Figure B–13 to the lowest-order C-diagram of Figure B–1.

It requires only a moment's thought to realize that the diagrammatic scheme for covariance functions involving $\hat{h}$, $\hat{q}$, and $\hat{T}_{ij}$ becomes extraordinarily complicated beyond the first few orders of perturbation theory, thereby defeating its original intent. For systems in which the two-body interaction is weak it is simply far better to write out by brute force the relevant terms in the expression for the covariance function—presuming there are only a few. For example, to calculate K_{hh} to leading order in the interaction we first recall Eq.(6–126) in which the energy-density operator is written

$$\hat{h}(\mathbf{x},t) = \hat{h}^0(\mathbf{x},t) + \hat{h}^1(\mathbf{x},t)\,, \tag{B–48}$$

and $\hat{h}^1$ contains a factor of the interaction potential. Then, through first order,

$$K_{hh} \simeq K^1_{hh} = K_{h^0 h^0} + K^1_{h^0 h^0} + K^0_{h^1 h^0} + K^0_{h^0 h^1}\,. \tag{B–49}$$

In the Boltzmann limit $K^0_{h^0 h^0}$ was evaluated in Eq.(6–157), and the form of $K^1_{h^0 h^0}$ is very similar to that for K^1_{nn}, Eq.(C–10). The third and fourth terms on the right-hand side of Eq.(B–49) are quite similar and, merely to satisfy curiosity, we record the formal expression in the Boltzmann limit:

$$K^0_{h^0 h^1}(\mathbf{k}, t-t') \simeq \tfrac{1}{2} z^2 (2S+1)^2 [v(0) + v(\mathbf{k})]$$
$$\times \int_0^\beta \frac{ds}{\beta} \frac{d^3 k_1}{(2\pi)^3} \int \frac{d^3 k_2}{(2\pi)^3} [\mathbf{k}_1 \cdot (\mathbf{k}_2 + \mathbf{k})]$$
$$\times e^{-\beta\epsilon(\mathbf{k}_1)}\, e^{-\beta\epsilon(\mathbf{k}_2 + \mathbf{k})}\, e^{(\tau-\tau')[\epsilon(\mathbf{k}_2) - \epsilon(\mathbf{k}_2 + \mathbf{k})]}\,. \tag{B–50}$$

REFERENCES

Beliaev, S.T.: 1958, 'Energy Spectrum of a Non-Ideal Bose Gas', *Sov. Phys. JETP* **7**, 299.

Fetter, A.L., and J.D. Walecka: 1971, *Quantum Theory of Many-Particle Systems*, McGraw-Hill, New York.

Galitskii, V.M.: 1958, 'The Energy Spectrum of a Non-Ideal Fermi Gas', *Sov. Phys. JETP* **7**, 104.

Gell-Mann, M., and K.A. Brueckner: 1957, 'Correlation Energy of an Electron Gas at High Density', *Phys. Rev.* **106**, 364.

Grandy, W.T., Jr., and F. Mohling: 1965, 'Quantum Statistics of Fully Ionized Gases', *Ann. Phys. (N.Y.)* **34**, 424.

Laughlin, R., and B.L. Scott: 1968, 'Off-Energy-Shell t-Matrix Elements for Local Potentials Containing Hard Cores', *Phys. Rev.* **171**, 1196.

Mohling, F.: 1961, 'Linked-Pair Expansions in Quantum Statistics', *Phys. Rev.* **122**, 1043.

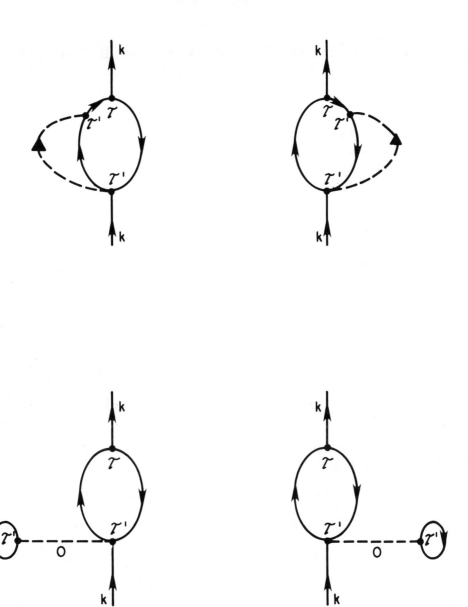

Fig. B–15. The four lowest-order modifications of the type shown in Figure B–13 to the lowest-order C-diagram of Figure B–1.

Appendix C

The Density Covariance Function

For a number of reasons it is of value to carry out a detailed calculation of the density-density covariance function in various orders of approximation. Not the least of these is the creation of a guide for evaluating other, more complicated correlations. With the presumption that the interparticle potential is weak, we evaluate K_{nn} in the Boltzmann limit and to leading order in the interaction. In terms of renormalized line factors, the contributing diagrams in this order are those of Figure C–1. The corresponding analytical expression is then

$$K_{nn}^1(\mathbf{k}, t-t') = \int_0^\beta \frac{ds}{\beta} \int \frac{d^3k_1}{(2\pi)^3} \sum_{\alpha,\beta} \varepsilon G_{\alpha\beta}^1(\mathbf{k}_1+\mathbf{k}, \tau'-\tau) G_{\beta\alpha}^1(\mathbf{k}_1, \tau-\tau')$$

$$-\frac{1}{2} \int_0^\beta \frac{ds}{\beta} \int_0^\beta d\tau_1 \sum_{\substack{\alpha,\beta \\ \mu,\lambda}} \int \frac{d^3k_1}{(2\pi)^3} \int \frac{d^3k_2}{(2\pi)^3} \left[v(\mathbf{k}) G_{\beta\lambda}^1(\mathbf{k}_1, \tau-\tau_1) \right.$$

$$\times G_{\lambda\beta}^1(\mathbf{k}_1+\mathbf{k}, \tau_1-\tau) G_{\alpha\mu}^1(\mathbf{k}_2+\mathbf{k}, \tau'-\tau_1) G_{\mu\alpha}^1(\mathbf{k}_2, \tau_1-\tau')$$

$$+ \varepsilon v(\mathbf{k}_2-\mathbf{k}_1) G_{\beta\lambda}^1(\mathbf{k}_1, \tau-\tau_1) G_{\lambda\alpha}^1(\mathbf{k}_2, \tau_1-\tau')$$

$$\left. \times G_{\alpha\mu}^1(\mathbf{k}_2+\mathbf{k}, \tau'-\tau_1) G_{\mu\beta}^1(\mathbf{k}_1+\mathbf{k}, \tau_1-\tau) \right]. \qquad (C-1)$$

Now substitute from Eq.(B–18) for G^1 and perform the τ_1-integration. The integrals are tedious, but straighforward, and the results are considerably simplified by use of the identity

$$e^{\beta(\epsilon_2+\epsilon_3-\epsilon_1-\epsilon_2)} \nu_2\nu_3(1+\varepsilon\nu_1)(1+\varepsilon\nu_4) = \nu_1\nu_4(1+\varepsilon\nu_2)(1+\varepsilon\nu_3), \qquad (C-2)$$

independent of whether or not the quantities are renormalized. It is also useful to remember that $\theta_C(\tau-\tau') = \theta_C(\tau-\tau) = \theta_C(\tau'-\tau') = 0$. After performing the spin sums as well we find that

$$K_{nn}^1(\mathbf{k}, t-t') = \frac{(2S+1)}{(2\pi)^3} \int_0^\beta \frac{ds}{\beta} \int d^3k_1 \, \nu^1(\mathbf{k}_1+\mathbf{k})[1+\varepsilon\nu^1(\mathbf{k}_1)]$$

$$\times e^{s[\epsilon^1(\mathbf{k}_1+\mathbf{k})-\epsilon^1(\mathbf{k}_1)]} e^{(i/\hbar)(t-t')[\epsilon^1(\mathbf{k}_1+\mathbf{k})-\epsilon^1(\mathbf{k}_1)]}$$

$$-\frac{(2S+1)}{(2\pi)^6} \int_0^\beta \frac{ds}{\beta} \int d^3k_1 \int d^3k_2 \, V_{\mathbf{k}}(\mathbf{k}_1, \mathbf{k}_2)\nu^1(\mathbf{k}_1+\mathbf{k})$$

$$\times [1+\varepsilon\nu^1(\mathbf{k}_1)][\nu^1(\mathbf{k}_2+!\mathbf{k}) - \nu^1(\mathbf{k}_2)]e^{s[\epsilon^1(\mathbf{k}_1+\mathbf{k})-\epsilon^1(\mathbf{k}_1)]}$$

$$\times e^{(i/\hbar)(t'-t)[\epsilon^1(\mathbf{k}_1+\mathbf{k})-\epsilon^1(\mathbf{k}_1)]}, \qquad (C-3a)$$

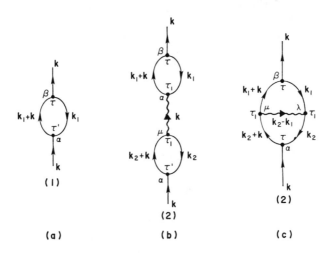

Fig. C–1. The three C-diagrams contributing to K_{nn} in leading order in the interaction.

with

$$V_{\mathbf{k}}(\mathbf{k}_1,\mathbf{k}_2) = \frac{(2S+1)v(k) + \varepsilon v(\mathbf{k}_2-\mathbf{k}_1)}{\epsilon^1(\mathbf{k}_2) - \epsilon^1(\mathbf{k}_2+\mathbf{k}) + \epsilon^1(\mathbf{k}_1+\mathbf{k}) - \epsilon^1(\mathbf{k}_1)}. \tag{C–3b}$$

No approximations have been made yet, but now we take the Boltzmann limit by approximating the functions $\nu(\mathbf{k})$ in the usual manner, omitting all exchange terms, and keeping only leading-order terms in the fugacity. [Note that at this point in the calculation the exchange term in Eq.(C–3) is not obviously smaller than the direct term. It will, however, contribute factors of $\hbar$.] We also retain only those terms linear in the interaction. Upon completing the Fourier transformation in time we find that in this approximation

$$K^1_{nn}(k,\omega) = z\frac{(2S+1)}{\lambda_T^3}\frac{(2\pi\beta m)^{1/2}}{k}e^{-\frac{1}{2}\beta m\omega^2/k^2}\left[1 - z\frac{(2S+1)}{\lambda_T^3}\beta v(0)\right]$$

$$- 2\beta m z^2\frac{(2S+1)^2}{\lambda_T^2}\frac{v(k)}{k^2}e^{-\frac{1}{2}\beta m\omega^2/k^2}\int_0^\infty ye^{-y^2}\,F(y;k,\omega)\,dy\,; \tag{C–4}$$

where

$$F(y;k,\omega) \equiv \frac{1}{\hbar}\int_{-k\lambda_T y/\sqrt{\pi}}^{k\lambda_T y/\sqrt{\pi}}\frac{1 - e^{-z}}{z + \beta\hbar\omega}. \tag{C–5}$$

This last integral exists as a Cauchy principal value for $\beta\hbar\omega \ll 1$, which is part of the Boltzmann limit, and we find that

$$F(y; k, \omega) \xrightarrow[\beta\hbar\omega \ll 1]{} 2^{3/2} \left(\frac{\beta}{m}\right)^{1/2} ky.$$ (C-6)

Hence,

$$K_{nn}^1(\mathbf{k}, \omega)_{\mathrm{B}} = \frac{(2\pi\beta m)^{1/2}}{k} e^{-\frac{1}{2}\beta m\omega^2/k^2} \left[z \frac{(2S+1)}{\lambda_T^3} - z^2 \frac{(2S+1)^2}{\lambda_T^6} \beta v(0) \right.$$
$$\left. - \beta z^2 \frac{(2S+1)^2}{\lambda_T^6} v(\mathbf{k}) \right].$$ (C-7)

In order to complete the calculation it is necessary to eliminate the fugacity in terms of the equilibrium particle-number density by means of the prescription

$$n = \frac{1}{V} z \frac{\partial}{\partial z} \ln Z, \qquad Z = \mathrm{Tr}\, e^{-\beta\hat{K}},$$ (C-8)

from the equilibrium grand canonical ensemble. This inversion is rather trivial and was carried out in detail in Chapter 8 of Volume I, where we found that

$$n \simeq z \frac{(2S+1)}{\lambda_T^3} - z^2 \beta v(0) \frac{(2S+1)^2}{\lambda_T^6},$$ (C-9a)

or

$$z \simeq \frac{n\lambda_T^3}{2S+1}[1 + \beta n v(0)].$$ (C-9b)

With this result we can return to Eq.(C-7) and write to leading order

$$K_{nn}^1(\mathbf{k}, \omega)_{\mathrm{B}} = n_0 \frac{(2\pi\beta m)^{1/2}}{k} e^{-\frac{1}{2}\beta m\omega^2/k^2} [1 - \beta n_0 v(\mathbf{k})].$$ (C-10)

It is interesting to conclude this calculation by recalling Eq.(6-76b) and evaluating the isothermal compressibility to this order. From the definitions of the inverse Fourier transforms we have

$$K_{nn}(\mathbf{r}) = \int \frac{d^3k}{(2\pi)^3} \int_{-\infty}^{\infty} \frac{d\omega}{2\pi} e^{i\mathbf{k}\cdot\mathbf{r}} K_{nn}(\mathbf{k}, \omega).$$ (C-11)

Substitution from Eq.(C-10) then yields

$$K_{nn}(\mathbf{r}) = \frac{n_0}{(2\pi)^3} \int d^3k\, e^{i\mathbf{k}\cdot\mathbf{r}} [1 - \beta n_0 v(\mathbf{k})]$$
$$= n_0 \delta(\mathbf{r}) - \beta n_0^2 V_2(\mathbf{r}).$$ (C-12)

Hence, to leading order,

$$\kappa_T = \frac{\beta}{n_0^2} \int K_{nn}(\mathbf{r})\, d^3r \simeq \frac{\beta}{n_0}[1 - \beta n_0 v(0)],$$ (C-13)

a result which could have been obtained by simpler methods, of course.

SUMMATION OF RING DIAGRAMS

When long-range forces are present in the many-body system, such as those described by the Coulomb interaction, it is mandatory that the sum over ring diagrams be carried out (e.g., as in Appendix B). If charged particles are to be considered one must construct a multicomponent theory, of course, but even for single-component systems with short-range forces the diagrammatic summation procedure may be of some value. In these latter cases the ring sum appears to be a high-density formulation, so that such a description might be of interest in studying moderately dense gases and liquids.

In order to construct the ring-sum description of K_{nn} we consider the five C-diagrams of Figure C–2, which contain the effective interaction U. Recall that the function U_τ was defined in Eq.(B–23) and represents a sum over ring diagrams, whereas $U^{(2)}$ denotes the second-order terms in which the δ-function term has been subtracted from $L_{\bar{\tau}}$ in Eq.(B–24). These particular diagrams arise because of the need to coordinate the ring sum with the self-energy analysis introduced earlier. Examination of the proper self-energy diagrams reveals that the ring-sum insertion is already incorporated into the diagram of Figure B–3b, say, as illustrated in Figure B–5d. But the diagram of Figure B–5e does *not* appear in $P_1(\mathbf{k})$, which explains the need for the two diagrams (b) and (c) in Figure C–2. The line factors in all the diagrams of Figure C–2 are then $G^1_{\alpha\beta}$, Eq.(B–18), with

$$\varepsilon P_1(\mathbf{k}) = -\int \frac{d^3 k_1}{(2\pi)^3} G^1_{\alpha\beta}(\mathbf{k}_1, 0)[(2S+1)v(0) + \varepsilon v(\mathbf{k} - \mathbf{k}_1)]. \tag{C–14}$$

In terms of these five diagrams the ring-sum approximation to K_{nn} is written

$$K^r_{nn}(\mathbf{k}, t-t') = K^a_{nn} + K^b_{nn} + K^c_{nn} + K^d_{nn} + K^e_{nn}, \tag{C–15}$$

where the corresponding analytic expressions will be studied presently. Owing to the symmetry problem, the evaluation of these terms is somewhat tedious, although the calculations are straightforward. In the Boltzmann limit, however, there is some simplification, and so we shall restrict the calculation to that region of the temperature-density plane. In that limit, and to leading order in the ring sum, we then have

$$K^r_{nn}(\mathbf{k}, t-t') \simeq K^a_{nn} + K^d_{nn}. \tag{C–16}$$

That is, owing to the presence of the function $U^{(2)}$ in diagrams (b) and (c) they are at least $O(v)$ smaller than the others. Moreover, diagram (e) is the exchange diagram to diagram (d).

The analytic expressions for the terms on the right-hand side of Eq.(C–16) are

$$K^a_{nn}(\mathbf{k}, t-t') = \int_0^\beta \frac{ds}{\beta} \int \frac{d^3 k_1}{(2\pi)^3} \sum_{\alpha,\beta} \varepsilon G^1_{\alpha\beta}(\mathbf{k}_1 + \mathbf{k}, \tau' - \tau) G^1_{\beta\alpha}(\mathbf{k}_1, \tau - \tau'),$$

$$\tag{C–17}$$

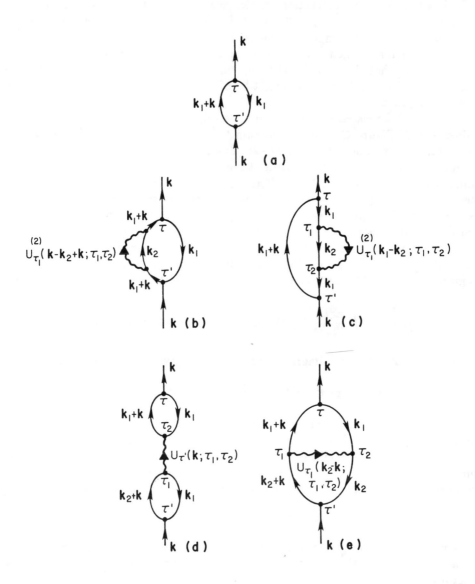

Fig. C–2. Leading-order contributions to the ring-sum approximation to K_{nn}.

$$K_{nn}^d(\mathbf{k}, t-t') = -\frac{1}{\beta}\int_0^\beta d\tau' \sum_{\substack{\alpha,\beta \\ \mu,\lambda}} \int \frac{d^3k_1}{(2\pi)^3} \int \frac{d^3k_2}{(2\pi)^3} \int_0^{\tau'} d\tau_2 \int_0^{\tau'} d\tau_1$$

$$\times U_{\tau'}(\mathbf{k};\tau_1,\tau_2) G_{\beta\lambda}^1(\mathbf{k}_1,\tau-\tau_2) G_{\lambda\beta}^1(\mathbf{k}_1+\mathbf{k},\tau_2-\tau)$$

$$\times G_{\alpha\mu}^1(\mathbf{k}_2+\mathbf{k},\tau'-\tau_1) G_{\mu\alpha}^1(\mathbf{k}_2,\tau_1-\tau') \,. \tag{C-18}$$

Note that in Eq.(C–18) we have employed the s-integral in order to properly account for symmetry numbers. This can be justified by deforming the contour C so as to make $s = \tau'$, effectively setting $t' = 0$. This scheme, however, makes no difference when the Fourier time transform is carried out.

Rather than substitute directly into these expressions from Eq.(B–23), it is instructive to pursue an alternative method by re-expanding the function $U_{\tau'}$ in K_{nn}^d. The reason for doing this is that the term K_{nn}^a is then more easily included in the ensuing sum. Thus, with reference to Figure B–8 and Eq.(B–23), one verifies that within the context of our approximations

$$K_{nn}^r(\mathbf{k}, t-t') \simeq \int_0^\beta \frac{ds}{\beta} \int \frac{d^3k_1}{(2\pi)^3} \sum_{\alpha,\beta} \varepsilon G_{\alpha\beta}(\mathbf{k}_1+\mathbf{k},\tau'-\tau) G_{\beta\alpha}(\mathbf{k}_1,\tau-\tau')$$

$$-\frac{1}{2}\int_0^\beta \frac{ds}{\beta} \int_0^\beta d\tau_1 \sum_{\substack{\alpha,\beta \\ \mu\lambda}} \int \frac{d^3k_1}{(2\pi)^3} \int \frac{d^3k_2}{(2\pi)^3} \varepsilon^2 v(\mathbf{k})$$

$$\times G_{\beta\lambda}(\mathbf{k}_1,\tau-\tau_1) G_{\lambda\beta}(\mathbf{k}_1+\mathbf{k},\tau_1-\tau)$$

$$\times G_{\alpha\mu}(\mathbf{k}_2+\mathbf{k},\tau'-\tau_1) G_{\mu\alpha}(\mathbf{k}_2,\tau_1-\tau')$$

$$+\frac{1}{3}\int_0^\beta \frac{ds}{\beta} \int_0^\beta d\tau_1 \int_0^\beta d\tau_2 \sum_{\substack{\alpha,\beta,\mu \\ \lambda,\gamma,\nu}} \int \frac{d^3k_1}{(2\pi)^3} \int \frac{d^3k_2}{(2\pi)^3} \int \frac{d^3k_3}{(2\pi)^3} \varepsilon^3 v^2(\mathbf{k})$$

$$\times G_{\beta\lambda}(\mathbf{k}_1,\tau-\tau_1) G_{\lambda\beta}(\mathbf{k}_1+\mathbf{k},\tau_1-\tau)$$

$$\times G_{\gamma\nu}(\mathbf{k}_2,\tau_1-\tau_2) G_{\nu\gamma}(\mathbf{k}_2+\mathbf{k},\tau_2-\tau_1)$$

$$\times G_{\mu\alpha}(\mathbf{k}_3,\tau_2-\tau') G_{\alpha\mu}(\mathbf{k}_3+\mathbf{k},\tau'-\tau_2)$$

$$- \cdots \,. \tag{C-19}$$

Although we have not included the superscripts explicitly, all quantities are to be interpreted as renormalized.

Next perform the spin sums and take the Boltzmann limit in each term by keeping only the leading-order contribution in z from each. One then obtains

$$K_{nn}^r(\mathbf{k}, t-t')_\text{B} = z(2S+1)\int_0^\beta \frac{ds}{\beta} \left\{ \int \frac{d^3k_1}{(2\pi)^3} I_1(\mathbf{k}_1) \right.$$

$$-\frac{1}{2}z(2S+1)v(\mathbf{k}) \int \frac{d^3k_1}{(2\pi)^3} \int \frac{d^3k_2}{(2\pi)^3} I_2(\mathbf{k}_1,\mathbf{k}_2)$$

$$+\frac{1}{3}z^2(2S+1)^2 v^2(\mathbf{k}) \int \frac{d^3k_1}{(2\pi)^3} \int \frac{d^3k_2}{(2\pi)^3} \int \frac{d^3k_3}{(2\pi)^3} I_3(\mathbf{k}_1,\mathbf{k}_2,\mathbf{k}_3)$$

$$\left. - \cdots \right\} , \tag{C-20}$$

where the functions $I_n(k_1, \ldots, k_n)$ are the results of doing the various integrals over the τ_i. By examining the first four functions in tedious detail one is able to discern the pattern and conclude that the general I_n is

$$I_n = n e^{(\tau - \tau')[\epsilon(k_1) - \epsilon(k_1 + k)]} e^{-\beta \epsilon(k_1 + k)}$$
$$\times \frac{[e^{-\beta \epsilon(k_2)} - e^{-\beta \epsilon(k_2 + k)}] \cdots [e^{-\beta \epsilon(k_n)} - e^{-\beta \epsilon(k_n + k)}]}{E_{12} \cdots E_{1n}}. \qquad \text{(C-21)}$$

Here we define

$$E_{1m} \equiv \epsilon(k_1) - \epsilon(k_1 + k) - \epsilon(k_m) + \epsilon(k_m + k), \qquad \text{(C-22)}$$

and recall that $\epsilon(k)$ has the renormalized form

$$\epsilon(k) = \left(\frac{\hbar^2 k^2}{2m} - \mu \right) - P_1(k), \qquad \text{(C-23)}$$

where

$$P_1(k) \equiv -\varepsilon \int \frac{d^3 k_1}{(2\pi)^3} G^1(k_1, 0)[(2S + 1)v(0) + \varepsilon v(k - k_1)]. \qquad \text{(C-24)}$$

With these results one can now write

$$K_{nn}^r(k, t - t')_{\mathrm{B}} = z(2S + 1) \int_0^\beta \frac{ds}{\beta} \int \frac{d^3 k_1}{(2\pi)^3} e^{(\tau - \tau')[\epsilon(k_1) - \epsilon(k_1 + k)]} e^{-\beta \epsilon(k_1 + k)}$$
$$\times \{ 1 - f(k_1) + f^2(k_1) - f^3(k_1) + \cdots \}$$
$$= z(2S + 1) \int_0^\beta \frac{ds}{\beta} \int \frac{d^3 k_1}{(2\pi)^3} e^{(\tau - \tau')[\epsilon(k_1) - \epsilon(k_1 + k)]} \frac{e^{-\beta \epsilon(k_1 + k)}}{1 + f(k_1)},$$
$$\text{(C-25)}$$

and

$$f(k_1) \equiv z(2S + 1)v(k) \int \frac{d^3 k_2}{(2\pi)^3} \frac{e^{-\beta \epsilon(k_2)} - e^{-\beta \epsilon(k_2 + k)}}{\epsilon(k_1) - \epsilon(k_1 + k) - \epsilon(k_2) + \epsilon(k_2 + k)}, \qquad \text{(C-26)}$$

Thus, we have succeeded in summing the series of ring diagrams.

Evaluation of the relevant integrals proceeds by means of Fourier transformation and use of the resulting δ-functions. In the Boltzmann limit

$$\beta \left(\frac{\hbar^2 k^2}{2m} \right) \ll \beta \hbar \omega \ll 1,$$

so that

$$K_{nn}^r(k, \omega)_{\mathrm{B}} \simeq z(2S + 1)\lambda_T^{-3} \frac{(2\pi \beta m)^{1/2}}{k} \frac{e^{-\frac{1}{2}\beta m \omega^2 / k^2}}{1 + f(k)} e^{\beta P_1(k)}, \qquad \text{(C-27)}$$

where as usual $\lambda_T \equiv (2\pi\hbar^2\beta/m)^{1/2}$ is the thermal wavelength. In the same limit evaluation of $f(\mathbf{k})$ yields [e.g., Eqs.(C–4)-(C–6)]

$$f(\mathbf{k}) \simeq z(2S + 1)\lambda_T^{-3}e^{\beta P_1(\mathbf{k})} \beta v(\mathbf{k}) . \qquad (C\text{–}28)$$

Thus, with

$$P_1(\mathbf{k}) \simeq -z(2S + 1)\lambda_T^{-3} v(0) , \qquad (C\text{–}29)$$

we have

$$K_{nn}^r(\mathbf{k},\omega)_B \simeq \frac{z(2S+1)}{\lambda_T^3} \frac{(2\pi\beta m)^{1/2}}{k} e^{-\frac{1}{2}\beta m\omega^2/k^2}$$
$$\times \frac{e^{\beta P_1(\mathbf{k})}}{1 + z(2S+1)\lambda_T^{-3}e^{\beta P(\mathbf{k}_1)} v(\mathbf{k})} . \qquad (C\text{–}30)$$

Then, when the fugacity is eliminated properly in terms of the density, one obtains the leading-order ring-sum form of K_{nn} in the Boltzmann limit:

$$K_{nn}^r(\mathbf{k},\omega)_B \simeq n\frac{(2\pi\beta m)^{1/2}}{k} \frac{e^{-\frac{1}{2}\beta m\omega^2/k^2}}{1 + \beta n v(\mathbf{k})} . \qquad (C\text{–}31)$$

LADDER SUM AND HARD SPHERES

When the two-body interaction contains a singular portion, such as an infinitely hard core, the formulation in terms of the Fourier transform $v(\mathbf{k})$ breaks down. As pointed out earlier, the potential need not even be singular for $v(\mathbf{k})$ to be meaningless. Thus, it becomes necessary to carry out the ladder sum in these cases so as to re-describe the theory in terms of scattering parameters, which are never singular.

We consider only the leading-order contributions from the ladder sum, meaning only the first term on the right-hand side of Eq.(B–40) need be considered, and we then must find the leading-order solution to the integral equation (B–41) for $\langle \mathbf{k}_1\mathbf{k}_2|T|\mathbf{k}_3\mathbf{k}_4\rangle$. If one performs the leading-order summation of ladder diagrams in each of the diagrams of Figure B–3—by means of the prescription (B–47) and the ensuing discussion—one finds that the total effect is merely to replace $v(\mathbf{k})$ with the quantity $\langle \mathbf{k}_1\mathbf{k}_2|T|\mathbf{k}_3\mathbf{k}_4\rangle$, with appropriate momentum labels. Consequently, it is still necessary to carry out the self-energy analysis and then consider the diagrams of Figure C–1 with ladder-sum insertions.

The next step is to study the integral equation (B–41). Because we shall be interested only in the Boltzmann limit we can approximate the Pauli factor $P(\mathbf{k}_5, \mathbf{k}_6)$ by unity, for the other terms will contribute additional factors of z, which are quantum corrections. One then recognizes this equation as being identical to that for the half-off-shell T-matrix of scattering theory. We take principal values in Eq.(B–41), so that the correspondence is actually to the standing-wave T-matrix.

Let us introduce relative two-body coordinates $\mathbf{K} \equiv \mathbf{k}_1 + \mathbf{k}_2$, $\mathbf{k} \equiv \frac{1}{2}(\mathbf{k}_1 - \mathbf{k}_2)$, and write

$$\langle \mathbf{k}_1' \mathbf{k}_2' | T(z) | \mathbf{k}_1 \mathbf{k}_2 \rangle = \langle \mathbf{k}' | t(z) | \mathbf{k} \rangle \delta_{\mathbf{K}'\mathbf{K}}, \tag{C-32a}$$

$$\langle \mathbf{k}_1' \mathbf{k}_2' | V | \mathbf{k}_1 \mathbf{k}_2 \rangle = \langle \mathbf{k}' | v | \mathbf{k} \rangle \delta_{\mathbf{K}'\mathbf{K}}. \tag{C-32b}$$

Equation (B–41) then becomes approximately the relative integral equation

$$\langle \mathbf{k}' | t(z) | \mathbf{k} \rangle = \langle \mathbf{k}' | v | \mathbf{k} \rangle + \sum_{\mathbf{q}}{}' \frac{\langle \mathbf{k}' | v | \mathbf{q} \rangle \langle \mathbf{q} | t(z) | \mathbf{k} \rangle}{z - \omega(\mathbf{k})}, \tag{C-33}$$

where $\omega(\mathbf{q}) = \hbar^2 q^2 / 2\mu$, $\mu \equiv m/2$, and half-on-shell $z \equiv \omega(\mathbf{k})$. With respect to Figure B–3 one then makes replacements such as

$$v(\mathbf{k}) \longrightarrow \langle \tfrac{1}{2}(\mathbf{k}_1 - \mathbf{k}_2 - \mathbf{k}) | t | \tfrac{1}{2}(\mathbf{k}_1 - \mathbf{k}_2 + \mathbf{k}) \rangle, \tag{C-34}$$

etc.

Because we are considering only spherically-symmetric interactions, we can make a partial-wave analysis by writing

$$\langle \mathbf{k}' | t(z) | \mathbf{k} \rangle = \sum_{\ell=0}^{\infty} (2\ell + 1) t_\ell(k', k; z) P_\ell(\mathbf{k}', \mathbf{k}), \tag{C-35}$$

and we retain only the s-wave contribution $t_0(k', k; z)$. For hard spheres of radius a, the half-on-shell solution for t_0 has been given explicitly by Laughlin and Scott (1968) as

$$\frac{4\pi^2 \mu}{\hbar^2} t_0^+(k', k; \omega_\mathbf{k}) = \frac{\sin(k'a)}{k'} e^{-ika}, \tag{C-36}$$

for outgoing scattered waves. But Eq.(C–33) corresponds to standing waves, so that the appropriate solution here is a superposition of the incoming and outgoing solutions: $t_0 = \frac{1}{2}(t_0^+ + t_0^-)$. Hence, for s-waves the appropriate T-matrix elements in Eq.(C–34) are just

$$\langle \mathbf{k}' | t | \mathbf{k} \rangle \simeq \frac{\sin(k'a)}{k'} \cos(ka). \tag{C-37}$$

In evaluating the Hartree-Fock self-energy of Eq.(B–14), we *first* do the sum over all ladder diagrams and *then* perform the self-energy analysis. This procedure is only valid in the leading-order ladder approximation. One then obtains the renormalized theory with

$$P_1(\mathbf{k}) \simeq \frac{\hbar^2(2S + 1 + \varepsilon)}{2m\pi^2} \int \frac{d^3 k_2}{(2\pi)^3} \nu(\mathbf{k}_2 + \mathbf{k}_1) \frac{\sin(k_2 a/2)\cos(k_2 a/2)}{k_2/2}, \tag{C-38}$$

to leading order. Note that here the exchange term contributes a quantity equal in magnitude to the direct term. In the Boltzmann limit, then,

$$P_1(\mathbf{k}_1) \simeq -\frac{z(2S + 1 + \varepsilon)}{4\pi^4 \beta k_1} e^{-\beta\omega(\mathbf{k}_1)} \int_0^{\infty} dk_2\, e^{-\beta\omega(\mathbf{k}_2)} \sin(k_2 a) \sinh(\beta\hbar^2 k_1 k_2 / m). \tag{C-39}$$

One then proceeds with the calculation of K_{nn} in this approximation in the same way as that leading to Eq.(C–4). The resulting formal expression is lengthy, and the integrals difficult to evaluate exactly, so we shall not record it here. But the discussion now provides a starting point for the reader interested in pursuing a more detailed evaluation of the covariance function for hard spheres.

$K_{nn}(\mathbf{k},\omega)_{\mathrm{B}}$ IN SECOND ORDER

In order to address some fundamental questions concerning covariance (and correlation) functions, it is necessary to extend the perturbation calculations beyond the leading terms. Hence, so as to provide a standard example, we analyze briefly the terms contributing to K_{nn} through second order in the interaction potential, and in the Boltzmann limit.

The sixteen different diagrams corresponding to the second-order terms are shown in Figures C–3, C–4, C–5, and C–6, and have been grouped in this particular way purposely. Figure C–3 contains the four second-order self-energy diagrams *not* included in the function $P_1(\mathbf{k})$ of Eq.(B–19). A preliminary examination of these terms reveals a difficulty similar to that which motivated the original self-energy analysis, and hence we see that it is necessary to extend that analysis in second order. That is, one must go beyond terms of the Hartree-Fock type and return to the full Dyson equation, Eq.(B–13). It does not suit our purpose to pursue that chore here, but the essential analysis is well understood (e.g., Fetter and Walecka, 1971).

As one might expect, the diagrams of Figure C–4 all yield expressions of order z^3, so that upon inversion the contributions are third order in the equilibrium density. This is consistent with the earlier result of Eq.(C–10) illustrating that the term first order in the interaction is second order in the density. In addition, each of these diagrams produces an analytical expression proportional to the characteristic factor $\exp[-\frac{1}{2}\beta m\omega^2/k^2]$—this is verified explicitly for the diagram of Figure C–4b in Eq.(C–27), when the latter is expanded through second order. One thus expects these contributions to vanish as $k \to 0$, so that no divergences are encountered in the hydrodynamic limit, say. All these diagrams appear to be well behaved in general.

The diagrams of Figures C–5 and C–6 present an entirely different problem, however, for their leading contributions are of order z^2. When the fugacity is eliminated we therefore obtain terms of order n^2, rather than of the expected order, n^3. Consequently, there is clear reason to examine these terms rather closely, for they may in some sense be 'different'.

We do not wish to go into excruciating detail regarding this point, so we shall examine only a small portion of the contribution from one of these diagrams— namely, that of Figure C–6c. When the Boltzmann limit is invoked and only terms of order z^2 retained, there arise four distinct contributions from this diagram. Two of these possess the characteristic factor $\exp[-\frac{1}{2}\beta m\omega^2/k^2]$, but the other two do not. Let us examine more closely one of these latter two terms, and denote it by $\overline{K}_{nn}^2$. Then, we find that

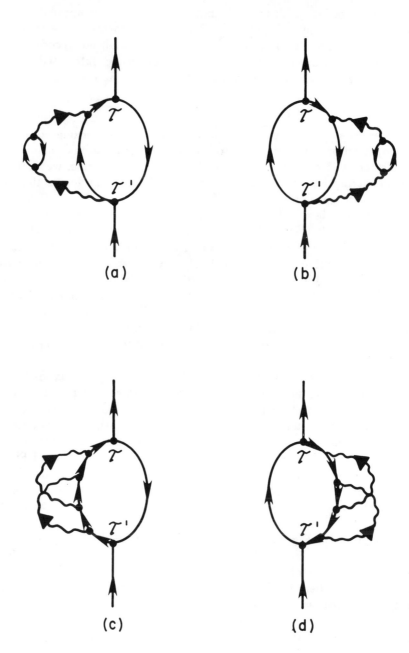

Fig. C–3. The four additional second-order self-energy diagrams contributing to K_{nn}.

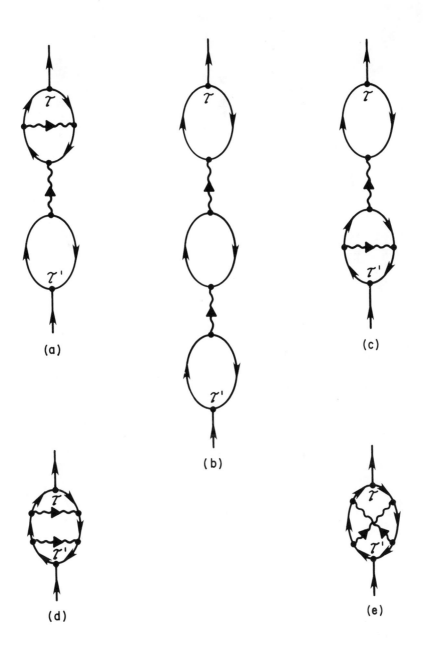

Fig. C–4. Five second-order diagrams contributing to K_{nn} which are completely well behaved.

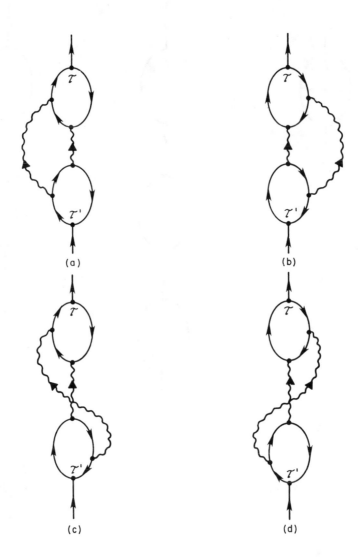

Fig. C-5. Four diagrams contributing to K_{nn} in second order which are anomalous.

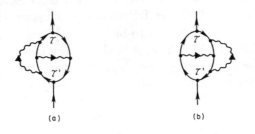

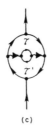

Fig. C–6. Three diagrams contributing to K_{nn} in second order which are anomalous.

$$\overline{K}^2_{nn}(\mathbf{k},\omega) = z^2(2S+1)\frac{4\pi m}{\beta\hbar^3}\frac{1-e^{-\beta\hbar\omega}}{\beta\hbar\omega}$$

$$\times \int \frac{d^3k_1}{(2\pi)^6}\frac{v^2(\mathbf{k}_1)}{k_1}e^{-\beta\omega(\mathbf{k}_1)-\frac{1}{2}\beta m\omega^2/k_1^2}\,e^{-\frac{1}{8}\beta\hbar^2 k^4/mk_1^2}\,e^{\beta\hbar\omega k^2/2k_1^2}$$

$$\times \int \frac{d^3k_2}{(2\pi)^2}\frac{e^{-\beta\omega(\mathbf{k}_2)}\,e^{-\beta\omega(\mathbf{k}_2)w_{12}^2}}{\left(\hbar\omega+\frac{\hbar^2 kk_2}{m}w_2+\frac{\hbar^2 k^2}{2m}\right)\left(\hbar\omega+\frac{\hbar^2 kk_2}{m}w_2+\frac{\hbar^2 k^2}{2m}+\frac{\hbar^2}{m}w_1\right)}$$

$$\times e^{-\beta\omega(\mathbf{k})k_2^2 w_2^2/k_1^2}\,e^{-\beta\hbar^2 k_1 k_2 w_{12}/m}\,e^{-\beta\hbar\omega(w_{12}-kw_2/k_1)k_2/k_1}$$

$$\times e^{-(\beta\hbar^2/m)(kk_2^2 w_2 w_{12}/k_1 + k_2 k^2 w_{12}/2k_1 - k^2 k_2 w_2/2k_1^2)}, \qquad (C\text{–}40)$$

where $w_1 \equiv (\mathbf{k}\cdot\mathbf{k}_1)/kk_1$, $w_2 \equiv (\mathbf{k}\cdot\mathbf{k}_2)/kk_2$, and $w_{12} \equiv (\mathbf{k}_1\cdot\mathbf{k}_2)/k_1 k_2$. Notice that here the characteristic factor mentioned above occurs in an integrand, and does not involve k. In the hydrodynamic limit,

$$\overline{K}^2_{nn}(\mathbf{k},\omega) \xrightarrow[k\to 0]{} z^2(2S+1)\frac{4\pi\beta^2 m}{\hbar^3}\frac{1-e^{-\beta\hbar\omega}}{(\beta\hbar\omega)^3}$$

$$\times \int \frac{d^3k_1}{(2\pi)^6}\frac{v^2(\mathbf{k}_1)}{k_1}e^{-\beta\omega(\mathbf{k}_1)}\,e^{-\frac{1}{2}\beta m\omega^2/k_1^2}\int \frac{d^3k_2}{(2\pi)^2}e^{-\beta\omega(\mathbf{k}_2)}$$

$$\times e^{-\beta\omega(\mathbf{k}_2)w_{12}^2}\,e^{-(\beta\hbar^2/m)k_1 k_2 w_{12}}\,e^{-(\beta\hbar\omega)k_2 w_{12}/k_1}. \qquad (C\text{–}41)$$

Although the integrals converge and the resulting expression obviously is finite, there *is* a difficulty. Both the Boltzmann and hydrodynamic limits require that $\beta\hbar\omega \ll 1$, for which $\overline{K}_{nn}^2$ is seen to become very large, diverging in the limit. The origin of this difficulty, of course, is in the energy denominators of Eq.(C–40): if we had invoked the limit $\beta\hbar\omega \to 0$ earlier, then the limit $k \to 0$ would not have existed. This calculation for the density-density covariance function is not directly applicable to the transport coefficients, but it does indicate that considerable care must be exercised in seeking density expansions for these coefficients. Such consequences are discussed further in Chapter 8.

REFERENCES

Fetter, A.L., and J.D. Walecka: 1971, *Quantum Theory of Many-Particle Systems*, McGraw-Hill, New York.

Laughlin, R., and B.L. Scott: 1968, 'Off-Energy-Shell *t*-Matrix Elements for Local Potentials Containing Hard Cores', *Phys. Rev.* **171**, 1196.

Index

This is both a subject and name index. The author has attempted to make it as comprehensive as possible, but only with respect to *significant* items. Matter that is mentioned only in passing, or which is so broad in meaning that indexing it is pointless, has been studiously omitted. Unless clearly called for, authors' names are not indexed to specific pages in the main text, but only to those pages on which a full reference is provided. In this latter case the page number is italicized.

Foundations
of
Statistical Mechanics

Volume I: Equilibrium Theory

by

Walter T. Grandy, Jr.

Department of Physics and Astronomy,
University of Wyoming, U.S.A.

This volume addresses the theory of many-body systems in thermal equilibrium from the viewpoint of the principle of maximum entropy. A sound basis in the theory of probability, based on the views of Laplace, Bayes, and Cox, is provided. This leads, in a clear and direct way, to the ensemble theory of Gibbs and eventually to quantum statistical mechanics. Much space is devoted to a detailed study of the fundamental structure of statistical mechanics, and to investigation of the reasons *why* the theory gives such an exceedingly accurate description of many-particle systems.

Numerous applications of the theory are presented, including consideration of both free-particle and interacting systems. A detailed study of the effects of external fields on such systems is also carried out.

This volume will be of particular interest to researchers in statistical physics and thermodynamics, and to theoretical physicists in general. It can also be recommended as a supplementary graduate textbook in this field.

ISBN 90–277–2489–X   FTP 19

Contents

Volume I: Equilibrium Theory

303